PATHOGEN AND MICROBIAL CONTAMINATION MANAGEMENT
IN MICROPROPAGATION

Developments in Plant Pathology

VOLUME 12

The titles published in this series are listed at the end of this volume.

Pathogen and Microbial Contamination Management in Micropropagation

Edited by

A.C. CASSELLS

KLUWER ACADEMIC PUBLISHERS
DORDRECHT / BOSTON / LONDON

Library of Congress Cataloging-in-Publication Data

ISBN 0-7923-4784-6

Published by Kluwer Academic Publishers,
P.O. Box 17, 3300 AA Dordrecht, The Netherlands.

Sold and distributed in the U.S.A. and Canada
by Kluwer Academic Publishers,
101 Philip Drive, Norwell, MA 02061, U.S.A.

In all other countries, sold and distributed
by Kluwer Academic Publishers,
P.O. Box 322, 3300 AH Dordrecht, The Netherlands.

Printed on acid-free paper

Printed in the Netherlands

CONTENTS

Overview

Pathogen and Contaminant Detection and Identification

Chemotherapy and Thermotherapy of Plants and Cultures

Laboratory Contamination Management

Disease at Microplant Establishment

PREFACE

This book is based mainly on invited and offered papers presented at the *Second International Symposium on Bacterial and Bacteria-like Contaminants of Plant Tissue Cultures* held at University College, Cork, Ireland in September 1996, with additional invited papers. The *First International Symposium on Bacterial and Bacteria-like Contaminants of Plant Tissue Cultures* was held at the same venue in 1987 and was published as *Acta Horticulturae* volume 225, 1988. In the intervening years there have been considerable advances in both plant disease diagnostics and in the development of structured approaches to the management of disease and microbial contamination in micropropagation. These approaches have centred on attempts to separate, spatially, the problems of disease transmission and laboratory contamination. Disease-control is best achieved by establishing pathogen-free cultures while laboratory contamination is based on subsequent good working practice. Control of losses due to pathogens and microbial contamination *in vitro* addresses, arguably, the most importance causes of losses in the industry; nevertheless, losses at and post establishment can also be considerable due to poor quality microplants or micro-shoots. In this symposium, a holistic approach to pathogen and microbial contamination control is evident with the recognition that micropropagators must address pathogen and microbial contamination *in vitro*, and diseases and microplant failure at establishment. There is increasing interest in establishing beneficial bacterial and mycorrhizal association with microplants *in vitro* and *in vivo*.

The contents are divided into five sections: overview, pathogen and contaminant detection and identification, chemotherapy and thermotherapy, laboratory contamination management and disease at microplant establishment. In each section there are papers that update contributions offered at the first symposium in 1987, for example, major advances have been made in microbial taxonomy and diagnostics based on advances in DNA-based techniques. Consolidation has occurred in therapy and laboratory contamination management. Novel, and arguably speculative, *in vitro* contamination control based on autotrophic cultures is described; and disease control at establishment using bacterial and mycorrhizal inoculants have also been discussed.

In the first two symposia in this series, microplant quality, in the sense of plant vigour, was not considered in the context of the production of high health status material. It is clear, however, from papers on the development of the industry, especially in Europe, and from papers on problems with the quality of microplants that aspects of microplant quality are important in the overall process of the production of good quality propagules of high health status. It was with this perspective in mind that the second symposium ended with the establishment of an ISHS working group on 'Quality Management in Micropropagation'.

(Further information on this working group can be obtained from the Editor at the above address or by email: a.cassells@ucc.ie)

ACKNOWLEDGEMENT

The Editor would like to thank Ms Regina McGarrigle of Mayo Editorial Services who is responsible for the technical editing and typesetting of this book.

PATHOGEN AND MICROBIAL CONTAMINATION MANAGEMENT IN MICROPROPAGATION — AN OVERVIEW

ALAN C. CASSELLS
Department of Plant Science, University College, Cork, Ireland

1. INTRODUCTION

The emphasis in this overview will be on disease and microbial contamination management in relation to the stages in micropropagation (Fig. 1); however, it should be recognized that micropropagation is an integrated process the objective of which is to produce, at an economic price, healthy propagules that develop into vigorous plants with normal or improved characteristics. In the production of clean microplants, attention must be paid to the selection of vigorous parental clones, to the establishment of cultures of appropriate health status, to multiplication with contaminant screening and to the production of vigorous propagules for establishment that will resist damping off disease and that will grow on to have normal or improved cropping characteristics.

Micropropagation is, for practical reasons, frequently described in terms of stages that reflect subculture to new media, etc. [1]. Initially, Murashige [2] proposed three stages: establishment of an aseptic culture, *in vitro* cloning and preparation for soil establishment. These stages were expanded by Debergh and Maene [3] in attempts to reduce contamination levels and also to improve establishment. The additional stages proposed were mother plant preparation and subdivision of the preparation for establishment (Fig. 1). Cassells [4,5] proposed expansion of Stage 0 to include genotype screening and stock plant indexing/therapy and addition of a further stage, i.e. monitoring for quality control, both with respect to health status and genetic/epigenetic stability (Fig. 1). Here, an integrated approach to the production of pathogen- and contaminant-free plants is considered.

2. THE NATURE OF MICROBIAL CONTAMINATION

A basic problem in addressing the broad issue of microbial contamination, including pathogen contamination, is, arguably, the impossibility of confirming axenic culture status in economic production. This is because of the problem of cryptic contamination. Nevertheless, an approach to the health certification of micropropagated plants can be made

A.C. Cassells (ed.), Pathogen and Microbial Contamination Management in Micropropagation, 1–13.
© *1997 Kluwer Academic Publishers. Printed in the Netherlands.*

(Fig. 2). The most likely sources of contamination are characterized pathogens of the crop, endophytic cultivable micro-organisms and laboratory contaminants. The risk is that infected plants may introduce a source of inoculum into the environment. This is especially important in relation to the international shipment of *in vitro* propagules where novel diseases may be introduced. The most problematic contaminants are latent forms of crop pathogens or latent micro-organisms that are potential pathogens of other crops, or of humans (Weller, this volume) or farm animals.

2.1. Contamination by crop pathogens

2.1.1. Major crops

The literature relating to diseases of major crops is well documented and increasingly accessible via computerized information retrieval systems (see e.g. CAB Abstracts (CAB International, Wallingford, UK) or the Internet web site of the American Phytopathological Society (http://www.scisoc.org)). However, such information may not be easily interpreted without local knowledge as regional variation in disease risk is common. Consultation with a crop specialist in the region of origin of the plant is recommended. If material is being imported, even if not mandatory under phytosanitary law, it is advisable to check with the local importing authority on perceived risks existing in the exporting country, and to obtain a phytosanitary certificate from authorities in the exporting region. This step should not be overlooked where *in vitro* cultures are introduced into the laboratory.

All material entering Stage 0 (Fig. 3) should be inspected, if possible, in the raiser's nursery during a growing season (some symptoms may be seasonally expressed) before introduction, and quarantined in the micropropagator's glasshouse during Stage 0. For quarantine purposes, stock plants should be grown individually in pots covered with Sunbags (Sigma Chemical Co., Poole, Dorset, UK) and the pots placed in individual trays to prevent movement of pathogens in run-off water, etc. The plants should be carefully inspected for symptoms. It has been the author's experience to import material with a phytosanitary certificate and to find that the material was infected with a pathogen new to the country. Introduction of disease with the planting material, like this, is the classic route of pathogen spread to a new growing region.

2.1.2. Minor crops

Unfortunately from the disease control perspective, micropropagators handle many minor crops whose pathology may be poorly documented or anecdotal. The first course of action should be to review the literature, and the guidelines indicated above should be followed. Diseases and disease risk can be anticipated by reference to the literature on diseases of related crops. As with major crops, APS Press (St. Paul, Minnesota, USA) has produced a series of plant disease compendia, by crop and by crop types.

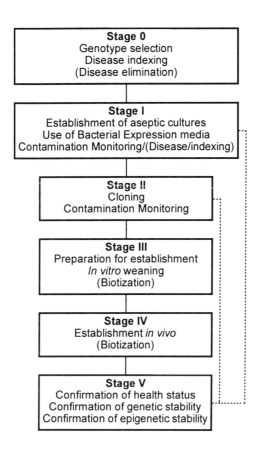

Figure 1. The stages in micropropagation expanded from Murashige [2] and Debergh and Maene [3]. It is recommended that genotype selection and disease indexing be carried out in Stage 0 (as Fig. 2) with the objective of identifying pathogen- and micro-organism-free individuals. The use of meristem explants on bacterial expression media is recommended in Stage I. If pathogen- and contaminant-free cultures are confirmed then the only requirement is culture indexing (as Fig. 3). *In vitro* weaning [17] is recommended in Stage III to increase disease resistance prior to Stage IV; biotization (i.e. mycorrhizal or beneficial bacterial inoculation may be carried out at this stage [21]). An additional stage, Stage V, is proposed as a quality control step, to confirm disease indexing results obtained *in vitro* and to confirm clone characters. Sample material from Stages I and II can be examined. Procedures in parentheses are optional.

2.1.3. Contamination in culture

It is taken as a truism that all micro-organisms restricted to the vascular system of the plant are eliminated by meristem culture, provided that the explant is taken from beyond the limit of differentiation of the phloem and xylem. Taking any other explant for Stage I is risky,

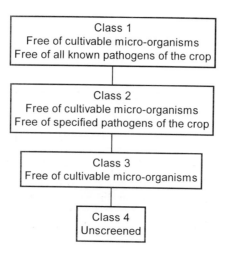

Figure 2. A proposed practical commercial plant health-certification scheme. Note that the 'axenic' category is not included.

particularly in view of concerns about detection of latent pathogens in cultures (see section 3; see also Holdgate and Zandvoort, this volume). If this assumption is valid, the main contaminants likely to enter cultures are the non-vascular, intracellular micro-organisms, i.e. viruses and viroids. In practice, this is too simplistic, as the survival potential of small explants may be low and, during excision, contamination may spread to the explant.

Bacterial contaminants introduced in/on the explant include cultivable or potentially cultivable bacteria. In the past these were assumed to be resident rhizosphere or phylloplane inhabiting bacteria that escaped surface sterilization treatment or had established themselves internally via wounds or natural openings. Increasingly, and alarmingly, these cultivable contaminants have been shown to include potential human pathogens ([6]; Zenkteler *et al.*, this volume). The presence of *Escherichia coli* in plant tissues [6] indicates a relatively nutrient-rich environment in the internal tissues which may explain the diversity of the *in vitro* flora associated with explants [7,8]. Some of the explant introduced organisms may not be expressed on the Stage I culture medium and may pass into clonal propagation to be expressed at later stages in the process, e.g. when the medium composition is changed [9], with serious economic consequences.

3. DETECTION OF CONTAMINANTS IN CULTURES

3.1. Specificity of diagnostics

Two aspects of the application of diagnostics in micropropagation are of concern, namely, specificity and sensitivity. Plant diagnostics, mainly serodiagnostics (Torrance, this

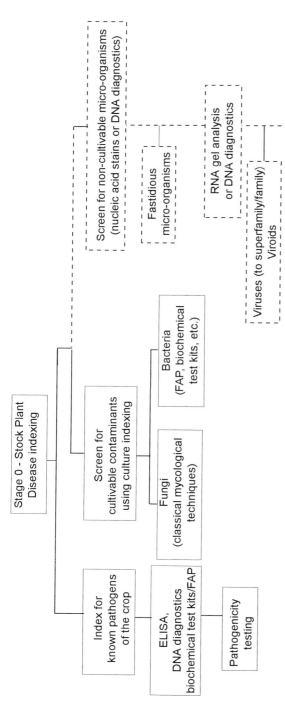

Figure 3. A summary chart of the disease-indexing requirements to produce high-quality health-status material *in vitro* based on indexation of the stock plant. It has been argued in the text that more reliable indexing may be obtained from mature plant material, albeit correct meristem excision procedure may eliminate pathogens. Indexing is divided into established protocols for characterized pathogens of the crop (confirmation of known diseases), indexing for cultivable endophytes (solid boxes) and screening for cryptic micro-organisms (open boxes). It is acknowledged that the latter is likely to be uneconomic and technically non-feasible for most non-specialist commercial laboratories. The steps in each pathway are based on detection and lead to protocols for identification/characterization. The latter may only be necessary where contaminant identification is essential.

volume) have been developed and validated for the detection of pathogen strains in mature plant tissues. It is usually important to confirm that the strains, particularly in the case of bacteria, are virulent, i.e. pathovars, by carrying out infectivity tests [10,11]. These, usually narrowly specific, diagnostics have a role in micropropagation where the client requires confirmation of freedom from specific disease strains. Micropropagators, however, especially when working with disease in less characterized crops, have a requirement for broad spectrum diagnostics, e.g. capable of detecting exotic pathogen strains and of simultaneous detection of micro-organisms of the major families of viruses, viroids and bacteria. Such diagnostics are being developed using mixtures of primers based on highly conserved genomic sequences. These approaches, currently, are largely restricted to basic applications but have the potential to transform contaminant detection (for more details see, for example, papers by Bove and Garnier and by Reeves, this volume).

3.2. Sensitivity of diagnostics

Given that diagnostics of appropriate specificity are available, there remains the problem of sensitivity of detection. There are two aspects of diagnostic sensitivity: firstly, the concentration or titer of the target organism in the tissue and secondly, the distribution of the micro-organism in the test tissues. In general, diagnostics have been developed for application to plants *in vivo* and for specific tissues at specific growth stages. It should not be assumed that the concentration of the contaminant and its localization in *in vitro* tissue will parallel that *in vivo*. Viruses have been shown to influence, and be influenced by, plant growth regulators *in vivo* (see e.g. [12]). *In vitro*, hormonal concentration in the medium may influence explant virus titer (and that of other micro-organisms: Cassells and McCarthy, in preparation). Secondly, the distribution of some viruses is irregular *in vitro* (Knapp *et al.*, this volume) and this may result in escape, i.e. false-negative test results. Suppression, rather than virus elimination, in *in vitro* treated material has been reported ([5,12] Leonhardt *et al.*, this volume). Matthews [13] has reviewed relevant aspects of plant virology.

Disease indexing and contaminant monitoring schemes for Stage 0, and for Stages I–III, respectively, are shown in Figs. 3 and 4.

4. ELIMINATION OF CONTAMINANTS

4.1. Treatment of Stage 0 plants

The merits of indexing plants at Stage 0 rather than *in vitro* have been discussed above. A classic approach has been to use diagnostics to locate disease escapes [5]. This could also be extended to selection against individuals heavily contaminated with cultivable endophytes [6]. It may be effective to apply heat therapy or antibiotics at this stage also, primarily to reduce contamination of the apex, so that explants may effectively escape disease [5,12].

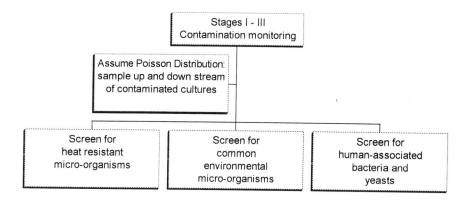

Figure 4. A strategy for contaminant monitoring in production. This scheme assumes that pathogen-free cultures were obtained in Stage I and emphasizes that contamination clusters may exist (see text for details).

4.2. Production of aseptic Stage I cultures

Serendipitous elimination of pathogens and contaminants may be a consequence of the good working practice of excising the minimum viable meristem tip explant for Stage I culture establishment. It is important also that media that facilitate the expression of bacterial contamination are used at Stage I [14]. Stage I cultures should then be increased to provide adequate material for testing before progression to Stage II. Where Stage 0 testing indicates contamination in the parent plant, the *in vitro* cultures should be re-screened for cultivable contaminants and for pathogens, albeit the above reservations regarding in vitro testing should be noted and, where possible, the plants should be randomly sampled and the sample grown on in isolation, to confirm the health status of the mature phenotype, i.e. to satisfy the Stage V requirement (Fig. 1). If disease/contamination is detected then *in vitro* chemotherapy or thermotherapy can be attempted and the progeny re-screened. The caution with any of these therapeutic strategies is that the possibility of contaminant suppression rather than elimination must be considered. Correct procedures must be followed [5,15].

4.3. Maintenance of aseptic cultures

This aspect of micropropagation has been systematically studied and clear guidelines provided ([16]; Leifert and Woodward, this volume). The main problems are entry of cultivable bacteria through system failure or poor aseptic technique and, secondly, active transmission of infection via insect infestation (Pype *et al.*, this volume). A consideration is that non-Poisson distribution of contamination may occur where aseptic technique breaks down in the case of the individual operative. It is important in this regard that production and shelving of cultures are linear, and traceable, such that when random sampling locates infection, up- and downstream tracking allows the infection cluster to be isolated and destroyed. Active transmission of infection, e.g. by micro-arthropods (see Pype *et al.,* this volume), can be avoided by the use of sealed vessels provided the vessels allow adequate gaseous exchange to avoid vitrification [17]. A novel strategy, based on photoautotrophic culture, has been proposed to prevent/suppress contamination (Long, this volume) or recover cultures that have become contaminated (Sayegh and Long, this volume). It remains to be confirmed whether this approach will gain widespread acceptance.

One aspect that has become clearer is the risk that cultures may become sources of human pathogens with potential risks for the micropropagation technicians and the industry. The risk that micropropagated plants may introduce human pathogens into hospitals via ornamental plants or into human or animal food/feed production must be of great concern and may merit special indexing measures for these markets.

4.4. Microplant resistance to disease at and post establishment

Previously, the issue of disease in micropropagation has been mainly restricted to issues related to the establishment and maintenance of disease- and contaminant-free cultures. Investigations on the establishment and weaning of microplants (or unrooted micropropagules) have recognized that poor microplant/propagule quality may influence performance. Vitrified microplants may be hypo-lignified, have thin walls and suffer a growth check on establishment, with loss of *in vitro* leaves and roots [5]. The latter are sources of nutrients for soil saprophytes which may become facultative pathogens. While beyond the scope of this article, parallel studies on *in vitro* weaning and improvement in microplant quality [18] will result in improved expression of constitutive disease resistance as the microplants mature more quickly, i.e. they will be less susceptible to young plant diseases [19].

Relevant here are investigations of microplant failure (Williamson *et al.* and Grunewaldt Stöcker, this volume). The development of measures based on bacterization and mycorrhization to protect microplants during the early stages of growth *in vivo* has been reported [20,21]. There is also evidence that these approaches may confer benefit on the growing crop and provide a value-added propagule for the micropropagation industry [22].

5. POTATO — A MODEL FOR THE INTEGRATION OF MICROPROPAGATION IN THE PRODUCTION OF CERTIFIED PLANT MATERIAL

Potato currently provides, arguably, the best example of good working practice for the exploitation of micropropagation in the production of planting material of high health status. The elements of a potato certification scheme based on that used in the Netherlands are shown in Fig. 5. This scheme exemplifies the points made above, in that Stage I cultures are produced by government regulated laboratories as disease- and contaminant-free ('axenic') mother cultures. These are multiplied in the micropropagation laboratory and the progeny (as microplants, micro- or mini-tubers) then returned to the field for disease indexing and certification. (*Note:* no official disease indexing or certification of health status is carried out on the cultures.)

In this scheme the parental material for clone selection in each cycle is taken from healthy 'basic seed' field crops. Individual plants are taken based on habit, vigour and freedom from disease symptoms. These are checked for specific pathogens by ELISA test on leaf samples from the field and by post-harvest indexing carried out on the tubers. 'Quarantine diseases' are monitored by State phytosanitary services, non-quarantine diseases by the potato inspection service. The latter are listed in Table 1. Disease-free clones are established in isolation glasshouses. Where necessary, thermotherapy may be used to eliminate pathogens and, subject to negative disease indexing, the material may be established. The progeny of the clonal selection is multiplied in isolation glasshouses and also may be used to establish aseptic mother cultures. These certified cultures may be multiplied by commercial micropropagators but for a limited period or number of subcultures, e.g. for a maximum of 2 years or 18 subcultures. In the conventional scheme, the basic seed is further multiplied in glasshouses and in the field, and then passed to commercial seed production companies for further multiplication prior to sale as certified grades. In all years the crop is inspected and graded on health status. Each year the crop is automatically down-graded to the grade specified by the disease tolerance levels (Table 1). The progeny from commercial micropropagation (as microplants, micro- or mini-tubers) may re-enter the certification scheme as older year Basic seed or in some schemes enter in later years, progeny being given the appropriate grade to their health status as determined by field/tuber indexing. A general principle of certification schemes is the recognition that health status declines with successive field multiplication cycles.

A certification scheme is in operation for strawberry where the same principles exist except that nuclear stock may be stored for an undefined period *in vitro* under slow growth condition and be used to initiate further propagation cycles subject to confirmation of its trueness to type (see Fig. 1, Stage V; [23]). For certification of grapevines see Leonhardt *et al.* (this volume).

6. CONCLUSIONS

Micropropagation is a substitution technology in the plant production industry and, as such, it brings both advantages and risks. It has the advantage that it is independent of ambient environmental conditions and has a low stock plant independence; thus, it can be exploited to rapidly clone elite genotypes, but there are risks. These include deviation from plant type and habit and, specifically of concern here, the risk of the release of clonally infected

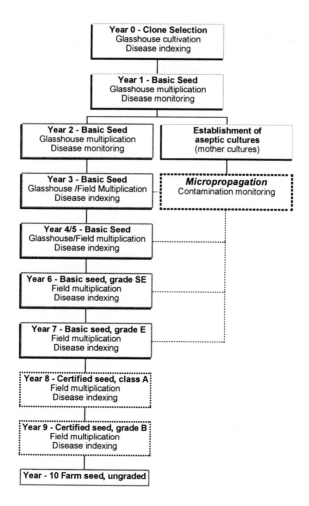

Figure 5. A consensus potato seed tuber certification programme using the Netherlands categories for the stages (after Netherlands General Inspection Service for Agricultural seeds and seed Potatoes (NAK), Ede, The Netherlands). Key features of such schemes are that definitive disease indexing is based on field material, i.e. mature tubers post harvest; and that 'certified' disease-free cultures are supplied to commercial laboratories for multiplication *in vitro*. There are restrictions on the number of subcultures permitted. *In vitro* material is not certified with respect either to health status or varietal purity. The progeny of the *in vitro* cultures, plants, tubers produced *in vitro* or mini-tubers produced in vector proof glasshouses, is subsequently certified. The *in vitro* progeny may re-enter the programme at not less than one grade lower than the mother cultures. Usually they are subject to several field multiplication cycles to reduce the tubers cost to comparability with material from conventional production. The objective of potato propagation is to introduce new varieties/produce seed of high health status, at an economic price, more rapidly than by conventional means.

Table 1. Field disease tolerance limits in potato seed crops in The Netherlands. Field results are based on symptom expression and sampling for ELISA. The field results for virus testing are confirmed by post-harvest sampling and further ELISA testing. The virus tolerances in the post-harvest test are: Class S and SE, no virus in 200 tuber sample; class E 1 in 200, class A 5 in 100 and class B 8 in 100. All crops are downgraded in successive seasons by at least one grade, or to the grade determined by the post-harvest disease indexing results. Tuber inspection is also carried out for *Rhizoctonia*, scab, wet rot, dry rot and *Phytophthora*

	S/SE	E	A	B
Severe mosaic/leafroll	0.03%	0.1%	0.25%	0.5%
Mild mosaic	0.03%	0.1%	2%	4%
Total max.	0.03%	0.1%	2%	4%
Verticillium wilt	2%	3%	4%	8%
Blackleg	0	0	0.03%	0.01%

propagules. Tissue culture methods have been used successfully to produce 'clean' plants based on disease escape through meristem culture and on good laboratory practice to keep out contamination during *in vitro* cloning [5]. Indeed, an important aspect of *in vitro* cloning is the low risk of pathogen contamination during the *in vitro* process due to the unlikelihood of exposure to vectors/pathogen inoculum in the laboratory. This is in contrast to multiplication of elite material in protected cropping or in the field, as reflected in the certification of potato seed tubers (Fig. 5; Table 1).

The risks in micropropagation of direct disease transmission or release of infected material which may introduce pathogen inoculum, relate mainly to the practice of relying on meristem culture to eliminate pathogens and contaminants from infected stock plants. These risks also relate to problems associated with disease indexing of *in vitro* material or, indeed, the absence of any indexing of production other than visual examination of cultures for signs of bacterial or fungal contamination (Fig. 2). These risks are especially great where micropropagation is used to facilitate international plant distribution (especially between quarantine zones). An issue for consideration is whether microplants should be axenic or whether free of 'quarantine' diseases (see [5]) should be the absolute standard.

In addition to the risks of plant disease, there is now evidence of risks to humans from potential 'food poisoning' organisms and human pathogens. These contaminants may pose a special threat to the health of the micropropagators, to hospital patients, to those whose immune systems are weak, or are after antibiotic treatment. Human consumption of contaminated plants produced in photoautotrophic ('microhydroponic') culture may also pose a risk.

At present, best practice may be to follow the application of micropropagation in the potato seed tuber certification industry where micropropagation is integrated into proven practice.

A criticism can be made of this application which restricts the use of certified 'mother cultures', e.g., to 2 years or 18 subcultures. Once certified, it should be possible, as in strawberry [23], to get agreement that these cultures retain their health certification status (trueness to type would have to be confirmed) as the risk of re-introducing pathogens is low. If this principle is accepted, it is advisable that the production of disease-indexed ('certified') mother cultures should be carried out under official regulation by, or in association with, qualified plant pathologists. The onus then on the micropropagator would be to exclude laboratory contaminants, i.e. cultivable organisms and to certify the freedom of their microplants from same.

In the absence of legal requirements, micropropagators are exposed solely to economic forces in deciding on acceptable quality standards for their products and, with price competition, economic factors will tend to lower standards (an anecdote has it that more companies with quality control procedures fail than those without). A reputation for poor quality, here poor health status, will however, tend to tarnish the whole industry and so it is in the interests of the industry to agree quality standards.

In spite of advances in diagnostics, their application in the micropropagation laboratory seems remote, both from the perspective of cost and technical competence. It has been suggested above that specialist help should be sought in establishing disease-indexed, clean mother cultures and that these should be the starting material for the micropropagator.

As a final comment, another aspect of microplant production, namely that aseptic microplants represent a 'biological vacuum', is also recognized ([21]; Hepton, this volume). This exposes microplants to the risk of promiscuous colonization by possible pathogenic micro-organisms; many environmental organisms may have this potential when released from competition from phylloplane/rhizosphere competitors. Phototrophic culture offers the opportunity to establish beneficial gnotobiotic relationships *in vitro* which may confer life-long beneficial effects on the microplant progeny also providing the industry with a potential novel value-added opportunity [21].

REFERENCES

1. George, E.F. (1993) Plant Propagation by Tissue Culture, Exegetics, Basingstoke, UK.

2. Murashige, T. (1974) Annu. Rev. Plant Physiol. 25, 135–166.

3. Debergh, P. and Maene, L. (1985) Acta Hortic. 166, 21–23.

4. Cassells, A.C. (1991) in Techniques for the Rapid Diagnosis of Plant Pathogens (Duncan J.M. and Torrance L., eds) pp. 179–191, Blackwell, Oxford, UK.

5. Cassells, A.C. (1997) in Biotechnology in Agriculture (Altman A., ed.) Marcel Dekker, New York, USA.

6. Cassells, A.C. and Tahmatsidou, V. (1996) Plant Cell Tissue Org. Cult. 47, 15–26.

7. Leifert, C., Waites, W.M. and Nicholas, J.R. (1989) J. Appl. Bacteriol. 67, 353–361.

8. Leifert, C., Waites, W.M., Nicholas, J.R. and Keetley, J.W. (1990) J. Appl. Bacteriol. 69, 471–476.

9. Cassells, A.C., Harmey, M.A., Carney, B.F., McCarthy, E. and McHugh, A. (1988) Acta Hortic. 225, 153–162.

10. Lelliot, R.A. and Stead, D.E. (1987) Methods for the Diagnosis of Bacterial Diseases of Plants, Blackwell, Oxford, UK.

11. Fox, R.T.V. (1993) Principles of Diagnostic Techniques in Plant Pathology, CAB International, Wallingford, UK.

12. Cassells, A.C. (1988) Prog. Med. Chem. 20, 119–156.

13. Matthews, R.E.F. (1991) Plant Virology, Academic Press, New York, USA.

14. Menard, D., Coumans, M. and Gaspar, Th. (1985) Med. Fac. Landbouww. Rijksuniv. (Gent) 50, 327–331.

15. Barrett, C. and Cassells, A.C. (1994) Plant Cell Tissue Org. Cult. 36, 169–175.

16. Leifert, C. and Waites, W.M. (1994) in Physiology, Growth and Development of Plants in Culture (Lumsden P.J., Nicholas J.R. and Davies W.J., eds) pp. 363–378, Kluwer, Dordrecht, The Netherlands.

17. Cassells, A.C. and Walsh, C. (1994) Plant Cell Tissue Org. Cult. 37, 171–178.

18. Lumsden, P.J., Nicholas, J.R. and Davies, W.J. (1994) (eds) Physiology, Growth and Development of Plants in Culture, Kluwer, Dordrecht, The Netherlands.

19. Isaac, S. (1992) Fungal–Plant Interactions, Chapman and Hall, London, UK.

20. Nowak, J., Asiedu, S.K., Lazarovits, G., Pillay, V., Stewart, A., Smith, C. and Liu, Z. (1995) in Ecophysiology and Photosynthetic *in vitro* Cultures (Carre F. and Chagvardieff P., eds) pp. 173–180, CEA, Aix-en-Provence, France.

21. Cassells, A.C., Mark, G.L. and Periappuram, C. (1996) Agronomie 16, 625–632.

22. Mark, G.L. and Cassells, A.C. (1996) Plant and Soil 185, 233–239.

23. Anonymous (1994) EPPO Bull. 24, 875–889.

STRATEGIC CONSIDERATIONS FOR THE ESTABLISHMENT OF MICRO-ORGANISM-FREE TISSUE CULTURES FOR COMMERCIAL ORNAMENTAL MICROPROPAGATION

D.P. HOLDGATE and E.A. ZANDVOORT

Integrated Plant Technology Applications, Ambledon House, 53 Church Road, Upper Tasburgh, Norfolk NR15 1ND, UK

1. INTRODUCTION

Plant tissue cultures have been used extensively since the mid-1960s for the creation of disease-indexed plants and mass micropropagation of ornamental and other horticultural plants. The first plants to be so propagated on a commercial scale were the orchids *Cymbidium* and *Cattleya* [1], followed by *Lilium* hybrids, gerbera and many foliage plants from the late 1960s [2]. A progressive expansion both in the range of species and the total numbers propagated has occurred world-wide through to the present day. Production in the major European centres, The Netherlands, France, Italy, Belgium and the UK, has experienced from moderate to severe reductions in the past 5–10 years as producers have transferred activity to nominally lower cost areas, or stopped production as more countries seek to supply the European growers at discounted prices. Production volume and costs have been major concerns for both producers and statisticians but, for the purchasing grower, plant quality represented by health status and uniformity are in reality paramount.

The micropropagation industry was initially built on the ability to generate, through *in vitro* culture, clonal progeny detectably free from a disease contaminating a parent plant. Strictly the concept of 'virus-free' and 'disease-free' should only be applied in respect of those specific viruses, and other pathogenic organisms, for which a diagnostic test has been applied and been shown to be negative. The term 'disease-indexed' has been extensively used to describe plants so analysed. Extending this application to a wide range of vegetatively propagated ornamentals, and subsequently to many other crops, provided a capability for rapid expansion of the ornamental industry, particularly in Europe. With rapid expansion, competitive markets and international prices to secure market share stimulated aggressive cost reductions. Frequently, cost reduction appears to have been achieved by ignoring the basic culture and diagnostic screening required to generate disease-indexed plants and to maintain cultures free from contaminating micro-organisms. However, by implication, the high health status, accredited to correctly initiated tissue cultures has been retained. Such strategic practice may well have proved self-defeating for many commercial operations in the longer term.

A.C. Cassells (ed.), Pathogen and Microbial Contamination Management in Micropropagation, 15–22.
© *1997 Kluwer Academic Publishers. Printed in the Netherlands.*

2. EFFECTS OF CONTAMINATION

It is generally accepted that micropropagation is high cost, with some 60% of this allocated to the intensive labour requirements, followed by costs of chemicals and/or energy, depending on country. However, plant losses due to systemic micro-organisms and contaminants, introduced during the culture handling procedures, can represent the highest cost element of any micropropagation operation when expressed in terms of labour, chemicals, energy and space charges, and especially when the recipient grower can claim compensation.

Unacceptable contamination can be identified at each subculture stage and discarded. In many instances contaminated cultures will not be removed, and attempts may be made to eliminate the micro-organisms by surface sterilisation of the cultures, employing fungicides, such as benlate, or antibiotics as potential remedies, irrespective of other difficulties which may be caused. Within a routine commercial production operation such actions do not eliminate the problem, although localised suppression of the micro-organism may mask the symptom. Systemic contamination reduces the net multiplication rate but can be most severely manifested at the rooting stage when failure to root *in vitro* or *in vivo* may be 100% due to the presence of bacteria or fungi. At the post-rooting acclimatisation stage losses can also be extremely high due to systemic pathogens surviving through the entire culture sequence. Losses have also been witnessed in high quality commercial plantings after the crop has been growing for some weeks or months. In the case of perennials, for example orchids, losses occur even years after transfer to the nursery due, generally, to viral infection, but occasionally to bacteria and fungi carried through the tissue culture.

The use of infected cultures, and general contamination problems, can cause high costs to the micropropagator and to the grower. Such costs are to a large extent avoidable.

3. ESTABLISHMENT OF MICRO-ORGANISM-FREE CULTURES

3.1. Non-viable activities

The carry-over of non-pathogenic endophytic bacteria and pathogenic micro-organisms at very low levels within a culture should be anticipated and a positive strategy adopted to eliminate the possibility. During the past 25 years there has been an attitude change from:

> – a tissue culture derived plant can be free from specified micro-organisms,

to one that:

> – tissue culture plants are (automatically) free from micro-organisms.

The latter is clearly not true. Even where an absolutely clean initial culture has been established, the standards of subculture must be at a very high level of competence in aseptic technique to maintain that status. Attempting to overcome contamination problems introduced during routine production programmes, or systemic fungal and bacterial contamination problems by incorporating antibiotics into the nutrient medium is not successful. Although micro-organism growth in the region of tissue contact with the nutrient medium can be inhibited, this cannot be equated with elimination from the entire culture. Masking of micro-organism contamination with antibiotics is a deplorable and dangerous practice which should be condemned when and wherever discovered.

3.2. Pre-treatment of mother stocks

To establish clean cultures of selections from ornamental crops for large-scale production it is desirable to develop a good pre-culture cultivation programme and, where possible, in an insect-free high quality growing environment, to reduce the chances of any pathogenic infections. Where rejuvenation procedures and quarantine periods are required, a series of isolation units or mechanisms may be invaluable. Plants destined to be used for culture purposes should be screened for infections prior to explant removal. This will determine the procedures to be employed to increase the chances of securing at least one disease-indexed culture suitable for establishing the production programmes. The initiation procedure should be based on the standard principle of meristem tip culture [3].

3.3. Meristem tip culture

It has been long established that culturing a shoot tip restricted to the apical dome plus one or, exceptionally, a maximum of two leaf primordia (the meristem tip explant) to initiate an *in vitro* culture, can result in the production of some cultures free from those infectious micro-organisms present in the parent plant [4–6]. For successful elimination of virus the size of the explant must be minimal, be carefully dissected from the shoot tip or bud, and grown in an *in vitro* culturing regime, preferably established as suitable for the particular genotype. Increasing the size of explants beyond the two leaf primordia stage will most likely result in the presence of the virus and/or other micro-organisms, thus making the culturing exercise a waste of resources. The success rate of meristem tip culture, as defined by the number of recovered disease-indexed cultures, tends to be low in percentage terms (2–40%). Success is dependent on the experience and skills of the operators, the tools available, the plant species and the type of organism to be eliminated, donor plant pre-treatment and size of the explant employed. The attention to detail and precision of application of these techniques in the 1960s was the foundation on which the substantial commercial micropropagation industry was created.

The practicality of dissecting out meristem tips in a viable state requires knowledge of the shoot tip or bud structure, the position of the meristem and the number of scale leaves which may be protecting the apical dome. Understanding of these and the phyllotaxis can be helpful and time saving. A major problem which can be experienced in fresh tissue is exudation from cut leaf bases, which can carry contaminant organisms from the older parts of the plant to the surface of the apical dome. In the experience of the authors, washing of the dome with sterile water immediately prior to excision has proven to be helpful. The use of broken pieces of razor blade mounted in an adjustable metal holder, glass rod or wood has proven to be superior to scalpel blades for excision as a better and more accurate and complete cut is made with the thinner sharper blade. In cases where phenolic oxidation and medium blackening occur, it is advantageous to make frequent transfers to new medium and to incorporate an antioxidant.

The use of meristem tip culture for routine initiation, even when the donor plant has been established as free from viral infection, improves the chances of obtaining cultures free from bacteria and fungi. However, attainment of this clean status should not be assumed. It is essential to undertake screening programmes covering several subculturing generations to establish the position. Where viral infected plants must be the source of explants for culture, the use of thermotherapy, in which the donor mother plant is treated at an elevated temperature (35–40°C) for periods of 34–38 days prior to dissection, can improve the overall result by reducing the spread of heat-labile viruses within the rapidly growing apical dome region. Although the debilitating effect of this treatment can reduce the survival rate of the dissected meristem tip explants, the net result of established disease-indexed cultures can be enhanced.

The general use of chemotherapy for the elimination of persistent endophytes is not a matter for routine application. The elimination of specifically identified micro-organisms by the application of antibiotics which have been experimentally established as lethal, and not merely biostatic, in combination with special longer term cultural procedures, may be achieved, albeit at low levels of efficiency. Plants derived from such programmes should be established as retaining the original genetic constitution and agronomic performance prior to propagation *en masse*.

Where an axillary or terminal bud is not available for culture initiation, as in the case of *Lilium* spp., adventitious bud formation may be employed with advantage. To secure the greatest benefit, the developing meristem tip, generally arising from the sub-epidermal level, should be excised from the parental structure at an early stage of development, thus prior to the migration of endophytic micro-organisms from the old tissues to these new growths. Since 1970, hundreds of millions of new virus-indexed lily plants have been produced by early isolation of adventitious meristem tips, which can be initiated in the sub-epidermal layers of bulb scales. Unfortunately, it is also true that large quantities of bulbs heavily infected with virus, bacteria and fungi have also been produced by those preferring to wait until the adventitious meristems have grown to easy-to-culture larger shoots or bulblets. This classical problem of infection of young bulblets faced by propagators was the major cause of the downfall and limitation of the lily business in Holland and the USA pre 1970. The linking of true disease indexing practices and micropropagation techniques, as demonstrated in orchid *in vitro* propagation, led to the start, in 1970, of mass propagation programmes of new cultivars of lilies and other bulb crops by a UK micropropagation operation in association with several Dutch breeders [7].

3.4. Disease indexing

The term 'disease-indexed' is used to indicate that the plant or culture being referred to has been shown to be detectably free from specified organisms through appropriate screening procedures. Freedom from disease-causing organisms should not be assumed, let alone stated or implied, without exhaustive screening for each organism of interest. To establish the freedom from specific infections, appropriate diagnostic procedures must be employed on cultures grown in a manner which would encourage the growth of the micro-organism, if present, and repeated at monthly to bimonthly intervals until at least two, but preferably three, genuine negative results have been obtained in sequence. All cultures providing positive and doubtful results should be discarded. The established elimination of one micro-organism does not prove the concomitant removal of any other organism.

The diagnostic tests used are variable, depending on the micro-organism and the level of sophistication developed and available for detection of that organism. Procedures used for detection and identification which have been reviewed recently [8] and are summarised in Table 1, are the subject of continuous upgrading and refinement. Within the application of diagnostic procedures a distinction must be made between those which can be employed for general screening, as in the case of nutrient broths, and those procedures used in combination for specific classification and identification purposes. General molecular probes for detecting bacterial DNA could make PCR a strong routine tool for establishing cultures to be free from bacterial infections and general non-pathogenic contamination, essential requirements for successful, cost-effective tissue culture activities. Biological screenings were standard procedures for detection of viruses prior to the introduction of the ELISA technique and, since it is the infective agent which expresses its own presence, are still advocated by many as the most reliable, albeit far more time consuming. The use of nutrient broths for detection of bacteria and fungi in macerated tissue samples are essential for the general screening required within the commercial tissue culture laboratory. Other procedures may be faster but they are subject to errors and false-negatives, especially when, for example, a tissue slice prevents direct contact between micro-organism and nutrient broth.

Regular sampling and biological screening should be a routine for all cultures maintained for the initiation of large-scale propagation programmes. A high standard and quality service requires that any culture showing any possibility of infection must be discarded. The adage "if in doubt throw it out" should prevail.

Contaminant-free, disease-indexed cultures have significant value and should be classified as elite stock. All micropropagation production programmes should be started from

Table 1. Tests for detection of infectious and contaminating micro-organisms in plant tissue cultures

Nature of test	Test type	Application to
Biological	Indicator plants	Virus, bacteria
	Nutrient broths	Bacteria, fungi
Physical	Electron microscopy	Virus
	Fluorescent microscopy	Bacteria
	Light microscopy	Fungi, bacteria
Molecular and chemical	Serological (ELISA, in particular)	Virus, bacteria
	PCR, RAPD	Virus, fungi, bacteria
	Fatty acid profiling	Bacteria

established elite cultures which can be maintained for long periods at reduced temperatures (2–14°C depending on the species) and regular but infrequent, subculturing (1 to 4 times per annum), to provide a constant source of quality explants for production. Although genetic stability of slow-growing juvenile material is high, routine biological screening and growing trials to ensure trueness to type should be part of the overall quality control activities.

During the propagation programme, and as a matter of routine, careful screening of containers and individual cultures for contaminants, prior to subculture, should be made by trained staff. All containers carrying a contaminated culture should be discarded. Without such actions ever-increasing losses are most likely to be experienced through transfer of fungal and bacterial contaminants, and total productive efficiency will be low and plants, as contaminated end-products, should be classified as substandard.

Despite the awareness of problems it is frequently stated, implied, and too generally accepted without question that all shoots derived from any tissue culture must be automatically free from all pathogens and micro-organisms. This is strictly and most emphatically NOT the case. With the increasing rate of international commercial trade in tissue culture produced plants and distribution of germplasm *in vitro* there is a case for phytosanitary standards and regulations to be more rigorously implemented.

The clean plants derived from micropropagation can become re-infected and, therefore, good husbandry practices should be employed for rooting and acclimatisation in the post-culture facilities, and in the field, to prevent, or at least delay, re-infection. A strict regime of replanting with clean stock derived from clean micropropagated shoots rooted in an environmentally controlled acclimatisation chamber has substantial benefits for the grower. However, application of resources may not be justified unless the grower has confidence that the micropropagator has done his total job — that is the elimination of micro-organisms from the cultures prior to and during the micropropagation production phase. Those micropropagators failing to employ clean culture strategies, including correct use of proper meristem tip culture and hygiene quality during the multiplication and rooting phases, will experience very high unit production costs, poor product quality and, ultimately, poor sales. A basic regime for production of contaminant-free and disease-indexed stocks is illustrated in Fig. 1.

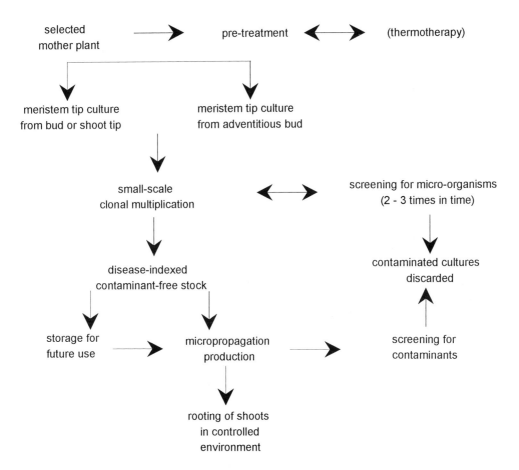

Figure 1. Strategy for the production of disease-indexed cultures.

4. CONCLUDING REMARKS

The original concept of commercial micropropagation based on the opportunities from mass clonal propagation of disease-indexed plants, has led to an enormous expansion of the applications and boosted productivity in a wide range of crops. Disease-indexed contaminant-free propagation, however, has not been universally sustained. Routine discarding of cultures at each transfer, due to bacterial or fungal contamination in the laboratory phases, and up to total losses at rooting and growing-on stages due to carry-over of pathogenic infections, or excessive presence of non-pathogenic micro-organisms which disrupt the immediate culture environment, generate major performance and cost difficulties.

The creation, maintenance and use of disease-indexed, contaminant-free cultures (Fig. 1) is the only reliable strategy for securing a sustainable micropropagation and tissue culture business. Furthermore, the disciplined aseptic technique required within the commercial micropropagation operation is not always accepted by new entrants to commercial production, but the real value of contaminant-free cultures is assessable from:
– reduced losses in the laboratory
– improved rooting rate and quality of shoots
– reduced losses in the weaning and hardening phases (acclimatisation)
– supply of quality plant products with superior growth rates, providing the customer with easier crop management, reliability and better yields.

REFERENCES

1. Marston, M.E. (1966) Hortic. Sci. 9, 80–86.

2. Holdgate, D.P., Aynsley, J., Fenwick, L., Hill, G.P., Krebs, H., Lynne, R., Rangan, T., Rothwell, A., Spurr, J., Stokes, M., Smith, S. and Thomas, E. (1965 to 1975) unpublished work during tenure at Twyford Laboratories Ltd., UK.

3. Morel, G. (1964) Rev. Cytol. Biol. Veg. 27, 307–314.

4. Morel, G. and Martin, C. (1952) Compt. Rend. 235, 1324–1325.

5. Quack, F. (1977) in Applied and Fundamental Aspects of Plant, Cell, Tissue, and Organ Culture (Reinert J. and Bajaj Y., eds) pp. 598–615, Springer-Verlag, Berlin.

6. Stone, O.M. (1973) Ann. Appl. Biol. 73, 45–52.

7. Holdgate, D.P. (1977) in Applied and Fundamental Aspects of Plant, Cell, Tissue, and Organ Culture, (Reinert J. and Bajaj Y., eds) pp. 18–43, Springer-Verlag, Berlin.

8. Reed, B.M. and Tanprasert, P. (1995) Plant Tissue Cult. Biotechnol. 1, 137–142.

IMPROVED VIRUS DETECTION IN ROSACEOUS FRUIT TREES *IN VITRO*

E. KNAPP[1], V. HANZER[1], D. MENDONÇA[2], A. da CÂMARA MACHADO[2], H. KATINGER[1] and M. LAIMER da CÂMARA MACHADO[1]

[1]*Institute of Applied Microbiology, University of Agriculture and Forestry, Nuszdorfer Lände 11, A 1190 Vienna, Austria*
[2]*Universidade dos Açores, Departamento de Ciências Agrárias, 9700 Angra do Heroísmo, Azores, Portugal*

1. INTRODUCTION

In vitro cultivation techniques are widely used for micropropagation and germplasm storage [1,2]. Moreover, *in vitro* culture techniques are the only effective tools for eliminating certain pathogens from elite germplasm. Many fruit tree viruses are well identified and characterised [3,4]. Conventional laboratory diagnostics, based on serology, have been developed and are routinely used for orchard surveys [5–8]. Drawbacks in the certification of fruit plants are, however, encountered since many viruses are either latent, irregularly distributed or highly localised. Thus, the risk of false results is increased especially with field plants [9,10]. Little has been published on the virus screening of *in vitro* cultures; however, optimisation of sampling of *in vitro* plants may be achieved if carried out in combination with studies on virus localisation using immuno-tissue printing [11–13].

2. MATERIALS AND METHODS

Fruit tree cultures were micropropagated on modified MS media supplemented with 0.36 mg/l benzylaminopurine and 0.01 mg/l indolebutyric acid. Thermotherapy of *in vitro* cultures was carried out at 38°C for 16 h and 36°C for 8 h. Meristems were excised from heat-treated cultures and further propagated for selection of healthy progeny plants [2,11,13]. Cultures that were subjected to the combined treatment of thermotherapy and meristem culture are referred to as 'treated' cultures in contrast to the non-treated cultures. Non-treated *in vitro* cultures were indexed by DAS or two-step ELISA [14,15], then immediately after thermotherapy to investigate the effects of elevated temperatures on viral antigen, and in the course of selecting healthy progeny plants, respectively. Sampling for ELISA was as described [11], unless otherwise stated. The threshold level was defined as the mean value of eight healthy replicates plus twice the standard deviation. OD values slightly above the threshold were considered dubious positives. Plum pox virus (PPV) was detected by immunocapture-PCR. Primers for the reverse transcriptase (RT) and

A.C. Cassells (ed.), Pathogen and Microbial Contamination Management in Micropropagation, 23–29.
© *1997 Kluwer Academic Publishers. Printed in the Netherlands.*

polymerase chain reaction (PCR) were based on the PPV NAT viral genome [16] and resulted in a 780 bp amplified fragment. The amplified products were separated by agarose gel electrophoresis and visualised by ethidium bromide staining.

3. RESULTS

3.1. Sampling for reliable large-scale screening

The localisation of plum pox virus (PPV), apple stem grooving virus (ASGV) and apple chlorotic leafspot virus (ACLSV) in host tissues was investigated using immuno-tissue printing [12,13]. These studies were undertaken in combination with the screening of cultures by ELISA and influenced the choice of the sample [17].

3.2. Influence of subculture intervals on ELISA results

In vitro plants undergo different growth phases (lag phase, active growth phase, stationary phase) after transfer to fresh media. Preliminary results with four apple cultures indicated that diagnosis was more reliable with samples from short subculture intervals. To confirm these results the apple cultivar 'Delbard Estival', infected with ASGV and ACLSV was chosen for a direct comparison of different sample sources. Leaf samples and the corresponding stem samples from single shoots, excluding the apical region, which was further propagated, were either investigated separately or mixed. Samples were taken after a subculture period of 65 days (stationary growth phase) and in the succeeding subculture interval 23 days later (active growth phase). Samples were stored at 20°C and the ELISAs were all performed at the same time. Ten to 17 samples were tested per source.

The reliability of the results of ASGV and ACLSV diagnosis in all samples taken after 23 days was either better than or equal to that in all samples taken after 65 days (Table 1), which confirmed the earlier results.

Table 1. The percentage of false test results occurring at different growth stages

Sample	ASGV		ACLSV	
	65 days	23 days	65 days	23 days
Leaf	14.2	14.2	14.2	7.1
Stem	11.1	0	28.5	14.2
Mixed	10.5	0	6.6	0

Although the reliability of diagnosis was higher in samples from actively growing cultures, the majority of samples from the stationary phase showed a higher virus titre, facilitating discrimination between positives and negatives (data not shown).

3.3. Distinguishing between virus suppression and virus elimination

It is known that antiviral therapy may depress the pathogen titre below the threshold level for detection [4]. In this respect the effects of thermotherapy on the titre of PPV, ASGV and ACLSV were investigated *in vitro* with three apricot and nine apple cultivars. PPV was detectable in samples from *in vitro* shoots of apricot immediately after termination of thermotherapy, in some but not in all of the samples tested. Thermotherapy was applied to the *in vitro* shoots for 17–20 days. PPV levels in heat-treated cultures were several times lower than in corresponding untreated cultures (data not shown). With apple cultures, irrespective of the length of thermotherapy, from 23 to 39 days, depending on the heat sensitivity of the plant material, virus titre of both ASGV and ACLSV decreased strongly following thermotherapy. ELISA readings were either negative or just slightly above the threshold level (data not shown).

To study the speed of re-accumulation of virus in clones treated by a combination of thermotherapy and meristem dissection, untreated apple cultures were tested as positive controls by ELISA. Treated plants showed a strong tendency for lower ACLSV level compared to the untreated cultures (Table 2).

Table 2. Comparison of ACLSV absorbance values in non-treated control and treated apple cultures several months after treatment

Apple cultivar		Month	$\bar{x}$	$s_{\bar{x}}$	n	Sample
'Jonagored'	non-treated		1.240	0.338	8	a
	treated	6	0.346	0.223	8	a
'Landsberger Renette'	non-treated		1.754	0.453	9	a
	treated	15	0.559	0.297	11	a
'Landsberger Renette'	non-treated		0.769	0.330	10	a
	treated	20	0.465	0.200	18	a
'Gelber Bellefleur'	non-treated		0.207	0.134	10	a
	treated	24	0.417	0.218	11	a
'Gelber Bellefleur'	non-treated		2.190	0.231	8	b
	treated	35	0.771	0.703	6	b

Month: tested after treatment; $\bar{x}$: mean of OD values of samples tested; $s_{\bar{x}}$: standard deviation; n: number of samples; a: mixed leaf from shoot bundle; b: base of stem plus leaf of further propagated shoot.

The behaviour of ASGV was investigated in the four apple cultures, 'Champagner Renette', 'Gelber Bellefleur', 'Roter Boskoop' and 'Summerred', as described for ACLSV. These cultures showed more variability than observed for ACLSV, as some of the heat-treated cultures had higher virus levels compared to the untreated cultures (data not shown).

3.4. The reliability of ELISA in comparison with PCR

Three ACLSV infected apple cultures were checked after treatment and no clear positives, but many dubious positives with OD values slightly above the threshold level, were found. The second test was undertaken 5 months later, the third 17 months later and the results were again negative or dubious positive. In another ACLSV infected apple culture the majority of samples tested positive after treatment and although negatives were selected for multiplication, they gave positive results 4 months later (data not shown).

A higher variability in the results was encountered in apple cultures infected with both ASGV and ACLSV. In three of four apple cultures tested more than 75% ASGV positives were found after the combined treatment. From these cultures only negative clones were subcultured, but again a high number of samples were positive for ASGV after more than 1 year *in vitro*. Negative results for ACLSV were confirmed. The apple cultivar 'Gelber Bellefleur' was chosen for an extensive test as two former tests did not contain ASGV positives. Two types of samples from shoot bundle were used for comparison. ASGV positives were compared to selected plants with either two negative or two dubious positive results (Table 3).

ELISA testing of apricot cultures 1 year after treatment provided evidence that thermotherapy and meristem culture were efficient in the elimination of PPV infection. However, when testing the apricot cultivar 'Ungarische Beste' a second time, samples were

Table 3. ASGV ELISA results of a third re-test of the treated apple cultivar 'Gelber Bellefleur'

Initial tests 'Gelber Bellefleur'	Sample Source	No.	Results as percentages		
			+	Dubious	−
Negatives	a	46	26	39.1	34.9
	b	15	46.6	26.6	26.8
Dubious positives	a	8	0	12.5	87.5
	b	8	0	12.5	87.5
Positives	a	6	100	0	0
	b	6	100	0	0

Sample source: a – base plus stem plus leaf of the further propagated main shoot; b – corresponding axillary shoots mixed.

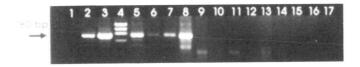

Figure 1. Agarose gel electrophoresis of IC–RT–PCR products from untreated and treated apricot cultivar 'Ungarische Beste'. 1, Virus-free *in vitro* apricot; 2, PPV infected *in vivo* apricot culture; 3, PPV containing plasmid; 4, Standard; 5–17, 'Ungarische Beste': 5–8, untreated, PPV infected; 9–12, treated, negative in ELISA; 13–15, treated, dubious positive in ELISA; 16–17, treated, weak positive in ELISA.

either negative, dubious positive or weak positive. Assays by IC–RT–PCR (immunocapture reverse-transcriptase polymerase chain reaction) confirmed that dubious and weak positive results in ELISA were false positives. The PPV specific amplified product of 780 bp was always absent in samples from the treated cultivar 'Ungarische Beste', when compared to the control (Fig. 1).

5. DISCUSSION

The combination of immuno-tissue printing and ELISA provides more reliable diagnostic results than ELISA alone. Above all the limitations of current ELISA based serological tests is the fact that they may not be sensitive enough to detect very low levels of virus. Virus titre can be a virus-specific trait, as is the case with ASGV, whose occurrence is extremely limited and highly localised in host tissues [5,13]. On the other hand, generally, high titre viruses like ACLSV and PPV show an irregular distribution in host plants [12,13]. Detailed knowledge about virus distribution patterns (decreasing from shoot base towards the apex, as in the case of ACLSV, or completely uneven, as in the case of PPV) helps to avoid sampling problems [13,17]. However, plant physiology and culture growth conditions exert further influences on virus accumulation, as was shown in our experiments. The appropriate phase of *in vitro* culture for sampling was determined with respect to maximum sensitivity and maximum reliability. Virus distribution was less localised in actively growing plants giving an increased reliability to the tests, whereas plants in the stationary phase accumulated higher virus levels thus allowing a better discrimination of infected plants, but being less reliable due to higher variability between samples.

It has been shown that thermotherapy depressed virus titre in a wide range of cultivars. Thermotherapy reduces viral protein and, as a consequence, serodiagnostics are of little value for early and reliable screening. We investigated whether thermotherapy merely suppressed rather than eliminated virus. A strong tendency for suppression could be observed with ACLSV, but not with ASGV, where a combination of thermotherapy and meristem culture failed. However, ACLSV was readily detected after re-accumulation above the threshold level whereas the well-known problems in reliable diagnosis of ASGV were encountered even after several tests of *in vitro* cultures [5,18,19]. IC–RT–PCR for PPV detection was introduced into the sanitation programme for final testing of well-characterised elite material. Sensitive nucleic acid based assays are still lacking, but are needed as routine procedures for the laboratory diagnosis of low titre pathogens like SGV in a system of certifying *in vitro* plants.

ACKNOWLEDGEMENTS

This research was supported by the Austrian Ministry for Agriculture and Forestry, project 853.

REFERENCES

1. Boxus, P., Quoirin, M. and Laine, J.M. (1977) in Applied and Fundamental Aspects of Plant Cell, Tissue, and Organ Culture (Reinert J. and Bajaj Y.P.S., eds.) pp. 130–140, Springer Verlag, New York.

2. Laimer da Câmara Machado, M., da Câmara Machado, A., Hanzer, V., Kalthoff, B., Weiss, H., Mattanovich, D., Regner, F. and Katinger, H.W.D. (1991) Plant Cell Tissue Org. Cult. 27, 155–160.

3. Németh, M. (1986) Virus, Mycoplasma and Rickettsia Diseases of Fruit Trees. Martinus Nijhoff/Dr W. Junk Publishers, Dordrecht, Boston, Lancaster.

4. Fridlund, P.R. (1989) Virus and viruslike diseases of pome fruits and simulating noninfectious disorders. Cooperative Extension College of Agriculture and Home Economics, Washington State University, Pullman, WA.

5. Fuchs, E., Grüntzig, M. and Al Kai, B. (1988) Nachrichtenbl. Pflanzenschutz DDR 42 (10), 208–211.

6. Varveri, C. and Bem, F. (1995) Acta Hortic. 386, 431–438.

7. Myrta, A., Di Terlizzi, B. and Digiaro, M. (1995) Acta Hortic. 386, 165–175.

8. Savino, V., Di Terlizzi, B., Digiaro, M., Catalano, L. and Murolo, O. (1995) Acta Hortic. 386, 169–175.

9. Casper, R. and Meyer, S. (1981) Nachrichtenbl. Dtsch. Pflanzenschutzdienstes (Braunschweig) 33 (2), 49–54.

10. Fridlund, P.R. (1983) Acta Hortic. 130, 85–87.

11. Knapp, E., Hanzer, V., Weiss, H., da Câmara Machado, A., Wang, Q., Weiss, B., Katinger, H. and Laimer da Câmara Machado, M. (1995) Acta Hortic. 386, 409–418.

12. Knapp, E., Hanzer, V., Weiss, H., da Câmara Machado, A., Wang, Q., Weiss, B., Katinger, H. and Laimer da Câmara Machado, M. (1995) Acta Hortic. 386, 187–194.

13. Knapp, E., da Câmara Machado, A., Pühringer, H., Wang, Q., Hanzer, V., Weiss, H., Weiss, B., Katinger, H. and Laimer da Câmara Machado, M. (1995) J. Virol. Methods 55 (2), 157–173.

14. Clark, M.F. and Adams, A.N. (1977) J. Gen. Virol. 34, 475–483.

15. Flegg, C.L. and Clark, M.F. (1979) Ann. Appl. Biol. 9, 61–65.

16. Maiss, E., Timpe, U., Brisske, A., Jelkmann, W., Casper, R., Himmler, G., Mattanovich, D. and Katinger, H.W.D. (1989) J. Gen. Virol. 70, 513–524.

17. Knapp, E. (1996) Ph.D. Thesis, University of Agriculture and Forestry, Vienna.

18. Gilles, G.L. and Verhoyen, M. (1992) I.R.S.I.A, Brussels. Viroses et maladie apparentees des arbres fruitiers et ornementaux assainissement et selection.

19. Knapp, E., Hanzer, V., Weiss, H., da Câmara Machado, A., Katinger, H. and Laimer da Câmara Machado, M. (1997) Proceedings of the 4th International EFPP Symposium on Diagnosis and Identification of Plant Pathogens, 9–12 September 1996, Bonn, Kluwer Academic Publishers, in press.

TRENDS IN EUROPEAN PLANT TISSUE CULTURE INDUSTRY FROM 1990 TO 1993

F. Ó RÍORDÁIN
Newbridge Research Centre, Celbridge, Co. Kildare, Ireland

1. INTRODUCTION

To date four surveys of the European plant tissue culture industry have been carried out by COST 87. This was a European co-operative research programme of "Plant *In Vitro* Culture". The surveys were organised by the Management Committee and show the changes that have taken place from 1982 to 1993. The directories were published by the European Commission [1–3], the latest in March 1994 [4]. This article discusses the additional information that was obtained in the survey and not included in the Directory. Details from earlier surveys have been already published [5,6].

Query forms were distributed by members of the Management Committee to laboratories in their countries and when completed they were returned for compilation and analyses to the author. The information was collected during 1993.

2. SIZE OF THE EUROPEAN INDUSTRY

Twenty-one countries are now included in the survey, although only 15 of these were in COST 87. The numbers of laboratories in the different countries covered by the survey are shown in Table 1. The numbers of laboratories, both commercial and official, in the four COST 87 surveys are shown in Fig. 1. Between 1990 and 1993 there has been a drop of 9% in the number of commercial laboratories in the countries that were represented in previous surveys.

3. PLANTS

The 20 most cited plants for commercial and official laboratories are shown in Tables 2A and 2B, respectively. With commercial laboratories the two most important, *Prunus* and *Ficus*, are the same as in 1990 and also in 1988. Next are *Spathiphyllum* and *Philodendron*, which have exchanged rank. In position five is *Calathea* which was not listed in 1990. With

A.C. Cassells (ed.), Pathogen and Microbial Contamination Management in Micropropagation, 31–37.
© *1997 Kluwer Academic Publishers. Printed in the Netherlands.*

the official laboratories *Solanum* has replaced *Prunus* in first place. They also show a high degree of interest in *Sorbus*, which is new to the list. Cereals have increased in importance. *Saintpaulia* no longer appears on the official list and is now of reduced importance to commercial laboratories.

The reported totals of plants produced in commercial laboratories are shown in Table 3. *Gerbera*, which was on the top in 1990, has been replaced by *Spathiphyllum*, with *Gerbera* second. Other interesting changes are the drop of 66% in the production of *Nephrolepis*, and the large decrease in *Rosa* production. Increases are shown by *Lilium*, while *Fragaria* and *Prunus* are similar to 1990 in spite of the recession.

Table 1. The numbers of laboratories in different countries in the 1993 survey

Country	Official	Commercial	Total
Austria	6	1	7
Belgium	21	12	33
Bulgaria	19	–	19
Czech Republic	18	2	20
Denmark	8	5	13
Finland	16	7	23
France	26	11	37
Germany	57	23	80
Greece	13	8	21
Hungary	14	8	22
Ireland	8	4	12
Italy	18	22	40
Netherlands	17	32	49
Norway	5	2	7
Portugal	7	–	7
Slovakia	9	3	12
Slovenia	4	2	6
Spain	18	11	29
Sweden	3	3	6
Switzerland	7	5	12
UK	35	11	46
Total	329	172	501

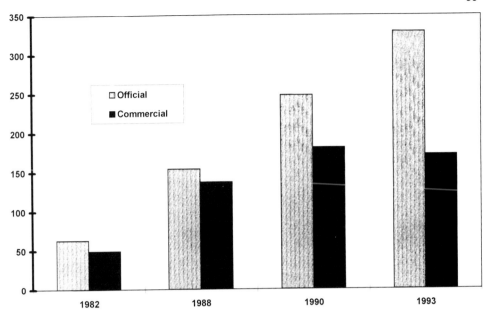

Figure 1. Numbers of laboratories in COST 87 surveys.

Table 2A. Plants with the numbers of commercial laboratories working on them

Genus	No. of laboratories	Genus	No. of laboratories
Prunus	87	*Pelargonium*	25
Ficus	61	*Solanum*	25
Spathiphyllum	55	*Brassica*	23
Philodendron	40	*Hydrangea*	23
Calathea	35	*Rhododendron*	23
Syngonium	35	*Betula*	22
Rosa	32	*Fragaria*	20
Nephrolepsis	29	*Rubus*	20
Malus	27	*Syringa*	19
Gerbera	26	*Actinidia*	18

Table 2B. Plant genera with the numbers of official laboratories working on them

Genus	No. of laboratories	Genus	No. of laboratories
Solanum	94	Rosa	37
Prunus	89	Rubus	33
Malus	78	Betula	32
Nicotiana	63	Populus	32
Sorbus	53	Lycopersicon	27
Brassica	49	Hordeum	26
Triticum	44	Allium	25
Vitis	42	Drosera	25
Quercus	38	Beta	21
Fragaria	37	Pyrus	20

Table 3. Production totals of more important plants (Millions)

Spathiphyllum	15.9	Syngonium	2.0
Gerbera	14.0	Rhododendron	1.7
Lilium	13.2	Rosa	1.7
Prunus	9.3	Begonia	1.6
Fragaria	6.2	Dipladenia	1.6
Nephrolepis	5.0	Calathea	1.5
Anthurium	4.2	Rubus	1.3
Ficus	4.0	Cordyline	1.2
Saintpaulia	3.7	Alstromeria	1.1
Bromeliaceae	2.7	Pyrus	1.1
Pogonatherium	2.5	Alpina	1.0
Solanum	2.3	Musa	1.0
Dahlia	2.1	Primula	1.0

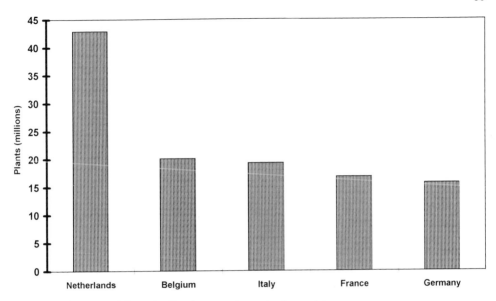

Figure 2. Commercial production in more important countries.

4. LABORATORY SIZE

4.1. Plants produced

Over 163 million plants were produced in total, 136 million by the commercial sector. National production figures for the main producing countries from commercial laboratories are shown in Fig. 2. All of these show a decrease on the last survey, except for Belgium which shows a small increase. The greatest production from a single laboratory was 10 million. There were 33 laboratories with a production of more than 1 million plants in 1992, 10 produced more than 5 million.

4.2. People employed

There were 3204 people employed in the 501 laboratories, very similar to that in 1990, in spite of being from more countries and laboratories. There has been a reduction in the mean level of employment in commercial laboratories from 10.5 to 7.8. The mean level in official laboratories has remained constant at 6.2. The percentage of laboratories with 1–3 persons has increased since 1990 to 37%. This is similar to the 1988 figure and the loss of the 1990 improvement. Now 24% of laboratories employ 1–2, which is again a loss of the 1990 improvement, when the number of laboratories in this smallest category fell to 13%. Those with 3–7 people have remained stable at 44% as against 47% in 1990 and 31% in 1988.

The largest laboratories are now those with 60. Since the 1988 survey a large number of laboratories with more then 60 employees have disappeared, including 7 laboratories employing up to 150. In 1988 there were 9 laboratories with 50 or more. In 1990 there were only 4 laboratories with more than 50 people and 6 laboratories in 1993 with more than 50 employees.

4.3. Clean work stations

The maximum number of work stations per laboratory was 58 as against 68 in 1990. The mean number of work stations has shown a slight decrease from 7.6 per laboratory to 7.2 with the commercial laboratories, 48% with 1–3 work stations. This is another example of the deteriorating position of the level of investment in the industry as this is worse than the 1988 position. There was no change in the mean number of work stations in official laboratories.

4.4. Efficiency

The mean number of plants produced per person for commercial laboratories is 99,000, which is an improvement on earlier figures, 86,000 and 85,000, in 1990 and 1988, respectively. The productivity per clean work station is down to 121,000 from 165,000 in 1988 and 126,000 in 1990 (Table 4). It is difficult to explain why the plant to employee ratio has risen while the plant/work station ratio has fallen; possibly percentage plant establishment is being increased at weaning.

Table 4. Productivity in the commercial sector (plants × 1000)

	1988	1990	1993
Plants/person	85	86	99
Plants/work station	165	126	121

4.5. Main interests

Rapid multiplication is the main interest of 74% of commercial and 36% of official laboratories. These figures are virtually the same as in 1990. Those laboratories that indicated Culture Storage and Plant Breeding as a primary interest in both sectors have also remained the same. The percentage of commercial laboratories that had a primary interest in the Elimination of Pathogens has increased from 5 to 10%. This may suggest an increasing concern about quality. There was an increased interest in Genetic Engineering. When first and second preferences are combined the interest in official laboratories increased to 50% from 20% in 1990. With commercial laboratories the interest remained at 10%. Another change is the increase in interest in Basic Research in official laboratories, from 28 to 36%. This mirrors the change in priorities of the research programme of COST 87. In the initial years it was very concerned with solving immediate practical problems in micropropagation. In the latter phase its aim was to achieve a better understanding of the processes involved in plant regeneration.

ACKNOWLEDGEMENTS

The author wishes to thank those in the 501 laboratories who took so much trouble to complete the query forms, and the national representatives on the COST 87 Management Committee who organised the survey in their respective countries.

REFERENCES

1. Ó Ríordáin, F. (1982) COST 87 Directory of European Plant Tissue Culture Laboratories. Commission of the European Communities, DG XII.

2. Ó Ríordáin, F. (1988) COST 87 Directory of European Plant Tissue Culture Laboratories. Commission of the European Communities, DG XII, 74 pp., ISBN 0 948321 50 4.

3. Ó Ríordáin, F. (1991) COST 87 Directory of European Plant Tissue Culture Laboratories 1990. Commission of the European Communities, DG XII, 124 pp., ISBN 0 948321 65 2.

4. Ó Ríordáin, F. (1994) COST 87 Directory of European Plant Tissue Culture Laboratories 1993. Commission of the European Communities, DG XII, 172 pp., ISBN 2 87263 113 5.

5. Ó Ríordáin, F. (1989) Proc. Natl. Ornamental Crops Conference, 1–11, Agricultural Institute, Dublin.

6. Ó Ríordáin, F. (1992) Agronomie, 12, 743–746.

COMMERCIAL MICROPROPAGATION LABORATORIES IN THE UNITED STATES

RICHARD H. ZIMMERMAN

US Department of Agriculture, Agricultural Research Service, Fruit Laboratory, 10300 Baltimore Avenue, Beltsville, MD 20705-2350, USA

1. INTRODUCTION

Micropropagation is the newest addition to the methods employed by horticulturists, foresters and agronomists for clonal multiplication of plants. Commercial use of this method dates back only about 30 years in the US. Commercial laboratories currently use micropropagation for (a) mass propagation of specific clones, (b) maintenance of pathogen-free (indexed) germplasm, (c) use as the initial step in a nuclear stock crop production system, (d) clonal propagation of parental stocks for hybrid seed production, and (e) year-round production of plants.

To determine the current status of commercial micropropagation in the US, an extensive survey of laboratories was made in March and April 1996. Individual laboratories were contacted and the laboratory manager or owner was interviewed by telephone to collect data on crops in culture and quantities produced. Information from laboratory operators and from research scientists was used to identify laboratories to ensure as broad a coverage of active laboratories as possible. All laboratories contacted and currently doing commercial production provided data. Some small laboratories were probably missed, but these are unlikely to have aggregate production of even 1% of the current estimated total.

2. HISTORICAL BACKGROUND

Orchid micropropagation, which developed as an offshoot of early successes in culturing meristem-tips for production of virus-free plants, was the key event in the founding of the micropropagation industry [1,2]. Success in applying meristem-tip culture to many other crops, combined with experience in orchid micropropagation and development of a widely adaptable tissue culture medium [3], led to successful *in vitro* propagation of numerous crops. These early results were summarized by Murashige [4], a key figure in the industry during this early stage. He did research on many commercially important crops, lectured widely, advised commercial growers, taught many training courses and advised several graduate students who made significant contributions in this area.

A.C. Cassells (ed.), Pathogen and Microbial Contamination Management in Micropropagation, 39–44.
© *1997 Kluwer Academic Publishers. Printed in the Netherlands.*

Early commercial use of micropropagation was made for foliage plants [5] and flower crops [6,7]. Many researchers contributed to the efforts to adapt micropropagation technology to woody plants. Anderson's [8,9] early developmental work on micropropagation of ericaceous plants was the foundation for commercial micropropagation of these crops in the US [10]. Micropropagation has been used on relatively few fruit and vegetable crops. Early applications were made to small fruit crops [11], which continues to be the main application, and several vegetable crops, particularly asparagus, potato and sweet potato [12]. It is currently applied on a large scale to production of certified potato plants to be used for minituber production in the greenhouse or seed tuber production in the field [13]. Efforts with forest tree species have lagged because the species of interest are often recalcitrant in culture and the cost per propagule must be very low to make the method economically feasible. However, protocols continue to be developed that broaden the range of plants for which commercial micropropagation is economic.

Commercial production of micropropagated plants has not grown steadily. The successes with many crops in the late 1970s and early 1980s led to a rapid increase in number and size of commercial laboratories by 1985. The requirement for immediate cash flow in many of these businesses quickly resulted in overproduction of certain crops. The resulting pressure to lower prices resulted in some laboratories closing and others consolidating. After several years of relatively stable growth, the industry, especially the foliage plant segment, was faced in 1992 with a rapid and large increase in low cost, imported tissue cultured plants. This situation caused further restructuring of the industry. Problems of quality control and crop scheduling with some imported material reduced the number of these imports and their impact starting in 1995. Domestic production once again seems to be increasing as demand continues to grow.

3. DESCRIPTION OF THE INDUSTRY

Commercial micropropagation laboratories are located in at least 26 states and most are situated near important production areas of the horticultural industries that they service. Florida leads in plants produced, followed by California, Washington and Oregon. California and Florida each have more than 15 laboratories; all other states have fewer than 10 each. Within states, the laboratories are often clustered in certain areas. In Florida, the heaviest concentration is near Apopka, where much of the foliage plant production is located. Californian labs are clustered mainly in coastal areas near San Francisco, Los Angeles and San Diego. The Pacific Northwest labs of Washington and Oregon are located west of the Cascade Range stretching from near the Canadian border to the south of Portland.

Some commercial laboratories are independent businesses whereas others are part of or owned by another horticultural enterprise, e.g. a nursery. Independent laboratories without greenhouse facilities sell only unrooted shoots or rooted plantlets that have not been acclimatized, but this type of production is decreasing in comparison to the situation 5–10 years ago. Laboratories with greenhouse facilities may still sell non-acclimatized shoots or plantlets, but are more likely to sell acclimatized plants, some of which may have been grown on for a period of weeks or months to obtain a higher value plant.

Some laboratories associated with nurseries or other production facilities produce only for internal use, so that the micropropagated plants are not offered for sale until they have been grown for several months to several years. Generally for these labs, part of the production is used within the company and the rest is sold on the open market. It is rare for a laboratory of this type to have sales only to other producers. Most laboratories sell at least part of their production on contract, but most production is sold on the open market.

Production of individual laboratories varies from a few thousand to tens of millions of plants per year. Small laboratories (<500,000 units per year) account for about 60% of the slightly more than 110 laboratories identified; 24 of these small labs produce only 50,000 units per year or fewer. About 30% are medium-sized laboratories (500,000–2,500,000 units per year); large laboratories (2,500,000–6,000,000 units) and very large laboratories (>6,000,000 per year) account for the remaining 10% .

Laboratory capacity is difficult to evaluate. Although an earlier estimate was 150 million units per year [14], currently capacity is probably closer to 175 million units, including some laboratories not currently propagating plants that could quickly re-open. Not included are several large new laboratories planned to open in 1997. The 11 laboratories currently having the largest production now have about 57% of the total capacity. Medium-sized labs account for about 27% and the small labs would have more than 10%, with the remainder in labs not currently in production.

4. PRODUCTION OF MICROPROPAGATED PLANTS

Total production of micropropagated plants is now more than 120 million plants per year based on a detailed survey conducted in March and April 1996 (Table 1). This quantity is considerably higher than earlier estimates of 61 to 75 million plants [14–16], but a much broader range of crops is now being micropropagated than 5–10 years ago. Also, this recent survey included a larger number of laboratories that were identified and contacted to obtain the figures reported here. The plants now being micropropagated can be grouped according to their uses or area of horticulture as shown in Table 1.

Foliage plants comprise the largest category, accounting for nearly 54% of the total production (Table 1). Ferns constitute the largest group of foliage plants, about 11% of all US micropropagated plants. *Spathiphyllum* production is nearly as large and other important genera include *Syngonium*, *Dieffenbachia*, *Ficus*, *Calathea* and *Philodendron*.

Table 1. Production of micropropagated plants in the United States by geographic region and type of crop. Numbers are thousands of plants

Crops	Eastern US[a]	Florida	West Central US[a]	Pacific NW[a]	CA & HI[a]	Total
Foliage plants	1,200	47,975	0	0	14,520	63,695
GH flowers	278	5,178	495	0	5,346	11,297
Perennials	4,388	1,030	1,320	2,080	630	9,448
Trees & shrubs	1,434	2,480	1,400	7,850	2,130	15,294
Vegetables	3,692	0	4,849	1,130	3,191	12,862
Fruits	1,431	10	50	1,970	260	3,721
Miscellaneous	25	975	1,715	1,270	560	4,545
Total	12,448	57,648	9,829	14,300	26,637	120,862

[a]Eastern US: states east of the Mississippi River; West Central US: states west of the Mississippi River except for Pacific coastal states; Pacific Northwest (NW): Washington and Oregon; CA and HI: California and Hawaii.

Orchids are about one-third of the greenhouse flower crop production (Table 1). *In vitro* orchid seed germination, done on a much larger scale than micropropagation, is not included. *Gerbera* is nearly half of the production in this category. Other important crops include *Anthurium* and bromeliads.

Herbaceous perennials are a rapidly increasing segment of the production (Table 1). *Hosta* and *Hemerocallis* are micropropagated in the largest quantities; other perennials include *Stokesia*, *Gypsophila*, *Heuchera*, *Leucanthemum* and *Rudbeckia*. Aquatic plants and species used for wetland restoration and other ecological purposes are a rapidly growing part and may account for as much as 30% of this category. Finally, some annual flowers are micropropagated to provide seed parents for hybrid seed production.

For trees and shrubs, ericaceous plants (*Rhododendron*, *Kalmia*, *Pieris*, *Leucothoe*) account for more than 23% of those now micropropagated (Table 1). Trees are another 22% and include both shade and ornamental trees (*Acer*, *Amelanchier*, *Betula*, *Magnolia*, *Malus*, *Prunus*, *Ulmus* and other genera) and trees grown for paper pulp or biomass (*Eucalyptus*, *Populus*). The remainder of the plants in this category are shrubs such as *Nandina*, *Syringa*, *Fothergilla*, *Hydrangea*, *Photinia*, *Viburnum* and numerous other genera.

Potatoes are about 90% of the micropropagated vegetable crops (Table 1) and are mainly grown in greenhouses for minituber production; a small proportion is planted in the field for elite seed potato production. Some microtubers are produced *in vitro*. Other crops propagated in tissue culture include asparagus, garlic, sweet potato (*Ipomoea batatas*) and some cabbage and cauliflower parent plants for hybrid seed production.

Fruit crop production is primarily that of small (soft) fruits with blueberry and raspberry the main crops (Table 1). Strawberry production is limited and is primarily to obtain stock plants for runner production; only a few laboratories are producing plants. Limited production of fruit tree rootstocks occurs, primarily for establishing layer beds.

Miscellaneous crops include a wide assortment of ornamental and tropical fruit crops and specialty crops (Table 1). Limited production of bulbs, corms and tubers occurs, e.g. *Lilium* and *Gladiolus*. Banana is the main tropical fruit produced, but most production is used for home gardens in southern states, especially Florida. A limited amount of *Citrus* rootstock is micropropagated. Other specialty crops include peppermint, spearmint, wasabii or Japanese horseradish (*Wasabia japonica*), and sugarcane.

5. EXPORT AND IMPORT OF MICROPROPAGATED PLANTS

Some micropropagated plants are exported from US laboratories, including foliage plants, woody ornamentals, fruit, vegetable and specialty crops. The total number is probably less than 5% of production. Imports have been mainly foliage plants (ferns, *Spathiphyllum*, *Syngonium*) and orchids. The total number may have been as high as 20 million units per year at the peak several years ago, but is now probably closer to 10 million.

6. CONCLUSIONS

A significant US commercial micropropagation industry has developed since 1965. At least 110 laboratories now produce more than 120 million ornamental and food crop plants per year. Foliage plants continue to be the leading product line, but other crops are growing in significance. Woody plants now account for more than 12% of production, but the dollar value of this segment of the industry is higher than 20% of the total. Perennials and vegetables are two additional crop types showing recent growth. The industry has overcome past problems from overproduction, excess laboratory capacity and strong competition from low-priced imports, and continues to grow.

REFERENCES

1. Morel, G. (1974) in The Orchids: Scientific Studies (Withner C., ed.) pp. l69–222, John Wiley, New York.
2. Griesbach, R.J. (1986) in Tissue Culture as a Plant Production System for Horticultural Crops (Zimmerman R.H. *et al.*, eds) pp. 343–349, Martinus Nijhoff Publishers, Dordrecht, The Netherlands.

3. Murashige, T. and Skoog, F. (1962) Physiol. Plant. 15, 473–497.

4. Murashige, T. (1974) Annu. Rev. Plant Physiol. 25, 135–166.

5. Hartman, R.D. and Zettler, F.W. (1986) in Tissue Culture as a Plant Production System for Horticultural Crops (Zimmerman R.H. *et al.*, eds) pp. 293–299, Martinus Nijhoff Publishers, Dordrecht, The Netherlands.

6. Oglevee-O'Donovan, W. (1986) in Tissue Culture as a Plant Production System for Horticultural Crops (Zimmerman R.H. *et al.*, eds) pp. 119–123, Martinus Nijhoff Publishers, Dordrecht, The Netherlands.

7. Stimart, D.P. (1986) in Tissue Culture as a Plant Production System for Horticultural Crops (Zimmerman R.H. *et al.*, eds) pp. 301–315, Martinus Nijhoff Publishers, Dordrecht, The Netherlands.

8. Anderson, W.C. (1975) Comb. Proc. Int. Plant Prop. Soc. 25, 129–135.

9. Anderson, W.C. (1978) Comb. Proc. Int. Plant Prop. Soc. 28, 135–139.

10. Briggs, B.A. and McCulloch, S.M. (1983) Comb. Proc. Int. Plant Prop. Soc. 33, 239–248.

11. Swartz, H.J. and Lindstrom, J.T. (1986) in Tissue Culture as a Plant Production System for Horticultural Crops (Zimmerman R.H. *et al.*, eds) pp. 201–220, Martinus Nijhoff Publishers, Dordrecht, The Netherlands.

12. Seckinger, G. (1991) in Micropropagation: Technology and Application (Debergh P.C. and Zimmerman R.H., eds) pp. 265–284. Kluwer Academic Publishers, Dordrecht, The Netherlands.

13. Jones, E.D. (1988) Am. Potato J. 65, 209–220.

14. Zimmerman, R.H. and Jones, J.B. (1991) in Micropropagation: Technology and Application (Debergh P.C. and Zimmerman R.H. eds), pp. 173–179. Kluwer Academic Publishers, Dordrecht, The Netherlands.

15. Jones, J.B. (1986) in Tissue Culture as a Plant Production System for Horticultural Crops (Zimmerman R.H. *et al.*, eds) pp. 175–182, Martinus Nijhoff Publishers, Dordrecht, The Netherlands.

16. Hartman, R.D. (1995) Greenhouse Prod. News 5(12), 10–14.

WALLED AND WALL-LESS EUBACTERIA FROM PLANTS: SIEVE-TUBE-RESTRICTED PLANT PATHOGENS

J.M. BOVÉ and MONIQUE GARNIER

INRA et Université Victor Segalen Bordeaux 2, Laboratoire De Biologie Cellulaire et Moléculaire, 71 Avenue Edouard Bourleaux, B.P. 81, 33883 Villenave D'Ornon Cedex, France

1. INTRODUCTION

The micro-organisms which form the object of this review are representatives of two groups of Eubacteria: (i) the *Mollicutes* (former mycoplasmas), i.e. a group of low guanine (G) plus cytosine (C) Gram-positive bacteria lacking the classic bacterial cell wall, and (ii) two members of the *Proteobacteria* (former purple bacteria), i.e. a bacterial group containing most of the Gram-negative bacteria. As shown in Fig. 1, some of these organisms are restricted to the sieve tubes of the infected plants and, as such, they are phytopathogenic. Others are found on the surfaces of plant organs, mainly flowers, and are non-phytopathogenic. Among the sieve-tube-restricted bacteria, only the spiroplasmas are available in culture; the phytoplasmas and the two proteobacteria have never been cultured. Properties distinguishing *Mollicutes* from other Eubacteria are listed in Table 1.

Since the first Symposium on Bacterial and Bacteria-like Contaminants of Plant Tissue Cultures in 1987, where a presentation on plant mollicutes was given [1], new developments have occurred not only in the field of the *Mollicutes*, but also in that of the sieve-tube-restricted, walled bacteria. Progress in the phylogeny, taxonomic characterization, identification and detection of the organisms covered in this review, has been due essentially to the availability of new or improved techniques in molecular biochemistry, biology and genetics: DNA cloning and sequencing, 16S ribosomal DNA sequence comparisons, pulsed-field gel electrophoresis for precise genome size measurements, testing UGA codon usage, i.e. its use as a stop codon or as a tryptophan codon, refined determination of cholesterol or Tween 80 requirements of mollicutes, testing for functional sugar phospho-transferase systems (PTS), and last, but not least, amplification of specific DNA fragments by Polymerase Chain Reaction (PCR) for sequencing and/or detection and characterization of bacterial agents, and many other purposes. Examples of these approaches will be provided in the following pages.

A.C. Cassells (ed.), Pathogen and Microbial Contamination Management in Micropropagation, 45–60.
© 1997 Kluwer Academic Publishers. Printed in the Netherlands.

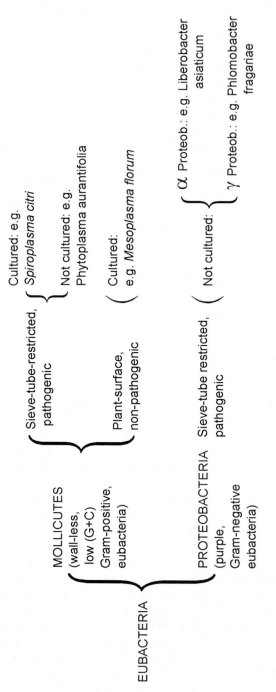

Figure 1. Walled and wall-less eubacteria from plants: sieve-tube-restricted plant pathogens and plant surface contaminants.

Table 1. Properties distinguishing mollicutes from other eubacteria[a]

Property	Mollicutes	Other eubacteria
Cell wall	Absent	Present
Plasma membrane	Cholesterol present in most species	Cholesterol absent
Genome size	580–2220 kbp	1450–>6000 kbp
G + C content of genome	23–41 mol%	25–75 mol%
No. of rRNA operons	1–2[b]	1–10
5S rRNA length	104–113 nucleotides	>114 nucleotides
No. of tRNA genes	30 (*M. capricolum*) 33 (*M. pneumoniae*)	51 (*B. subtilis*) 78 (*E. coli*)
UGA codon usage	Tryptophan codon in *Mycoplasma, Ureaplasma, Spiroplasma, Mesoplasma (Entomoplasma)*	Stop codon in *Acholeplasma*
RNA polymerase	Resistant to rifampicin	Rifampicin sensitive

[a]Adapted from Razin [39] and Bove [40].
[b]Three rRNA operons in *Mesoplasma lactucae* [40].

2. CONFIRMATION OF THE EUBACTERIAL ORIGIN OF THE MOLLICUTES

Comparison of the 16S ribosomal DNA sequences of representative members of the *Mollicutes* with those of other bacteria has shown that the mollicutes represent a branch of the phylogenetic tree of the Gram-positive eubacteria [2,3]. The mollicutes are now seen as having been derived by regressive evolution (loss of genes, genome size reduction) from an ancestor of the Gram-positive bacteria with low (G + C) in the genome. Their closest walled, eubacterial relatives are two low (G + C) Gram-positive, bacterial species: *Clostridium ramosum* and *Clostridium innocuum*. Like these clostridia, the phylogenetically "early" mollicutes, i.e. the anaeroplasmas and asteroleplasmas, are still obligate anaerobes, suggesting that anaerobiosis has been inherited from the bacterial ancestor. Rifampin insusceptibility of the mollicutes (see Table 1) as well as of the two clostridial species has probably also been acquired from the bacterial ancestor, and so has the low (G + C) content of the DNA. Figure 2 shows the phylogenetic tree of the mollicutes. The following points should be made. (i) The tree shows five phylogenetic groups: (1) the spiroplasma group with the new *Mesoplasma* and *Entomoplasma* genera but also certain *Mycoplasma* sp., such as *M. mycoïdes*; (2) the *Mycoplasma pneumoniae* group; (3) the *Mycoplasma hominis* group; (4) the anaeroplasma-acholeplasma group where the phytoplasmas cluster, and (5) the asteroleplasma group with only one species. (ii) The

48

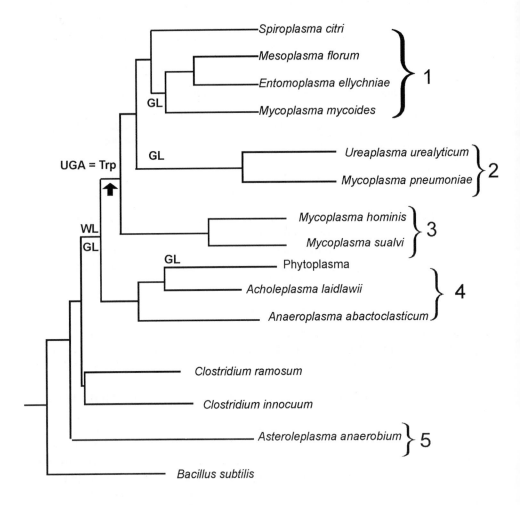

Figure 2. Phylogenetic tree of the *Mollicutes*. From [3], [5], [38]. GL: Gene loss; WL: Wall loss; 1, 2, 3, 4, 5: Phylogenetic groups.

Mollicutes represent a coherent phylogenetic branch of the low (G + C) Gram-positive tree, and loss of cell wall has occurred early, probably only once (WL on Fig. 2) as a result of gene loss (GL on Fig. 2) during regressive evolution. On Fig. 2, the branching of the asteroleplasma group is indicated as having occurred prior to loss of cell wall. This is probably wrong and shows that the exact branching of this group has not yet been established. (iii) In the universal genetic code, there is only one codon for tryptophan (Trp):

5′UGG 3′. The mollicutes, as low (G + C) organisms, have managed to develop a new Trp codon with less G: 5′UGA 3′ in which the 3′A replaces the 3′G of UGG. This is indicated on Fig. 2 as "UGA = Trp". To do that, they have evolved a new transfer RNA with anticodon 5′UCA 3′ capable of reading not only UGA but also UGG because of wobble (see [4]). Hence, the evolutionary "late" mollicutes: spiroplasmas, entomoplasmas, mesoplasmas, mycoplasmas and ureaplasmas use UGA as a Trp codon, while the "early" mollicutes (anaeroplasmas, asteroleplasmas, acholeplasmas, phytoplasmas) use UGA as a stop codon, i.e. as it is used normally. (iv) Even though the phytoplasmas are not available in culture, it could be shown that they cluster close to the acholeplasmas (see section 5.2) and are genuine mollicutes [5].

3. MOLLICUTE TAXONOMY AT THE GENUS LEVEL

The definition of mollicute genera is based on several criteria such as anaerobiosis, helical morphology, requirement for cholesterol or Tween 80, optimum growth temperature, hydrolysis of urea.

Genome size was previously considered an important property for genus differentiation. With the development of pulsed-field gel electrophoresis (PFGE), a much more accurate technique than the early renaturation kinetics, it became evident that within several genera there were wide genome size ranges.

Cholesterol requirement was considered for a long time a major criterion in establishing high taxonomic groupings within the *Mollicutes*. Recent findings appear to weaken the high status of cholesterol requirement in mollicute classification. Several spiroplasma species (*S. floricola*, *S. apis*, *S. diabroticae* and *S. chinense*) have been shown to grow in the absence of cholesterol (and Tween 80) [6]. This implies either that the family Spiroplasmataceae be split at the generic level, with the provision of a new genus (*Helicoplasma*) for helical mollicutes which do not require sterol for growth, or that cholesterol requirement be abandoned as an important character at higher levels of mollicute classification. For the present, the latter alternative has been preferred [7].

Finally, the order Entomoplasmatales (*Entomoplasma*, *Mesoplasma*, *Spiroplasma*) is based on habitat. Indeed, most organisms in this order have insect or other arthropod hosts. In the case of the phytopathogenic spiroplasmas, the insect host is the leafhopper vector. Table 2 lists the distinctive properties of the three genera of the Entomoplasmatales.

A schematic approach to laboratory differentiation of major *Mollicute* genera is indicated in Table 3 [10].

Table 2. Taxonomy and characteristics of the order Entomoplasmatales

Property	Entomoplasmatales		
	Entomoplasmataceae		Spiroplasmataceae
	Entomoplasma	*Mesoplasma*	*Spiroplasma*
Morphology	Non-helical	Non-helical	Helical
Number of species	5	12	46
G + C content (mol%)	27–29	27–30	25–30
Genome size (kbp)	790–1140	870–1100	780-2400
Cholesterol requirement	Yes	No	Yes No
Tween 80 requirement (0.04%)	No	Yes	No
Habitat	Insects Plant-surface	Insects Plant-surface	Insects Plant surface, Phloem
Phytopathogenic	No	No	Yes No
Optimum growth temperature (°C)	30–32	30–32	30–32[a]

[a]*Spiroplasma mirum*: 37°C.

Finally, species differentiation is accomplished essentially by serological techniques. Molecular and Diagnostic Procedures in Mycoplasmology have recently became available [8,9].

4. PATHOGENIC, SIEVE-TUBE-RESTRICTED WALLED BACTERIA: LIBEROBACTER AND PHLOMOBACTER (PROTEOBACTERIA)

4.1. Citrus greening disease and the Liberobacters

The micro-organism associated with citrus greening disease was first observed in 1970 [11] in the phloem of affected sweet orange leaves. It was initially thought that the greening organism was a mycoplasma-like organism (MLO), but the organism was soon found to be enclosed by a 25-nm-thick envelope, which was much thicker than the unit membrane envelope characteristic of MLOs (thickness, 7–10 nm). These properties suggested that the

Table 3. Differentiation of major *Mollicute* genera[a]

A.	PRELIMINARY CHARACTERISTICS	
	Growth on liquid and solid medium	
	Passage through bacterial filters	
	Growth in the presence of penicillin and no reversion in its absence	
B.	MORPHOLOGY (dark-field microscopy)	
	Helical → SPIROPLASMA	
	Non Helical → OTHER MOLLICUTE → C	
C.	GROWTH IN SERUM-FREE MEDIUM	
	Yes → ACHOLEPLASMA	
	No → OTHER MOLLICUTE → D	
D.	GROWTH IN SERUM-FREE MEDIUM CONTAINING TWEEN 80 (0.04%)	
	Yes → MESOPLASMA	
	No → OTHER MOLLICUTE → E	
E.	OPTIMAL GROWTH TEMPERATURE (°C)	
	30–32 → ENTOMOPLASMA	
	35–37 → OTHER MOLLICUTE → F	
F.	HYDROLYSIS OF UREA	
	Yes → UREAPLASMA	
	NO → MYCOPLASMA	

[a]From [10].

greening organism was a walled bacterium and not a mycoplasma. Organisms similar to the greening agent occur in plants other than citrus and are involved in more than 20 different diseases. As far as is known, these organisms are always restricted to the sieve tubes within the phloem tissue. None of them has been obtained in culture. By analogy with MLOs, these organisms have been called bacterium-like organisms (BLOs); they have also been inappropriately called rickettsia-like organisms.

Greening is one of the most severe diseases of citrus. It has a large geographic distribution because it is transmitted by two psyllid insect vectors, *Diaphorina citri* in Asia and *Trioza erytreae* in Africa. Symptoms of greening in Asia occur even when temperatures are well above 30°C, while in Africa the disease is present only in cool regions. These temperature effects have been reproduced under phytotron conditions. In addition, when the greening BLO was experimentally transmitted from citrus to periwinkle plants by dodder [12], the greening reaction in periwinkle was the same as that observed in citrus. Therefore, the African BLO is heat sensitive and the Asian BLO is heat tolerant, and this suggests that the two BLOs are somewhat different. Characterization has been slow and difficult because the BLOs have not been cultured.

In order to determine the phylogenetic position of the greening BLO and the evolutionary distance between African and Asian BLOs, we have PCR-amplified the 16S ribosomal DNA (rDNAs) of an Asian strain and an African strain of the greening BLO, using the universal primers described [13]. The 16S rDNA amplicons of the two BLO strains were cloned and sequenced. Comparisons with sequences of 16S rDNAs obtained from the GenBank data base revealed that the two BLOs belong to the α subdivision of the class Proteobacteria [14]. Even though their closest relatives are members of the α-2 subgroup, the BLOs are distinct from this subgroup as there is only 87.5% homology between the 16S rDNAs examined. Therefore, the two BLOs represent a new lineage in the α subdivision of the Proteobacteria.

Bacteriologists have had, hitherto, a conservative attitude in refraining from giving Latin binomial names to non-cultured organisms. However, with the development of PCR and DNA sequencing, it is now possible to characterize such organisms on the molecular and phylogenetic level. On the basis of such considerations, the designation "Candidatus" has been proposed as an interim taxonomic status to provide a proper record of sequence-based potential new taxa at the genus and species level [15]. We have used this possibility in the case of the greening organisms by naming the African greening BLO *Candidatus* Liberobacter africanum and the Asian greening BLO *Candidatus* Liberobacter asiaticum [14].

From the 16S rDNA sequences of the liberobacters, we have designed primers for the specific amplification of their 16S rDNA in plant extracts. With both liberobacters, amplicons close to 1160 bp are obtained. *Xba*I digestion of the L. asiaticum amplicon yields two fragments (640 bp and 520 bp), and that of L. africanum gives three fragments (520 bp, 506 bp and 130 bp), permitting easy distinction between the two species [16].

Two DNA probes, In-2.6 and AS-1.7, containing genes for ribosomal proteins have been produced respectively for L. asiaticum and L. africanum. In dot-blot hybridization assays, In-2.6 detects all Asian strains tested but not African strains, while AS-1.7 detects the African but not the Asian strains [17–19].

4.2. Phlomobacter

Leaf marginal chlorosis of strawberry is a new important disease of strawberry in France since 1988. A BLO was detected by electron microscopy in infected plants [20].

Work similar to that described above for the greening liberobacter has very recently resulted in PCR-amplification and sequencing of the 16S rDNA of the strawberry BLO. Sequence comparisons have shown this organism to be a representative of the γ subdivision of the Proteobacteria [21]. We propose to designate the strawberry BLO as *Candidatus* Phlomobacter fragariae.

The γ-Proteobacteria are divided in three main subgroups: (i) mainly photosynthetic organisms of the purple sulfur type, e.g. *Chromatium*, (ii) species associated with legionnaires disease, e.g. *Legionella*, and (iii) a mixture of non-photosynthetic genera from the enterics (e.g. *Escherichia coli*), vibrios, fluorescent pseudomonads, and also endosymbionts of ants, aphids (*Buchnera*), tsetse-flies, whiteflies, or parasites of leafhoppers (BEV), wasps (*Arsenophonus nasomiae*). It is interesting to find the Phlomobacter in the same subdivision as these symbionts or parasites of insects, as the phlomobacter has probably a similar symbiotic and/or parasitic association with its putative insect vector.

On the basis of the 16S rDNA sequence of the strawberry phlomobacter, a PCR assay has been developed which, for the first time, permits detection of the strawberry agent in infected plants [21] and will be undoubtedly useful in identifying the insect vector.

In summary, even though the sieve-tube-restricted micro-organisms associated with citrus greening and strawberry leaf marginal chlorosis are not available in culture, the 16S rDNAs could be obtained by PCR, sequenced, and compared to the rDNAs of other organisms. This work has shown that the former BLOs are true eubacteria, and more precisely Proteobacteria. The greening Liberobacters represent a new lineage in the α subdivision of the Proteobacteria, while the strawberry Phlomobacter is a new lineage of the γ-Proteobacteria. Both the α and γ subdivisions are known to contain bacteria associated with insects. It is, therefore, not a surprise to find the insect-transmitted liberobacter and phlomobacter agents within these groups of Eubacteria.

5. PATHOGENIC, SIEVE-TUBE-RESTRICTED, WALL-LESS BACTERIA: SPIROPLASMAS AND PHYTOPLASMAS

As opposed to the walled eubacteria examined in the previous chapter, spiroplasmas and phytoplasmas are mollicutes, i.e. wall-less eubacteria. Table 4 lists the major properties which distinguish the spiroplasmas from the phytoplasmas. The spiroplasmas are helical and available in culture, the phytoplasmas are non-helical and have never been cultured. The spiroplasmas represent a relatively "late" phylogenetic group with UGA coding for tryptophan; the phytoplasmas are part of the phylogenetically "early" acholeplasma branch with UGA being a stop codon. However, the phytopathogenic spiroplasmas and all the phytoplasmas have the same habitat in the plants: the sieve-tubes of the phloem tissue, and they have two hosts: plants and insects (mainly leafhoppers).

Table 4. Pathogenic, sieve-tube-restricted *Mollicutes*

Property	Spiroplasmas	Phytoplasmas
Morphology	Helical	Non-helical
Cultured	Yes	No
UGA codon	Trp	Stop
Functional sugar PTS	Yes	(No)[a]
Evolutionary relationship	*Spiroplasma* branch	*Acholeplasma* branch
Spiroplasma species (named)	31	–
Phytopathogenic *Spiroplasma* spp.	3	–
Characterized phytoplasmas	–	51
Phytoplasma groups	–	14
Plant diseases	*S. citri:* Citrus stubborn, many others *S. kunkelii:* Corn stunt *S. phoeniceum:* Periwinkle yellows	Over 300 diseases in 98 plant families

[a](No): probably no.

5.1. Spiroplasmas

The spiroplasmas were discovered through the study of two diseases of plants: corn stunt and citrus stubborn. The stubborn agent was the first mollicute of plant origin to be obtained in culture in 1970, and shown to have, unexpectedly for a wall-less organism, a helical morphology *in vitro* as well as *in situ* (phloem sieve tubes). The stubborn organism is known as *Spiroplasma citri* since 1973, and is the first spiroplasma to have been cultured and characterized as the result of an intense international collaborative effort [22]. The corn stunt agent was cultured in 1975 and fully characterized by 1986 as *Spiroplasma kunkelii* [23]. The third and only other sieve-tube-restricted phytopathogenic spiroplasma is *Spiroplasma phoeniceum* cultured in 1983 and 1984 from naturally infected periwinkle plants in Syria [24]. The host range of the three phytopathogenic spiroplasmas has been reviewed earlier [1].

Many other spiroplasmas have been discovered since the early work on *S. citri* and *S. kunkelii*. The great majority of spiroplasmas has been cultured from insects. Even the sieve-tube-restricted plant spiroplasmas have insect hosts: the leafhopper vectors through which the spiroplasmas are transmitted from infected plants to healthy ones. As the spiroplasmas are culturable, many studies have been devoted to these organisms from 1970 on. After 25 years of work, *Spiroplasma citri* is probably the best understood spiroplasma and is now available for genetic analysis (see section 6).

5.2. Phytoplasmas

Doi *et al.* [25] observed in the sieve-tubes of plants affected by yellows diseases, micro-organisms that resembled morphologically and ultrastructurally animal mycoplasmas. On the basis of this resemblance, the plant agents were called Mycoplasma-Like Organisms (MLOs). Today over 300 different plant species from 98 families have been found to be infected with MLOs. In spite of intensive efforts, the MLOs have never been obtained in culture, and their true nature, mycoplasmal or not, could not be determined for many years. Only when specially adapted molecular biology technics could be applied to the MLOs, did the characterization work progress quickly. Today, it is demonstrated that the MLOs are indeed members of the Class Mollicutes, for the following reasons [5]: (i) the $(G + C)$ content of their DNA, 25–30%, is similar to that of the culturable mollicutes; (ii) their genome size, as determined by pulsed-field gel electrophoresis, is small, 600–1240 kbp, well within the range characteristic of mollicute genomes; (iii) DNA extracted from leaves infected with a given MLO was used to PCR-amplify the 16S rDNA of the MLO, using universal primers for 16S rDNA of Eubacteria. The MLO 16S rDNA could be cloned and sequenced. Such work was carried out for several MLOs (for references see [26]). Sequence comparisons showed the MLOs to be phylogenetically close to the *Acholeplasma/Anaeroplasma* group (Fig. 2); (iv) The evolutionary relationship with the acholeplasmas was confirmed by the fact that MLOs use UGA as a stop codon, not as a tryptophan codon [27]. This suggests that the MLOs, like the acholeplasmas, are phylogenetically "early" mollicutes, as opposed to the "later" spiroplasmas.

Specific primers for PCR amplification of phytoplasma 16S rDNA and 16S/23S spacer region have been designed. The amplified rDNA of a given phytoplasma can be sequenced and used for phylogenetic placement of the phytoplasmas within the phytoplasma tree; it can also serve for restriction fragment length polymorphism (RFLP) analyses for additional phylogenetic data.

The Subcommittee on the Taxonomy of Mollicutes recognized, in 1994, that the MLOs are members of the Class Mollicutes and adopted the trivial name phytoplasma to replace MLO [28]. The subcommittee, in collaboration with the phytoplasma working team of the International Research Program on Comparative Mycoplasmology (IRPCM), has also recommended that the *Candidatus* designation [15], be used for the major phylogenetic groups (subclades) of the phytoplasmas, each group representing a distinct *Candidatus* phytoplasma species. Fourteen groups have been derived from 16S rDNA sequence analysis (Table 5). The phytoplasma associated with witches' broom disease of lime (WBDL, group 14) is the first phytoplasma to have been described as a *Candidatus* species: *Candidatus* Phytoplasma aurantifolia. Its description is based on 16S rDNA sequence, 16S/23S spacer region sequence, genome size, Southern hybridization profiles obtained with

Table 5. The 14 Phytoplasma groups (*Candidatus* species)

1.	Aster Yellows group
2.	Apple Proliferation group
3.	X-disease group
4.	Rice yellow dwarf group
5.	Flavescence dorée group
6.	Coconut Lethal Yellowing group
7.	Stolbur group
8.	Pigeon Pea group
9.	Ash Yellows group
10.	Clover Proliferation group
11.	Loofah Witches' broom group
12.	Lethal decline of coconut group
13.	Peanut Witches' broom group
14.	Lime Witches' broom[a] group

[a]First published *Candidatus* species: *Candidatus* Phytoplasma aurantifolia [29].

WBDL-phytoplasma specific probes, and genomic similarities with other phytoplasma groups [29].

Regarding this symposium, it might be interesting to mention the behavior of phytoplasmas in plant tissue cultures. Phytoplasmas associated with apple proliferation (AP) disease of apple trees could be maintained in their micropropagated natural host plant, *Malus pumila*, since 1985 [30]. Different isolates of this pathogen could thus be studied *in vitro*. Amplification of a pathogen-specific DNA fragment by PCR confirmed the presence of AP phytoplasmas in the diseased plants even after 10 years of *in vitro* propagation. RFLP analysis of the amplified chromosomal DNA fragments revealed no genetic difference between the AP phytoplasma isolates. Growth parameters, symptom expression and phytoplasma concentration were examined to compare the *in vitro* behaviour of four different AP phytoplasma isolates and to compare different subculture conditions. A comparison of these data obtained after 2 or 8 years of micropropagation revealed no essential differences. Eight years after culture initiation, diseased shoots still exhibited typical symptoms such as witches' broom, small leaves with large stipules and stunted growth. However, when the AP phytoplasma was maintained for 8 years on a micropropagated non-natural host, *Pyronia veitchii*, no symptoms and no significant differences could be observed between healthy and infected *P. veitchii* plants, even though phytoplasmas were present in all diseased plantlets tested [31]. *In vitro* micropropagation of phytoplasmas infecting poplar, chrysanthemum, *Gladiolus, Hydrangea, Rubus,* periwinkle, eggplant and *Prunus marianna* have also been reported [32–35].

Leaf-tip cultures of the evening primrose (*Oenothera hookeri*) have been obtained on media for leaf-tip propagation, from surface sterilized viviparous plantlets (embryos) taken on field-infected plants [36]. The leaf-tip cultures were maintained by subculturing every 3 weeks. The cultures from the infected plants (aster yellows phytoplasma) were slightly chlorotic, with narrow strap-like leaves, and had a more frequent initiation and proliferation of lateral shoots. Unlike the control leaf-tip cultures, they often appeared spindly, sending out thin, stem-like shoots, with occasional die-back at the tips, even on fresh media. Electron microscopy revealed abundant phytoplasmas in the sieve-tube elements. Analogous to the application of tetracycline in field conditions, remission of symptoms of the *Oenothera* leaf-tip cultures could be accomplished readily by adding low levels of tetracycline (12.5 μg/ml) to the plant medium. However, if plants were removed after only short exposures to the antibiotic, symptoms returned at a high frequency. However, effective curing required several months exposure to tetracycline, during vigorous growth of the plant cultures. Erythromycin and streptomycin also accomplished some curing, but they were not as efficient as tetracycline. A constant heat treatment (32–34°C) under continuous light for 4 months was totally ineffective.

6. CONCLUSION: TOWARDS UNDERSTANDING INTERACTIONS BETWEEN *SPIROPLASMA CITRI*, THE LEAFHOPPER VECTOR AND THE HOST PLANT

This review has focused on identification, characterization and phylogeny of walled and wall-less bacterial agents associated with plant disease and has tried to show the important developments that have occurred in the last 10 years. The evolutionary relationships between the *Mollicutes* and the Gram-positive eubacteria with low (G + C) have been confirmed. New genera of mollicutes (*Entomoplasma, Mesoplasma*) have been created to accommodate organisms that were improperly classified as *Mycoplasma* or *Acholeplasma* species. The former sieve-tube-restricted, non-cultured Mycoplasma-Like Organisms (MLOs) have been shown to be phylogenetically related to the acholeplasmas and to be, indeed, genuine mollicutes, now called phytoplasmas. Similarly, the sieve-tube-restricted, non-cultured Bacteria-Like Organisms (BLOs) could also be characterized and shown to belong to the α and γ subdivisions of the Proteobacteria. These developments have been summarized in Table 2. As seen in this review, the work accomplished was essentially devoted to the study of the bacterial agents themselves. Little has been done so far to understand the interactions between the agents (and especially the phytopathogenic agents) and their plant and insect hosts. The following work on *Spiroplasma citri* is presented to show that the time has come for such studies.

S. citri is a plant pathogen. A convenient experimental host plant is periwinkle (*Catharantus roseus*). In nature, infection of a plant can only be achieved by insect vectors. The leafhopper *Circulifer haematoceps* is the major vector in the Mediterranean countries and Western Asia. Thus, *S. citri* has two hosts in which it multiplies: the leafhopper and the plant. We have been interested in the genes involved in the interactions between the spiroplasma and its two hosts. Classically, such genes can be identified by mutations and adequate screening procedures to detect the mutants. We have now developed a technique for *S. citri* mutagenesis by random insertion of transposon Tn 4001 into the *S. citri* genome. This technique is the successful outcome of intensive studies devoted to the construction of gene vectors for *S. citri* [37]. The first vector used was the replicative form (RF) of *S. citri* virus SpV1, an Inoviridae such as *E. coli* phage M13. However, the RF vector turned out to be unstable, the DNA insert being quickly deleted. This phenomenon has led to the demonstration that homologous recombination (HR) was involved in deletion formation, even though the recA protein, normally required for HR, was deficient in all five *S. citri* strains tested. A second approach was to use the origin of *S. citri* DNA replication (*oriC*) to construct a number of artificial plasmids, with or without the colE1 replication origin functioning in *E. coli*, and containing various antibiotic resistance determinants (*tet* M, *cat*, *aacA-aph*D). These plasmids have been successfully used as cloning vectors. Those with the colE1 sequences function as shuttle vectors between *E. coli* and *S. citri*. Some behave as extrachromosomal plasmids, others integrate into the spiroplasmal genome at *oriC*. With these plasmids, the spiralin of *S. phoeniceum* could be introduced and expressed at high levels in *S. citri*. They have also been important to show that only some *S. citri* strains can easily be transformed. *S. citri* strain GII3 was chosen for Tn 4001 mutagenesis precisely because it can be readily transformed and also because it is efficiently transmitted by the leafhopper *C. haematoceps* to periwinkle plants.

Over 1000 Tn 4001 insertion mutants of *S. citri* have now been obtained. Mutant 553 grows well in the insect, is transmitted to the periwinkle plant, and reaches high titers in the plant, but it does not induce symptoms as long as there is no reversion to the wild-type spiroplasma by loss of the transposon. Mutant 470 does not multiply in the leafhopper and is not transmitted to the plant. A third mutant has lost motility. The mutant genes in which the transposon is inserted have been identified. In the non-phytopathogenic mutant 553, the affected gene is within the fructose operon. Fructose cannot be transferred into, and metabolized by, the spiroplasmal cells. How absence of fructose utilization results in absence of symptoms remains to be understood.

It is hoped that these studies not only contribute to our understanding of host–parasite interactions but will also offer new approaches for the control of plant diseases.

REFERENCES

1. Bové, J.M. (1988) Acta Hortic. 225, 215–222.

2. Woese, C.R. (1987) Microbiol. Rev. 51 (2), 221–271.

3. Weisburg, W.G., Tully, J.G., Rose, D.L., Petzel, J.P., Oyaizu, H., Yang, D., Mandelco, L., Sechrest, J., Lawrence, T.G., Van Etten, J., Maniloff, J. and Woese, C.R. (1989) J. Bacteriol. 171 (12), 6455–6467.

4. Citti, C., Maréchal-Drouard, L., Saillard, C., Weil, J.H. and Bové, J.M. (1992) J. Bacteriol. 174 (20), 6471–6478.

5. Sears, B.B. and Kirkpatrick, B.C. (1994) Am. Soc. Microb. News 60 (6), 307–312.

6. Rose, D.L., Tully, J.G., Bové, J.M. and Whitcomb, R.F. (1993) Int. J. Syst. Bacteriol. 43 (3), 527–532.

7. Bové, J.M., Carle, P., Tully, J.G., Whitcomb, R.F. and Laigret, F. (1994) IOM Lett. 3, 449–450.

8. Razin, S. and Tully, J.G. (eds.) (1995) Molecular and Diagnostic Procedures in Mycoplasmology, Academic Press, Vol. I.

9. Razin, S. and Tully, J. G. (eds.) (1996) Molecular and Diagnostic Procedures in Mycoplasmology, Academic Press, Vol. II.

10. Tully, J.G. (1996) in Molecular and Diagnostic Procedures in Mycoplasmology, Vol. II, pp. 1–21, Academic Press.

11. Laflèche, D. and Bové, J.M. (1970) C. R. Acad. Sci. Paris 270, 1915–1917.

12. Garnier, M. and Bové, J.M. (1983) Phytopathology 73 (10), 1358–1363

13. Weisburg, W.G., Barns, S.M., Pelletier, D.A. and Lane, D.J. (1991) J. Bacteriol. 173 (2), 697–703.

14. Jagoueix, S., Bové, J.M. and Garnier, M. (1994) Int. J. Syst. Bacteriol. 44 (3), 397–386.

15. Murray, R.G.E. and Schleifer, K.H. (1994) Int. J. Syst. Bacteriol. 44(1), 174–176.

16. Jagoueix, S., Bové, J.M. and Garnier, M. (1996) Mol. Cell. Probes, 10, 43–50.

17. Villechanoux, S., Garnier, M., Renaudin, J. and Bové, J.M. (1992) Curr. Microbiol., 24, 89–95.

18. Villechanoux, S., Garnier, M., Laigret, F., Renaudin, J. and Bové, J.M. (1993) Curr. Microbiol. 26 (3), 161–166.

19. Planet, P., Jagoueix, S., Bové, J.M. and Garnier, M. (1995) Curr. Microbiol. 30 (3), 137–141.

20. Nourrisseau, J.G., Lansac, M. and Garnier, M. (1993) Plant Dis. 77 (10), 1055–1059.

21. Garnier, M., Zreik, L. and Bové, J.M. (1996) unpublished.

22. Saglio, P., Lhospital, M., Laflèche, D., Dupont, G., Bové, J.M., Tully, J.G. and Freundt, E.A. (1973) Int. J. Syst. Bacteriol. 23 (3), 191–204.

23. Whitcomb, R.F., Chen, T.A., Williamson, D.L., Liao, C., Tully, J.G., Bové, J.M., Mouches, C., Rose, D.L., Coan, M.E. and Clark, T.B. (1986) Int. J. Syst. Bacteriol. 36 (2), 170–178.

24. Saillard, C., Vignault, J.C., Bové, J.M., Tully, J.G., Williamson, D.L., Fos, A., Garnier, M., Gadeau, A., Carle, P. and Whitcomb, R.F. (1986) Int. J. Syst. Bacteriol. 37 (2), 106–115.

25. Doi, Y., Teranaka, M., Yora, K. and Asuyama, H. (1967) Ann. Phytopathol. Soc. Jpn. 33, 259–266.

26. Schneider, B., Seemueller, E., Smart, C.D. and Kirkpatrick, B.C. (1995) in Molecular and Diagnostic Procedures in Mycoplasmology (Razin S. and Tully J.G., eds) pp. 369–380, Vol. 1, Academic Press.

27. Toth, K.F., Harrison, N. and Sears, B.B. (1994) Int. J. Syst. Bacteriol. 44, 119–124.

28. Tully, J.G. (1995) Int. J. System. Bacteriol. 45 (2), 415–417.

29. Zreik, L., Carle, P., Bové, J.M. and Garnier, M. (1995) Int. J. Syst. Bacteriol. 45 (3), 449–453.

30. Jarausch, W., Lansac, M. and Dosba, F. (1996) Plant Pathol. 45, 778–786.

31. Lansac, M., Jarausch, W. and Dosba, F. (1995) Adv. Hortic. Sci. 9, 140–143.

32. Cousin, M.T., Roux, J., Millet, N. and Michel, M.F. (1990) J. Phytopathol. 130, 17–23.

33. Bertacci, S., Davies, R. E. and Lee, I. M. (1992). Hortic. Sci. 27, 1041–1043.

34. Raj Bhansali, R. and Ramawat, K.G. (1993) J. Hortic. Sci. 68, 25–30.

35. Jarausch, W., Lansac, M. and Dosba, F. (1994) Acta Hortic. 359, 169–176.

36. Sears, B.B. and Klomparens, K.L. (1989) Can. J. Plant Pathol. 11, 343–348.

37. Renaudin, J. and Bové, J.M. (1995) in Molecular and Diagnostic Procedures in Mycoplasmology (Razin S. and Tully J.G., eds), Vol. I, chapter B6.

38. Tully, J.G., Bové, J.M., Laigret, F. and Whitcomb, R.F. (1993) Int. J. Syst. Bacteriol. 43 (2), 378–385.

39. Razin, S. (1995) in Molecular and Diagnostic Procedures in Mycoplasmology (Razin S. and Tully J.G., eds) pp. 1–25, Vol. 1, Academic Press.

40. Bové, J.M. (1993) Clin. Infect. Dis. 17 (Suppl. 1), S10–S31.

MODERN METHODS FOR IDENTIFYING BACTERIA

DAVID E. STEAD, JUDY HENNESSY and JUDITH WILSON
Central Science Laboratory, Sand Hutton, York, YO4 1LZ, UK

1. INTRODUCTION

Bacteria found as contaminants of plant tissue cultures are diverse and belong to a range of ecological groups. They include plant pathogens, epiphytes, endophytes and accidental contaminants, e.g. from air or from humans during handling. Bacteria from all these broad ecological niches can cause problems in plant tissue cultures. However, perhaps paradoxically, plant endophytes and pathogens may often cause less obvious symptoms than saprophytic and other contaminant bacteria not normally considered pathogens of plants. The effects of antibiotics and nutrients in culture media also influence bacterial growth. Thus, pathogenicity in plant tissue cultures is a complex issue. Identification of the causal agent is often important since this may determine the appropriate control measures.

Consulting a standard text on diagnosis of plant pathogenic bacteria may be insufficient since a much wider range of bacteria may be involved. Methods are required that give as much taxonomic information as possible. In recent years there has been a move away from the traditional nutritional and physiological methods to more cost-effective methods. However, many of the more modern methods are designed to compare two bacteria and ask the question — does this bacterium belong to species X or not?, rather than — to which species/genus/family does this bacterium belong?

This paper reviews some of the modern methods available with particular reference to contaminants of plant tissue cultures and takes into account accuracy, speed and cost. It also reviews methods for which no data are presented.

2. MATERIALS AND METHODS

2.1. Bacterial cultures

All strains were either from the National Collection of Plant Pathogenic Bacteria (NCPPB) housed at CSL or from the CSL diagnostic clinic, to which bacterial contaminants of plant

A.C. Cassells (ed.), Pathogen and Microbial Contamination Management in Micropropagation, 61–73.
© *1997 Kluwer Academic Publishers. Printed in the Netherlands.*

tissue cultures are regularly submitted for identification. All strains were cultured on appropriate media [1,2] and were maintained either by freeze drying or by storage at -80°C [3].

2.2. Identification by fatty acid profiling

All strains were cultured on trypticase soy agar at 28°C for either 24 or 48 h. Fatty acid profiles were prepared according to standard protocols [2] using the Microbial Identification System based on software available from MIDI (Newark, DE, USA). Fatty acid profiles were compared with libraries available commercially (MIDI) or self-generated using MIDI library generation software.

2.3. Identification by protein profiles

All strains were cultured on nutrient agar at 28°C for either 24 or 48 h. Protein profiles were prepared largely according to the methods outlined by Stead [4] with occasional small unpublished modifications. In brief, after appropriate extraction and purification, proteins were separated by polyacrylamide gel electrophoresis at 6°C on 1 mm thick gels for 3 h at 20 mA per gel. Gels were stained with Coomassie blue before drying overnight on a simple air drying frame. Gels were scanned either by laser densitometry or a flat bed scanner and the gel images normalised according to calibration standards using Gelcompar software (Applied Maths, Kortricht, Belgium). Normalisation accounted for inter- and intra-gel differences in electrophoretic rates. Profiles were compared using a range of methods available in the Gelcompar software package, primarily by unweighted pair group matching analysis and principal component analysis.

2.4. Genetic fingerprints

Several PCR-based methods were used. Randomly amplified polymorphic DNA assays (RAPD-PCR) were done using the primers and methods of Maki-Valkama and Karjalainen [5]. All results presented are for *Erwinia chrysanthemi*.

Repetitive sequence PCR (rep-PCR) assays were based on the use of primers to REP, ERIC and BOX repetitive DNA sequences according to the methods of Louws *et al.* [6] with occasional as yet unpublished modifications depending on the primer set and genus of bacterium used.

PCR products were separated by agarose gel electrophoresis and DNA bands stained by ethidium bromide before visualising and photographing under ultraviolet light. Profiles were scanned on a flat bed scanner and the gel images normalised and compared as for proteins.

3. RESULTS

3.1. Fatty acid profiles

Fatty acid profiles contained much useful taxonomic information. Tables 1 and 2 list the key features of several genera commonly found as contaminants of plant tissue cultures. In general, Gram-negative genera have unique patterns of hydroxy fatty acids and rarely have large quantities of branched fatty acids. Gram-positive genera rarely have hydroxy fatty acids but most have large quantities of branched fatty acids.

It is less easy to differentiate Gram-positive genera according to presence or absence of specific fatty acids than for Gram-negative genera but Table 1 shows significant differences between the profiles of *Bacillus, Rhodococcus, Staphylococcus* and *Streptococcus*.

Table 2 lists the key acids of taxonomic value in the Gram-negative bacteria. Most of these are hydroxy fatty acids which are excellent chemotaxonomic markers often at the genus level although this is not the case for example in the Enterobacteriaceae. Most fatty acids in the Gram-negative bacteria have even numbers of carbon atoms. Branched acids with odd numbers of carbon atoms are common in Gram-positive bacteria but are uncommon in Gram-negative bacteria. Cyclopropane acids are almost exclusively found in the Gram-negative bacteria. Unsaturated acids are found in most bacteria and most of these are mono-unsaturated, *cis*-forms although tri-saturated and *trans*-isomers are occasionally found.

In addition to providing useful taxonomic markers, fatty acid profiles of strains under test were compared with libraries of profiles. The comparison took into account both types and amounts of fatty acids in the profile using covariance matrix, principal component analysis and pattern recognition components of the MIDI software. Table 3 shows the accuracy of identification primarily of plant pathogenic bacteria based on use of a self-generated library representing most strains in the NCPPB.

Accuracy of identification was generally 100% at genus and species levels. In those cases where it was less than 100%, it is likely that some strains had been wrongly classified on accession or were from taxa with as yet unresolved taxonomic problems such as in the *Pantoea–Erwinia* complex. At subspecies and biovar levels, accuracy of identification varied from 84 to 100% and 87 to 100%, respectively. At pathovar level, accuracy of identification varied widely, but was generally higher for *Xanthomonas campestris* pathovars than for *Pseudomonas syringae* pathovars (results unpublished). Within *P. syringae*, accuracy of identification was as low as 12% for pv. *syringae* itself but this reflects the fact that this pathovar contains strains from many hosts which have not been classified in other pathovars because no pathovar has been proved or described for a particular host.

Table 1. Selected fatty acids in Gram-positive bacteria

Fatty acid	Genus									
	1	2	3	4	5	6	7	8	9	10
14:0	+	+	+	(+)	(+)	+	(+)	+	+	+
15:0	(+)	+	(+)	(+)	-	+	-	(+)	-	-
14:0 ISO	+	+	+	(+)	-	-	+	-	-	-
15:0 ISO	+	+	+	+	(+)	-	+	-	-	-
15:1 ANTEISO A	-	-	-	-	+	-	-	-	-	-
15:0 ANTEISO	+	+	+	+	+	-	+	-	-	-
17:0 ISO	+	+	+	(+)	(+)	-	+	-	-	-
17:0 ANTEISO	+	+	+	+	+	-	+	-	-	-
17:1 ω9c	-	(+)	+	-	-	-	-	-	-	-
16:1 ω7c	+	(+)	(+)	(+)	-	-	+	-	+	+
18:1 ω9c	+	(+)	-	-	-	-	+	-	+	-
Hydroxy acids	(+)	-	-	-	-	-	-	(+)	-	(+)
10 methyl 18:0	-	-	-	-	-	+	-	-	-	-

-: normally absent in strains from most species of the genus; +: present in most strains within species of the genus; (+): present in most strains within some species of the genus. 1, *Bacillus*; 2, *Arthrobacter*; 3, *Micrococcus*; 4, *Curtobacterium*; 5, *Clavibacter*; 6, *Rhodococcus*; 7, *Staphylococcus*; 8, *Streptococcus*; 9, *Enterococcus*; 10, *Lactobacillus*.

3.2. Protein profiling

The major problem with SDS–PAGE protein profiling was found to be lack of reproducibility. A minimum of three calibration standards per gel was required to allow inter- and intra-gel normalisation. Since the actual protein bands are not identified, proteins cannot be used as chemotaxonomic markers as effectively as with fatty acids. There were clear differences between profiles of unrelated species, but although clear differences between pathovars were sometimes found, they were often too small for reliable identification at the pathovar level. Fig. 1 shows normalised reconstructed gel images for several plant pathogenic bacteria taken from several different gels following stepwise

Table 2. Key fatty acids in genera of Gram-negative bacteria

Fatty Acid	Genus																		
	1	2	3	4	5	6	7	8	9	10	11	12	13	14	15	16	17	18	19
10:0 3OH	+	+	+	-	-	-	(+)	-	-	-	-	-	-	+	-	+	-	-	-
12:0 2OH	-	-	(+)	-	-	-	(+)	-	-	-	-	-	(+)	+	-	+	-	-	-
12:0 3OH	-	-	+	+	-	-	(+)	-	-	-	-	-	+	-	-	+	-	-	-
14:0 3OH	-	-	(+)	-	+	+	+	-	-	(+)	(+)	+	(+)	-	+	-	+	+	+
14:0 2OH	-	-	(+)	-	(+)	(+)	(+)	(+)	+	-	+	-	-	-	-	-	-	-	-
16:0 2OH	-	-	-	-	-	-	-	(+)	-	-	+	+	-	-	-	-	-	+	-
16:0 3OH	-	(+)	(+)	-	-	-	-	+	+	+	+	+	-	-	(+)	-	+	(+)	+
16:1 2OH	-	(+)	-	-	-	-	-	(+)	-	(+)	-	-	-	-	-	-	-	(+)	-
18:1 2OH	-	-	-	-	-	-	-	-	-	-	-	+	-	-	(+)	-	-	+	+
11:0 ISO 3OH	-	-	-	+	-	-	-	-	-	-	-	-	-	-	-	-	-	-	-
13:0 ISO 3OH	-	-	-	+	-	-	-	-	-	-	-	-	-	-	-	-	-	-	-
15:0 ISO 3OH	-	-	-	-	-	-	-	+	+	+	-	-	-	-	-	-	-	-	-
17:0 ISO 3OH	-	-	-	-	-	-	-	+	+	+	-	-	-	-	-	-	-	-	-
12:0	+	+	+	-	+	(+)	+	-	-	-	-	-	-	+	+	-	+	-	-
14:0	+	+	(+)	+	+	+	+	(+)	+	+	+	(+)	(+)	-	-	(+)	-	+	-
15:0	(+)	+	(+)	+	(+)	(+)	(+)	(+)	-	+	-	-	(+)	-	-	-	-	-	-
16:0	-	-	-	-	-	-	-	-	-	-	-	-	-	-	-	-	-	-	-
17:0 CYCLO	-	(+)	(+)	(+)	+	(+)	+	-	-	-	-	-	-	+	-	-	-	+	-
19:0 CYCLO	-	-	(+)	-	(+)	(+)	+	-	-	-	+	-	-	-	(+)	-	+	+	+
15:0 ISO	-	-	-	+	-	-	-	+	+	+	-	-	-	-	-	-	-	-	-
15:0 ANTEISO	-	-	-	+	-	-	-	(+)	(+)	+	-	-	-	-	-	-	-	-	-
18:1 ω7c	+	+	+	(+)	+	+	+	-	-	-	+	+	(+)	+	+	+	+	+	+
18:1 ω9c	-	-	-	(+)	-	-	-	-	-	-	-	-	-	+	-	-	-	-	-
20:3 ω6,9,12c	-	-	-	-	-	-	-	-	-	-	-	-	-	-	(+)	-	-	-	-

1, *Acidovorax;* 2, *Comamonas;* 3, *Pseudomonas;* 4, *Xanthomonas;* 5, *Enterobacter;* 6, *Erwinia;* 7, *Serratia;* 8, *Flavobacterium;* 9, *Sphingobacterium;* 10, *Cytophaga;* 11, *Acetobacter;* 12, *Gluconobacter;* 13, *Acinetobacter;* 14, *Janthinobacterium;* 15, *Methylobacterium;* 16, *Flavimonas;* 17, *Agrobacterium;* 18, *Burkholderia;* 19, *Phyllobacterium.*

comparison to these calibration standards on each gel and finally to a unique reference calibration standard. As can be seen, the normalisation process was not perfect. The natural variation between runs of the same strain was similar to that between type strains of closely related pathogens (data not shown).

However, protein profiling is an excellent taxonomic tool for classification of bacteria. This work has been pioneered by Vauterin and his colleagues for plant pathogenic bacteria and differences in profiles have been used extensively in providing phenotypic information to

Table 3. Accuracy of identification of plant-associated bacteria at different levels based on fatty acid profiling

Taxon	Rank	% strains listed as 1st choice	No. strains tested
Agrobacterium	Genus	100	20
Agrobacterium biovar 1	Biovar	87	15
Burkholderia	Genus	100	50
B. cepacia	Species	100	14
Ralstonia	Genus	99	155
R. solanacearum	Species	98	150
Acidovorax	Genus	100	48
A. avenae subsp. *avenae*	Subspecies	98	35
Comamonas	Genus	100	10
C. testosteroni	Species	100	5
Erwinia	Genus	99	307
E. amylovora	Species	97	76
E. herbicola	Species	82	17
E. carotovora subsp. *carotovora*	Subspecies	98	57
E. chrysanthemi	Species	100	21
E. chrysanthemi bv. 3	Biovar	100	7
Pantoea	Genus	100	43
P. agglomerans	Species	100	6
Pseudomonas	Genus	100	185
P. syringae	Species	100	175
P. syringae pv. *syringae*	Pathovar	12	16
P. syringae pv. *berberidis*	Pathovar	100	8
P. aeruginosa	Species	100	11
Stenatrophomonas maltophilia	Species	100	12
Xanthomonas	Genus	100	48
X. campestris pv. *campestris*	Pathovar	97	33
X. hyacinthi	Species	100	15
Bacillus	Genus	100	21
B. pumilus	Species	100	6
Clavibacter michiganensis	Species	100	41
C. michiganensis subsp. *michiganensis*	Subspecies	84	31
Curtobacterium	Genus	95	48
C. flaccumfaciens	Species	95	38
C. flaccumfaciens pv. *flaccumfaciens*	Pathovar	90	21
Rhodococcus fascians	Species	100	24

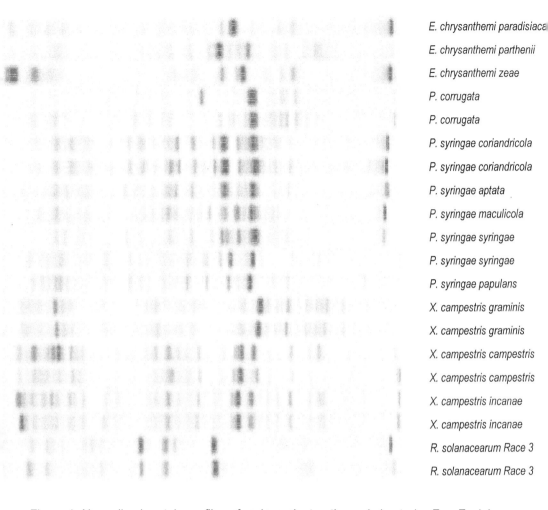

Figure 1. Normalised protein profiles of various plant pathogenic bacteria. *E. = Erwinia*; *P. = Pseudomonas*; *X. = Xanthomonas*; *R. = Ralstonia*.

support DNA/DNA homology results used as the basis for reclassification of *Xanthomonas* into some 20 species [7]. The use of SDS–PAGE protein profiling for general identification is of value down to species and possibly subspecies level, but because it is more costly, time-consuming and slower than fatty acid profiling, we have not developed it as a routine identification technique.

3.3. Nutritional kits

Results are not presented here but kits are discussed. The kits contain a battery of dehydrated reagents. Addition of a standardised aqueous inoculum of the test bacterium initiates the reaction. This is expressed either by growth seen as visual turbidity or biological activity visualised by an indicator. Nutritional kits are based on the traditional methods for identification and often have the same limitations. For example, utilisation of a particular substrate may not be a highly conserved feature within strains of a given taxon. Closely related bacteria may have similar utilisation patterns. A particular pattern of utilisation usually gives little useful taxonomic information except for comparison with other profiles. They can give fairly accurate identification down to species level but usually fail at pathovar level. However, despite their somewhat limited use in identification, they are of great value in classification, offering a wide range of useful phenotypic characters.

3.4. Genetic fingerprints

There are now a wide range of methods for genetic fingerprinting. They fall into four groups:

(i) Restriction methods. These rely on harvesting fairly large quantities of DNA, purifying it, cleaving it into a large number of fragments with restriction endonuclease enzymes derived from bacteria e.g. *Eco*R1, separating the fragments by gel electrophoresis before staining and comparing profiles (fingerprints). It is often difficult to obtain clear band separation and this can be overcome either by using a rare-cutting restriction endonuclease to reduce the number of fragments or by blotting the profile onto a membrane and hybridising with a labelled probe such as a pathogenicity gene or rRNA. If using a rare-cutting enzyme, it is necessary to separate fragments by pulsed-field gel electrophoresis because of the presence of extremely large and small fragments in the sample.

These techniques are often referred to as restriction fragment length polymorphisms (RFLP) [8–14].

(ii) Polymerase chain reaction (PCR)-derived genetic fingerprints. These rely on the use of oligonucleotide primers annealing at intervals along the DNA molecule and the use of PCR to amplify the sequences between annealed primers. Primers can be very short oligonucleotides, about 10 base pairs long which under non-stringent, low temperature conditions (e.g. 42°C), anneal wherever there is sufficient homology. A range of such random primers is commercially available. The PCR products are separated by gel electrophoresis and stained, for example, with ethidium bromide to obtain a fingerprint. This technique is often referred to as randomly amplified polymorphic DNA assay (RAPD or RADP–PCR). However, fingerprints are not always reproducible because of the small changes which affect the annealing process [4,15].

In recent years, several families of repetitive sequences have been found in bacterial DNA, some of which are common to all bacteria. These sequences are believed to have a function in cell division. Three commonly used sequences are the repetitive extragenic palindromic (REP), enterobacterial repetitive intergenic consensus (ERIC) and BOX elements. Primers flanking these regions are commercially available. Since these primers have high homology with bacterial DNA, more stringent annealing temperatures can be used and the fingerprints obtained tend to be more reproducible.

Collectively these repetitive sequence PCRs are termed rep-PCR which should not be confused with one of them (REP–PCR) [5,16,17].

(iii) Amplified Fragment Length Polymorphism (AFLP). This method combines elements of (i) and (ii) and relies on the use of one or two restriction endonucleases to cleave the DNA into fragments. Ligation of one or more unique DNA fragments to the cut ends and subsequent use of primers to the unique ligated fragments allows amplification of the restriction fragments. This technique is a relatively new development [18].

(iv) PCR–RFLP. Conserved DNA regions amplified by PCR are restricted using restriction endonucleases and the products separated by gel electrophoresis. This technique is often used to confirm that a PCR product has the same sequence as expected, but apart from this use in confirming detection it can be used to classify closely related bacteria and identify at the pathovar level, for example, within *Erwinia chrysanthemi* [19].

3.4.1. Randomly Amplified Polymorphic DNA (RAPD–PCR) fingerprints

Figure 2 shows great variation in the RAPD–PCR fingerprints of strains of *Erwinia chrysanthemi*. This variation is linked to host and to geographic distribution rather than to the biovar determination based on nutrition. Almost all European potato strains had banding patterns similar to that of *Dianthus* strains. All *Dieffenbachia* strains had a unique major band. By contrast, three maize strains all belonging to biovar 3 had different fingerprints. All the profiles shown in Fig. 2 were carried out at the same time. Unfortunately, repeat assays of the same strains at daily intervals gave different banding patterns (data not shown). This lack of reproducibility for RAPD–PCR is due to several factors, e.g. annealing temperature, and limits the use of RAPD–PCR in identification especially where comparison of a fingerprint with a library of fingerprints is required.

3.4.2. Repetitive Extragenic Palindromic PCR (REP–PCR)

REP–PCR profiles consistently allowed differentiation of strains at species and often at infraspecific levels such as pathovars. Figure 3 shows the REP–PCR profiles for strains representing three pathovars of *Pseudomonas syringae*. All pv. *syringae* strains from lilac had very similar profiles but the strains from pear fell into two distinct clusters. Most pv. *lachrymans* strains fell into a single cluster although the type strain, known to be atypical, had a clearly different profile. Five pv. *morsprunorum* strains from three different hosts had very similar profiles.

These results, if substantiated, could have significance for the future of the current pathovar system within *P. syringae*. They also indicate that rapid, cheap, accurate identification at an infraspecific level, certainly comparable to the pathovar level, may be possible. Early indications are that these techniques are reasonably reproducible but that differences in

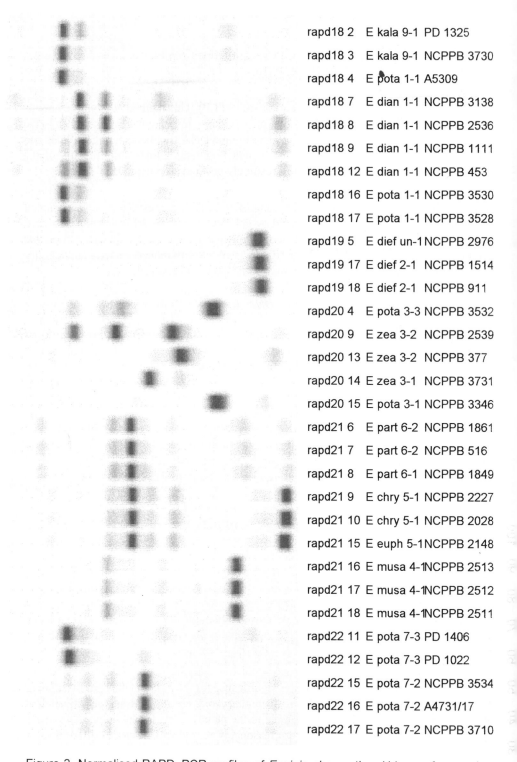

rapd18 2	E kala 9-1	PD 1325
rapd18 3	E kala 9-1	NCPPB 3730
rapd18 4	E pota 1-1	A5309
rapd18 7	E dian 1-1	NCPPB 3138
rapd18 8	E dian 1-1	NCPPB 2536
rapd18 9	E dian 1-1	NCPPB 1111
rapd18 12	E dian 1-1	NCPPB 453
rapd18 16	E pota 1-1	NCPPB 3530
rapd18 17	E pota 1-1	NCPPB 3528
rapd19 5	E dief un-1	NCPPB 2976
rapd19 17	E dief 2-1	NCPPB 1514
rapd19 18	E dief 2-1	NCPPB 911
rapd20 4	E pota 3-3	NCPPB 3532
rapd20 9	E zea 3-2	NCPPB 2539
rapd20 13	E zea 3-2	NCPPB 377
rapd20 14	E zea 3-1	NCPPB 3731
rapd20 15	E pota 3-1	NCPPB 3346
rapd21 6	E part 6-2	NCPPB 1861
rapd21 7	E part 6-2	NCPPB 516
rapd21 8	E part 6-1	NCPPB 1849
rapd21 9	E chry 5-1	NCPPB 2227
rapd21 10	E chry 5-1	NCPPB 2028
rapd21 15	E euph 5-1	NCPPB 2148
rapd21 16	E musa 4-1	NCPPB 2513
rapd21 17	E musa 4-1	NCPPB 2512
rapd21 18	E musa 4-1	NCPPB 2511
rapd22 11	E pota 7-3	PD 1406
rapd22 12	E pota 7-3	PD 1022
rapd22 15	E pota 7-2	NCPPB 3534
rapd22 16	E pota 7-2	A4731/17
rapd22 17	E pota 7-2	NCPPB 3710

Figure 2. Normalised RAPD–PCR profiles of *Erwinia chrysanthemi* biovars from various hosts.

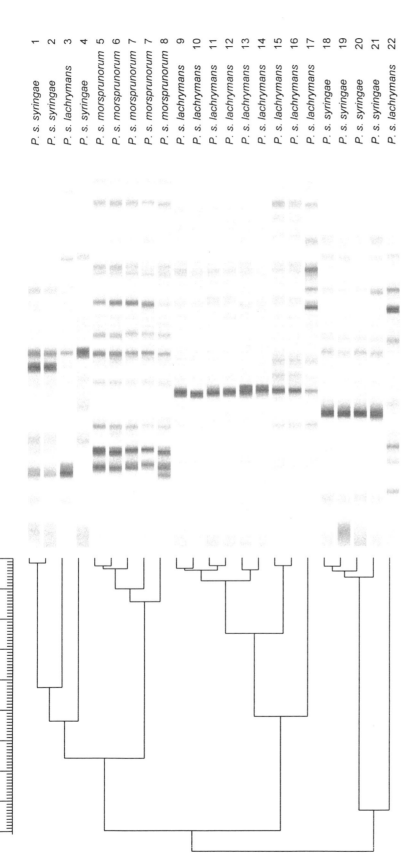

71

Figure 3. Normalised REP-PCR profiles of *Pseudomonas syringae* pathovars. 1, NCPPB 1077 pear; 2, NCPPB 1072 pear; 3, NCPPB 3544 cucumber; 4, NCPPB 1085 pear; 5, NCPPB 1095 plum; 6, NCPPB 1459; 7, NCPPB 2787 cherry; 8, NCPPB 2427 apricot; 9, NCPPB 542 cucumber; 10, NCPPB 541 cucumber; 11, NCPPB 2754 cucumber; 12, NCPPB 1097 cucumber; 13, NCPPB 467 cucumber; 14, NCPPB 277 cucumber; 15, NCPPB 1096 cucumber; 16, NCPPB 540 cucumber; 17, NCPPB 2916 melon; 18, NCPPB 1070 lilac; 19, NCPPB 1205 lilac; 20, NCPPB 524 lilac; 21, NCPPB 93 lilac; 22, NCPPB 537 cucumber.

minor bands may be accounted for by problems with reproducibility of annealing and gel normalisation.

Similar classifications with these and many other bacteria (unpublished results) have been obtained with ERIC–PCR and BOX–PCR. The optimum method probably varies with the genus.

4. DISCUSSION

Most of the techniques allowed rapid, fairly accurate identification or differentiation of bacteria. Protein profiling is perhaps the least cost-effective method. It takes several days and has a relatively large labour cost. Fatty acid profiling gave results in less than 60 h and usually in approximately 30 h from receiving a pure culture. Although the costs of capital equipment and software are high, it is a cheap method to run because labour costs are low (approximately 15 min per culture). Even in the absence of a good match, fatty acid profiling gave much useful taxonomic information, unlike other profiling methods, most of which are based on bands in gels.

Nutritional profiling is cheap to set up but fairly expensive to operate. Also nutritional profiles may vary within strains of a given taxon.

Genetic fingerprints all have their uses but the REP–PCR methods appear to be more reproducible than RAPD–PCR and are thus of greater use in identification based on library development but problems with reproducibility and normalisation still need to be addressed. At present, no relevant commercial libraries are available.

In general, the amplification techniques are more rapid and cheap than the restriction techniques such as RFLP. Amplified fragment length polymorphism (AFLP), a hybrid of the two, has not been evaluated in this study, but also appears to have good potential.

In summary, fatty acid analysis is the method of choice for identification down to species level. One or more of the repetitive PCR profiling techniques is likely to allow differentiation of subspecies, biovars, pathovars and possibly races within species.

REFERENCES

1. Lelliott, R.A. and Stead, D.E. (1987) Methods for the Diagnosis of Bacterial Diseases of Plants. pp. 216, Blackwell Scientific Publications, Oxford, UK.

2. Stead, D.E. (1992) Int. J. Syst. Bacteriol. 42, 281–295.

3. Stead, D.E. (1990) in Methods in Phytobacteriology (Klement Z., Rudolph K. and Sands D.C., eds) pp. 275–278, Akademiai Kiado, Budapest, Hungary.

4. Stead, D.E. (1992) in Techniques for Rapid Diagnosis of Plant Pathogens (Duncan J. and Torrance L., eds) pp. 76–111, Blackwell Scientific Publications, Oxford, UK.

5. Maki Valkama, T. and Karjalainen, R. (1994) Ann. Appl. Biol. 125, 301–309.

6. Louws, F.J., Fulbright, D.W., Stephens, C.T. and de Bruijn, F.J. (1994) Appl. Environ. Microbiol. 60, 2286–2295.

7. Vauterin, L., Hoste, B., Kersters, K. and Swings, J. (1995) Int. J. Syst. Bacteriol. 45, 472–489.

8. Nassar, A., Bertheau, Y., Dervin, L., Narcy, J.P. and Lemattre, M. (1994) Appl. Environ. Microbiol. 60, 3781–3789.

9. Cook, D., Barlow, E. and Sequira, L. (1989) Mol. Plant-Microbe Interact. 2, 113–121.

10. Cook, D. and Sequira, L. (1994) in Bacterial Wilt : the Disease and its Causative Agent, *Pseudomonas solanacearum* (Hayward A.C. and Hartman G.L., eds) pp. 77–93 , CAB International, Wallingford, UK.

11. Gillings, M.R. and Fahy, P. (1994) in Bacterial Wilt : the Disease and its Causative Agent, *Pseudomonas solanacearum* (Hayward A.C. and Hartman G.L., eds) pp. 95–112, CAB International, Wallingford, UK.

12. Cooksey, D.A. and Graham, J.H. (1989) Phytopathology 79, 745–750.

13. Berthier, Y., Verdier, V., Guesdon, J.L., Chevrier, D., Denis, J.B., Decous, G. and Lemattre, M. (1993) Appl. Environ. Microbiol. 59, 851–859.

14. Lazo, G.R., Roffey, R. and Gabriel, D.W. (1987) Int. J. Syst. Bacteriol. 34, 214–221.

15. Williams, J.G.K., Kubelik, A.R., Livak, K.J., Rafalski, J.A. and Tingey, S.V. (1990) Nucleic Acids Res. 18, 6531–6535.

16. Welsh, J. and McClelland, M. (1990) Nucleic Acids Res. 18, 7213–7218.

17. Versalovic, J., Schneider, M., de Bruijn, F.J. and Lupski, J.R. (1994) Methods Mol. Cell. Biol. 5, 25–40.

18. Janssen, P., Coopman, R., Huys, G., Swings, J., Bleeker, M., Vos, P., Zabeau, M. and Kersters, K. (1996) Microbiology 142, 1881–1893.

19. Nassar, A., Darrasse, A., Lemattre, M., Koutoujansky, A., Dervin, C., Vedel, R. and Bertheau, Y. (1996) Appl. Environ. Microbiol. 62, 2228–2235.

DEVELOPMENTS IN SEROLOGICAL METHODS TO DETECT AND IDENTIFY PLANT VIRUSES

L. TORRANCE
Scottish Crop Research Institute, Invergowrie, Dundee, DD2 5DA, Scotland, UK

1. INTRODUCTION

Several recent publications have reviewed serological methods for the detection and diagnosis of plant viruses [1–4]. Consequently, this paper will focus on some developments to methods that may be relevant to the detection of contaminants in plant tissue cultures, particularly developments that improve the sensitivity of virus detection. Although antibodies have been used to facilitate highly sensitive assays based on the polymerase chain reaction [5,6] these are considered primarily nucleic acid based methods and will not be included in this article. The last section will describe the production of novel antibody-like proteins by recombinant DNA methods, and their future potential for application to virus detection will be discussed.

2. ENZYME IMMUNOASSAYS

Although there are many different antibody-based methods available to detect and identify relationships between viruses, enzyme immunoassays (EIA) have superseded most of them, and EIA are now used widely for the detection and diagnosis of virus diseases [7,8]. These tests are well suited to the work of routine testing laboratories because they are very sensitive yet they need little input of skilled manpower or specialist equipment. The microtitre plate format, together with commercially available ancillary equipment for automatic processing such as sample extractors, dispensers, plate washers, and data recorders with the facility for computerised data manipulation and storage [1,2,9], make EIA well suited for throughput of thousands of samples. EIA methods give reliable results over a wide range of antigen concentrations, and are largely unaffected by sample matrix.

The performance of the EIA depends largely on the quality of the antibodies incorporated. Antibodies are made with preparations of plant viruses that are purified from the leaves of infected plants. Each antiserum must be tested for suitability for use in EIA because some may contain antibodies specific for plant proteins and will give false positives with control samples. Also, different batches of antisera can be a source of variation. The problems are

A.C. Cassells (ed.), Pathogen and Microbial Contamination Management in Micropropagation, 75–82.
© *1997 Kluwer Academic Publishers. Printed in the Netherlands.*

particularly evident with viruses which are present in low concentration (e.g. potato leafroll luteovirus) or are unstable and difficult to purify (e.g. potato mop-top furovirus).

It was thought that the production and use of monoclonal antibodies would overcome the problems of quality and continuity of supply, and since the early 1980s monoclonal antibodies have been produced to over 60 plant viruses in 20 virus groups [4]. However, few commercially available tests incorporate monoclonal antibodies. This is partly because monoclonal antibodies are expensive to prepare and partly because some monoclonal antibodies do not work when employed in different test formats, e.g. they lose reactivity when used to coat plates or when chemically coupled to enzymes [3]. However, many monoclonal antibodies work well as detecting antibodies in EIA when antigen is coated directly on plates or when trapped by a layer of coating polyclonal antibodies (together with a generic anti-mouse or anti-rat enzyme conjugate to reveal bound monoclonal antibody). The monoclonal antibodies which are commercially available, e.g. those specific for barley yellow dwarf virus strains, potato viruses X, V, mop-top and leafroll, and a general potyvirus detector (Adgen and Agdia) are sold for use in these formats.

Several companies sell plant virus antisera including Adgen, Agdia, Bioreba, Sigma, and Loewe. Although the source and provenance of the antibody preparations are not always revealed, they have usually been obtained from research laboratories because they are surplus to requirements, and they can be of variable quality and specificity. Sigma Chemical Company sell a range of pathogen detection kits for testing plants produced in commercial greenhouse, nursery and plant tissue culture businesses. The range is based on horseradish peroxidase (HRP) enzyme-conjugated virus-specific antibodies and the substrate o-phenylenediamine (OPD). Previously it has been shown that extracts from plant roots can cause non-specific oxidation of another HRP substrate, tetramethylbenzidine [10]. Although the factors responsible for the non-specific oxidation were not determined, it may be caused by the presence of endogenous enzyme in some plant tissues and appropriate controls (i.e. sample wells containing no enzyme-conjugated antibodies) should be included in the tests. For this reason alkaline phosphatase conjugates are preferred for tests on plant samples.

Closteroviruses have been detected by EIA in grapevines grown in the greenhouse and in plants grown in tissue culture [11]. The viruses were found to be irregularly distributed in infected tissue and several samples were taken from each plant to ensure reliable detection. EIA was used in a survey of plants growing in commercial greenhouses in British Columbia; 2600 samples were tested for the lettuce and impatiens strains of tomato spotted wilt virus (TSWV), and TSWV was detected in samples from 25 of the 38 sites tested [12]. Similarly, EIA was used to assess the incidence of four viruses in orchids growing in commercial farms and nurseries in Hawaii [13]. Cymbidium mosaic virus (CyMV) was detected in only 4% of the seed-propagated hybrids that were less than 3 years old, but the incidence of CyMV increased in older hybrids, up to 94% on some orchid farms. Also, CyMV was found in 45% of 330 cloned orchids tested.

3. IMPROVEMENTS IN SENSITIVITY

The direct double antibody sandwich (DAS) microtitre plate format of EIA first introduced to plant virology by Clark and Adams in 1977 [14] is still widely used. In this test, virus-specific polyclonal antibodies are used to coat the wells of a microtitre plate, and an antibody–alkaline phosphatase conjugate preparation for virus detection. Conversion of the chromogenic substrate p-nitrophenyl phosphate (colourless) to *p*-nitrophenol (yellow in alkaline solution) reveals the presence of bound enzyme. This assay can be very sensitive, detecting viruses at concentrations of 1–10 ng/ml. Also, composite samples of tissue from several plants (10–100 leaves) can be tested [15]. DAS EIA is suitable for the detection of many viruses. However, there are occasions when it may not be sufficiently sensitive, e.g. detection of very low concentrations of virus present in vector insects [16], or potato virus Y-infected dormant potato tubers produced on plants infected during the growing season. Also polyclonal antibody-based DAS EIA may fail to detect serologically related strains because of the rather narrow strain specificity it exhibits [17]. Some viruses are latent in plants grown in tissue culture [18]. It was proposed that *in vitro* cultures of some plant species may be more resistant to viruses, perhaps because of the increase in pathogenesis related proteins produced in response to growth regulators incorporated in the tissue culture medium [18]. Therefore, other more sensitive EIA methods may be needed for their detection.

Various alternative EIA formats and substrates have been published over the years which claim to improve the limit of detection and broaden strain specificity [7,17]. Incorporation of monoclonal antibodies has provided improvements, for example, by decreasing non-specific background reactions and providing more accurate discrimination between strains of barley yellow dwarf virus [19]. The biotin–avidin modification [20] in which the high affinity of avidin for biotin is exploited, has proved versatile and sensitive. Also, it was shown that biotin-labelled antibodies specific for tobacco mosaic virus detected serologically related tobamoviruses, thus broadening the range of serologically related virus isolates detectable by DAS EIA [21]. Nowadays, it is more usual to use streptavidin (from *Streptomyces avidinii*) to decrease possible problems of non-specific binding of avidin [20]. In these assays the primary (virus-specific) antibodies are biotinylated, and biotin bound antibodies revealed by reaction with a universal streptavidin–enzyme conjugate. Recently, an assay which incorporates biotinylated antibodies, and conjugates comprising streptavidin coupled to homopolymers of HRP improved detection sensitivity by 12–25 times over a monoclonal antibody-based DAS EIA in assays to detect barley stripe mosaic virus (BSMV) [22] and was used to detect BSMV in seeds. Increases in sensitivity of 4- to 16-fold have been obtained with fluorescent substrates [23] or by enzyme cycling [16]. Also, chemiluminescent substrates have been promoted as providing increases in sensitivity of one or more orders of magnitude. In tests to detect potato leafroll luteovirus (PLRV) particles we compared the chromogenic pNPP with the 1,2-dioxetane chemiluminescent CSPD®. An indirect EIA was done where microtitre plates were coated with an anti-PLRV polyclonal antibody preparation, and MAb SCR3 was used to detect PLRV particles. We found that the CSPD$^{®}$ substrate improved detection sensitivity; the limit of detection was approx. 1 ng/ml, compared with 6 ng/ml obtained with pNPP (L. Torrance and G.H. Cowan, unpublished).

The increases in sensitivity with fluorogenic or chemiluminescent substrates are obtained because it is possible to accurately measure very small amounts of the products. Often the increases in sensitivity obtained in practice in these assays are not as great as expected, and to obtain the best results much greater care has to be taken to minimise background non-specific reactions; otherwise the increase in specific signal will be accompanied by a corresponding increase in non-specific signal. For example, assays should be optimised with respect to dilution of antibody preparations, blocking agents and incubation times for use with the particular substrate. Antibody preparations with little or no titre against plant proteins such as monoclonal antibodies should be used to minimise background noise, and antibody–enzyme conjugates should be prepared with heterobifunctional cross-linking reagents (Pierce) instead of glutaraldehyde, and purified by HPLC, for controlled chemical coupling and to eliminate aggregates.

Another consideration is that the tests with chemiluminescent substrates are approximately 20 times more expensive than chromogenic methods. However, the additional expense is probably justified in tests on plants that will be the starting material for the propagation of very large numbers of progeny in tissue cultures, e.g. plants treated to eliminate viruses, such as material propagated from meristem tip cultures, or after thermotherapy, or other explants.

4. DOT-IMMUNOBINDING ASSAY

Dot-immunobinding assays are useful alternatives to microtitre plate assays [7,24]. In these tests, plant sap is spotted directly onto nitrocellulose (Schleicher & Schuell) or polyvinylidene difluoride (PVDF; Millipore) transfer membranes, and the membranes are incubated with antibody–enzyme conjugates. The enzyme substrate is insoluble and precipitates at the enzyme reaction site (Fig. 1). However, because there is no antibody capture step, they do not work reliably with viruses present in low concentration such as PLRV or other luteoviruses. Six potato viruses were readily detected by EIA and dot-immunobinding assays of single leaf discs (6 mm) taken from tissue culture plants. The results of the dot-immunobinding assay could be quantified using densitometry and the readings were found to be comparable with absorbance values obtained in EIA [25].

Use of the membranes as support matrices has allowed development of fast portable EIA test kits and dip-stick tests to be devised [26–28]. These test kits are self-contained, and tests can be completed quickly in the field or office on small numbers of samples [26].

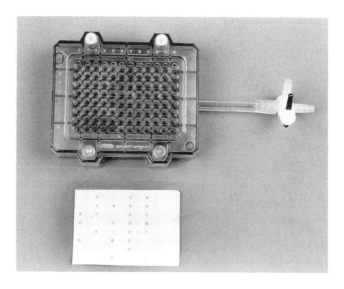

Figure 1. Dot immunobinding assay. Bio-dot apparatus (BioRad) and membrane after immunodetection of narcissus latent virus from samples of *Narcissus* leaves (courtesy of W.P. Mowat and S. Dawson).

5. ANTIBODIES FROM PHAGE DISPLAY LIBRARIES

Key developments in immunology research such as the application of polymerase chain reaction to amplify antibody genes, together with the expression of functional fragments of antibody molecules on the surface of filamentous phage (phage display) [29] have allowed new approaches to be made to produce novel antibodies.

Fragments of antibodies that comprise the heavy and light chain variable (antigen binding) domains of the antibody molecules linked together by a short sequence of about 15 amino acids to form a single polypeptide chain (scFv), can be expressed fused to the minor coat protein pIII of filamentous bacteriophage particles (Fig. 2). Antibody genes from hybridomas, as well as diverse repertoires of antibody genes from immunised and non-immunised donors have been displayed in this way [30]. Specific scFv can be obtained from a population of phages carrying many different scFv by binding to and then eluting from antigen. The selected phage preparation can then be enriched for binders by re-infecting *Escherichia coli* with the eluted phage and repeating the procedure. In this way genetically pure populations of phage which encode the scFv can be obtained after several repeated cycles. Large combinatorial phage display libraries have been produced containing $>10^8$ different clones and it has been shown that a range of recombinant immune reagents with diverse specificities can be isolated from such libraries [31]. Selection from phage

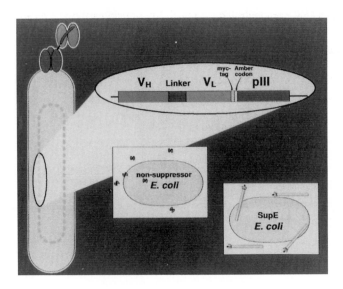

Figure 2. The display of antibodies (scFv) on phage and the strategy for expression in *Escherichia coli*. The scFv gene (V$_H$–linker–V$_L$) is fused via an amber stop codon to pIII, the minor coat protein of filamentous phage; when phage infects *E. coli* strains that suppress the amber codon (supE) the scFv is expressed on phage but when grown in non-suppressor strains soluble scFv is secreted.

display libraries yields specific antibody fragments without the need either to immunise animals or to use hybridoma technology, and should, therefore, overcome several difficulties in the production of conventional polyclonal sera or monoclonal antibodies such as poor immunogenicity, toxicity of the antigens or high production costs. In addition, the scFv can be fused to a range of reporter molecules such as alkaline phosphatase and the fusion proteins expressed in bacterial cultures [32,33].

We used the MRC human synthetic library [31] to obtain scFv against cucumber mosaic cucumovirus and potato leafroll luteovirus [33,34]. Although the absorbance values obtained in EIA with the scFv were weaker than those with the polyclonal antibody preparations, the scFv were used without any modification after only four rounds of selection. Several methods have been used to improve the binding affinity of selected scFv, e.g. mutagenesis by error-prone PCR or chain shuffling [30,35,36]. Also, scFv are monovalent molecules and it is possible to produce bivalent molecules by shortening the linker peptide which should increase the avidity [37] and, therefore, improve performance in EIA. Moreover, the library we used incorporated only one kind of V$_L$ chain, and antibodies with affinities in the nanomolar range have recently been selected directly from a larger more diverse library [38].

Our work at SCRI has demonstrated the feasibility of using antibody gene libraries to select scFv specific for plant viruses. We obtained virus-specific scFv quickly (within a few weeks), and without recourse to animal immunisations, from a phage-antibody library. However, the binding affinity of such scFvs must be improved before they are incorporated into routine diagnostic tests. We think that further work to improve these reagents is worthwhile because bacterial expression of antibody-like proteins for the detection and diagnosis of plant pathogens would produce standardised, reproducible assays at a fraction of the costs of production of monoclonal antibodies. Furthermore, costly methods for storage of hybridoma cell lines would be avoided which is an advantage in countries where supplies of liquid nitrogen are unreliable.

ACKNOWLEDGEMENTS

I thank the Scottish Office of Agriculture, Environment and Fisheries Department for financial support.

REFERENCES

1. Torrance, L. (1992) in Techniques for the Rapid Detection of Plant Pathogens (Duncan J.M. and Torrance L., eds) pp. 1–33, Blackwell Scientific Publications, Oxford, UK.

2. Torrance, L. (1992) Neth. J. Plant Pathol. 98 (Suppl. 2), 21–28.

3. Torrance, L. (1995) Eur. J. Plant Pathol. 101, 351–363.

4. van Regenmortel, M.H.V. and Dubs, M.-C. (1993) in Diagnosis of Plant Virus Diseases (Matthews R.E.F., ed.) pp. 159–214, CRC Press, London.

5. Nolasco, G., de Blas, C., Torres, V. and Ponz, F. (1993) J. Virol. Methods 45, 201–218.

6. Brandt, S. and Himmler, G. (1995) Vitis 34, 127–128.

7. Cooper, J.I. and Edwards, M.L.(1986) in Developments and Applications in Virus Testing (Jones R.A.C. and Torrance L., eds) pp. 139–154, Association of Applied Biologists, Wellesbourne, UK.

8. Matthews, R.E.F. (1991) Plant Virology, third edition. Academic Press Inc., New York.

9. Torrance, L. and Jones, R.A.C. (1981) Plant Pathol. 30, 1–24.

10. Jones, A.T. and Mitchell, M.J. (1987) Ann. Appl. Biol. 111, 359–364.

11. Monis, J and Bestwick, R.K. (1996) Am. J. Enol. Viticulture 47, 199–205.

12. Bitterlich, I. and Macdonald, L.S. (1993) Can. Plant Dis. Surv. 73, 137–142.

13. Hu, J.S., Ferreira, S., Wang, M. and Xu, M.Q. (1993) Plant Dis. 77, 464–468.

14. Clark, M.F. and Adams, A.N. (1977) J. Gen. Virol. 34, 475–483.

15. Torrance, L. and Dolby, C.A. (1984) Ann. Appl. Biol. 104, 267–276.

16. Torrance, L. (1987) J. Virol. Methods 15, 131–138.

17. Koenig, R. (1981) J. Gen. Virol. 55, 53–62.

18. Leifert, C., Morris, C.E. and Waites, W.M. (1994) Crit. Rev. Plant Sci. 13, 139–183.

19. Torrance, L., Pead, M.T., Larkins, A.P. and Butcher, G.W. (1986) J. Gen. Virol. 67, 549–556.

20. Kohen, F., Amir-Zaltsman, Y., Strasburger, C.J., Bayer, E.A. and Wilchek, M. (1988) in Complementary Immunoassays (Collins W.P., ed.) pp. 57–69, John Wiley and Sons, London.

21. Zrein, M., Burckard, J. and van Regenmortel, M.H.V. (1986) J. Virol. Methods 13, 121–128.

22. Sukhacheva, E., Novikov, V., Plaskin, D., Pavlova, I. and Ambrosova, S. (1996) J. Virol. Methods 56, 199–207.

23. Torrance, L. and Jones, R.A.C. (1982) Ann. Appl. Biol. 101, 501–509.

24. Mowat, W.P., Dawson, S. and Duncan, G.H. (1989) J. Virol. Methods 25, 199–210.

25. Singh, R.P., Boucher, A., Somerville, T.H. and Coleman, S. (1996) Am. Potato J. 73, 101–112.

26. Miller, S.A., Rittenburg, J.H., Petersen, F.P. and Grothaus, G.D. (1992) in Techniques for the Rapid Detection of Plant Pathogens (Duncan J.M. and Torrance L., eds), pp. 208–221, Blackwell Scientific Publications, Oxford, UK.

27. Dewey, F.M., MacDonald, M.M., Phillips, S.I. and Priestley, R.A. (1990) J. Gen. Microbiol. 136, 753–760.

28. Thornton, C.R., Dewey, F.M. and Gilligan, C.A. (1993) Plant Pathol. 42, 763–773.

29. Winter, G., Griffiths, A.D., Hawkins, R.E. and Hoogenboom, H.R. (1994) Annu. Rev. Immunol. 12, 433–455.

30. Marks, J.D. (1995) in Antibody Engineering, second edition (Borrebaeck C.A.K., ed.) pp. 53–88, Oxford University Press, New York, USA.

31. Nissim, A., Hoogenboom, H.R., Tomlinson, I.M., Flynn, G., Midgley, C., Lane, D. and Winter, G. (1994) EMBO J. 13, 692–698.

32. Kerschbaumer, R.J., Hirschl, S., Schwager, C., Ibl, M. and Himmler, G. (1996) Immunotechnology 2, 145–150.

33. Ziegler, A., Harper, K. and Torrance, L. (1996) in BCPC Symposium Proceedings No 65: Proceedings Diagnostics in Crop Production. pp. 35–38, University of Warwick, Coventry, UK 1–3 April 1996. British Crop Protection Council.

34. Ziegler, A., Torrance, L., Macintosh, S.M., Cowan, G.H. and Mayo, M.A. (1995) Virology 214, 235–238.

35. Thompson, J., Pope, T., Tung, J.-S., Chan, C., Hollis, G., Mark, G. and Johnson, K.S. (1996) J. Mol. Biol. 256, 77–88.

36. Schier, R., Balint, R.F., McCall, A., Apell, G., Larrick, J.W. and Marks, J.D. (1996) Gene 169, 147–155.

37. Holliger, P., Prospero, T. and Winter, G. (1993) Proc. Natl. Acad. Sci. USA 90, 6444–6448.

38. Vaughan, T.J., Williams, A.J., Pritchard, K., Osbourn, J.K., Pope, A.R., Earnshaw, J.C., McCafferty, J., Hodits, R.A., Wilton, J. and Johnson, K.S. (1996) Nature Biotechnol. 14, 309–314.

MOLECULAR DIAGNOSTICS FOR PATHOGEN DETECTION IN SEEDS AND PLANTING MATERIAL

J.C. REEVES
NIAB, Huntingdon Road, Cambridge, CB3 0LE, UK

1. INTRODUCTION

Parallels between seed health testing and the screening of plant production for disease during micropropagation are not superficially obvious but brief consideration will show that they, nonetheless, exist. Both procedures have the ultimate objective of contributing to the production of healthy plants and are multiplication processes where infection or contamination at an early stage can lead to the proliferation of micro-organisms throughout production. These micro-organisms can be overtly pathogenic or otherwise interfere to reduce final yield and quality either in the field or in the tissue culture laboratory [1].

There are a number of other similarities between the two procedures. For example, during planting material multiplication procedures, it is possible and desirable to reduce or even eliminate infection in the early stages of multiplication. The benefit from early elimination of infection before it proliferates as multiplication proceeds is patent. However, reinfection can occur or subliminal infection undetected at an early stage can become manifest so that, in seed production and in micropropagation, there is a continued need for monitoring of health status throughout. More attention to sampling is required as the lot size increases during multiplication.

Because inoculum levels of pathogens are frequently low on seeds sensitive tests are extremely important for effective disease monitoring. This may also be true in micropropagation, particularly if tissue culture media contain components which suppress growth of micro-organisms whose presence only subsequently becomes apparent. In addition, diseases are frequently asymptomatic on seed or latent and there is, therefore, a need for a definitive test which can provide a diagnosis unsupported by a typical syndrome.

Testing also needs to be fast if the speed of availability of data is to be of any use in disease management during the progression of multiplication from one stage to the next. For example, in the production of winter cereal seed there is only a very short window of opportunity for testing between the harvest of a seed crop and its sowing. If seed-borne

A.C. Cassells (ed.), Pathogen and Microbial Contamination Management in Micropropagation, 83–95.
© *1997 Kluwer Academic Publishers. Printed in the Netherlands.*

disease data are not available to influence disease management choices then prophylactic seed treatment becomes necessary. This is costly and undesirable if not justified by the presence of seed-borne disease [2]. An analogous situation may apply during micropropagation.

Nevertheless, there are some differences between seed health testing and disease monitoring during micropropagation. Seed health testing deals exclusively with crop, and sometimes human, pathogens. In micropropagation, other organisms, and not exclusively or necessarily pathogens, can be important through a direct effect on the production media [1]. Although seed health testing takes place against a background of saprophytic organisms carried on the seed, these do not generally affect ultimate crop production but are capable of interfering with a seed health test. These organisms can be of greater importance in tissue culture where sterility of cultures is paramount. Because of this, tolerances in the level of infection of tissue cultures may be inappropriate in a similar way that in testing for some quarantine organisms no level of infection is acceptable. In seed health testing this is not the case and for some pathogens a certain level of infection is tolerable on the grounds that it is not expected to cause significant losses when present below that level.

Clearly, therefore, there are sufficient similarities between disease monitoring in seed and in micropropagation for the experience gained in the development of molecular diagnostics for seed health testing to have some relevance to, and application in, plant production through tissue culture. Accordingly, this paper aims to present some background to the development of nucleic acid based molecular diagnostic techniques for use in seed health testing.

2. SEED HEALTH TESTING

Seed-borne diseases are caused by pathogens of various kinds. In the UK most seed health testing concentrates on the detection of fungal pathogens but bacteria, viruses and nematodes can also be important. Testing for and monitoring of these pathogens is a component of seed quality control. Seed certification is a formal and statutory means of ensuring the flow of quality seed to the market place. Paradoxically, these certification schemes contain very little requirement for seed health testing for various historical reasons. Instead, other quality attributes such as germination, purity and trueness to type are emphasised. Nevertheless, some seed health testing for certain named pathogens is included and the importance of seed health status is recognised. This may not necessarily be only achieved by laboratory tests and field plot inspection may be applied. There are other reasons for seed health testing which include the conduct of surveys for pathogen incidence, the investigation of reasons for poor germination and the detection of disease organisms for quarantine. A major further incentive for determining the presence of a seed-borne pathogen is in order to make decisions about the need to apply a seed treatment which may not be required in the absence of infection [2]. All these methods of control for seed-borne diseases have as their basis the need for seed health testing.

3. SEED HEALTH TESTING REQUIREMENTS

The most common techniques used in testing for seed-borne pathogens are not recent developments. To be effective in routine use these methods require that a distinct pathogen with well-defined diagnostic attributes is the target for detection and should be simple, fast and cheap, giving repeatable results between samples within the limits of sampling error [3].These general requirements apply in most diagnostic contexts including tissue culture. Sheppard [4] has given these criteria in more detail and draws a distinction between factors of importance in screening for pathogens and those factors important in confirmatory tests (Table 1).

Table 1. The relative importance of laboratory test factors (from Sheppard [4])

Factor	Screening	Confirmation
Simplicity	+++	+
Mass application	+++	+
Cost	+++	+
Precision	+	++++
Accuracy	+	+++
Diagnostic sensitivity	++++	+++
Diagnostic specificity	+	++++

+: minor.
++: moderate.
+++: major.
++++: crucial.

Sheppard [4] also identifies a difference between analytical specificity and sensitivity and diagnostic specificity and sensitivity. The former measure the accuracy with which a test identifies a specific target and the minimum level of target detectable, respectively. The latter are more complex. Diagnostic sensitivity measures the ability of a test to identify correctly a sample known *a priori* to be diseased and is given by the formula:

$$\text{Diagnostic sensitivity} = \frac{TP}{TP + FN} \times 100$$

where TP denotes true-positive results and FN false-negative results.

The obverse of this is the diagnostic specificity which measures the ability of a test to identify correctly as healthy a sample known *a priori* not to be diseased. This is given by:

$$\text{Diagnostic specificity} = \frac{\text{TN}}{\text{TN} + \text{FP}} \times 100$$

where TN denotes true-negative results and FP false-positive results.

Although apparently rarely used, these criteria could be valuable in the evaluation of new test techniques in comparison with existing test methods and should be borne in mind during the development process.

4. SEED HEALTH TESTING METHODS

There are a number of methods used in seed health testing. The simplest in principle of these is the growing-on test where samples of the potentially infected seed lot are grown and scored for the development of symptoms. This has a number of disadvantages, for example the seed has to be of germinable quality for sufficient plants to be obtained from a reasonably sized sample and a large number of plants are required to detect low levels of infection. A test of this nature also requires that disease symptom expression is unambiguous, easily identified, uncomplicated by the presence of other diseases, which may or may not be seed borne, or by the presence of other disorders and which preferably is manifest early in the development of seedlings. This is necessary to avoid an excessively lengthy test time in a method which is intrinsically time consuming, relying as it does on the growth of the test plants. For these reasons this technique is not suited to high throughput routine applications and is rarely used. Another simple method involves the inoculation of indicator plants with washings taken from a potentially infected seed sample. This may require fewer plants but is still reliant on the development of diagnostic symptoms, which can take time.

Most test methods are based on direct observation of some kind. This can take the form of examination of seed washings under a microscope for the presence of fungal spores, as in the case of testing for bunts, or for the presence of hyphal elements or bacteria. Direct observation of dry seed is rarely of any value. However, direct observation of moistened seed on agar plates or blotters is the most common approach adopted. This involves the placement of the seed, usually surface sterilised, onto an appropriate agar medium and incubating under conditions suitable for the development of the diagnostic characteristics of the pathogen. This typically takes a number of days before the agar plate is examined by an observer skilled in the identification of fungal pathogens. Paper blotters can be used in a similar way and identification in some cases, for example in tests for *Pyrenophora* spp., can use the development of a colour reaction characteristic of the genus.

These techniques can be extremely effective for a number of pathogens but are not without their limitations. Because they rely on growth of the pathogen they inevitably show the presence of live inoculum but this also implies a time delay of several days and the tests are labour intensive, needing the attention of skilled personnel. These are major disadvantages where high throughput is paramount. This is illustrated by the current trend for autumn sowing of cereals. The majority of winter cereal seed is prophylactically treated with fungicides without any knowledge of the disease burden of that seed. This costs UK farmers some £23 million per annum. If farmers wished to have information about seed-borne diseases before making a management decision to treat the seed only if justified by the presence of a pathogen at a sufficiently high level to threaten a disease outbreak in the subsequent crop, it is unlikely that the traditional seed health tests would be suitable to cope with the expected demand. This is because of the short amount of time available between harvesting the seed and its resowing [2]. It is probable that similar constraints would apply in the multiplication of plantlets during micropropagation.

There are also other disadvantages to these traditional test methods. Inoculum levels on seeds can be extremely low but nevertheless able to produce a significant disease outbreak. This is particularly true of bacterial diseases. Consequently, sensitive (in the analytical sense) tests are desirable. Because saprophytes are often present at high levels on seed, particularly when harvested under poor conditions and despite surface sterilisation of the seed, they can interfere with the recording of the test by masking or even inhibiting the growth of the pathogen. Moreover, there can be confusion, even for skilled observers, between saprophytes and the target organism or between the target and closely related organisms. For example, it requires much skill to consistently distinguish between *Pyrenophora graminea* and its close relative *Pyrenophora teres* on barley seed. Accuracy of diagnosis can, therefore, also be compromised, especially in the absence of symptoms to assist identification.

These problems have focused attention on the improvement of test methods. For bacterial pathogens advancements in the use of selective and semi-selective media have been of value and a number of biochemical methods are now available, for example the use of different carbon sources to effect an identification. Fatty acid methyl ester analysis (this volume) has also been used with considerable success to identify bacterial cultures. Notable also has been the development of effective immunodiagnostic techniques. These are also covered elsewhere in this volume.

Over the past decade considerable advancement has taken place in the development of nucleic acid technology and this promises to have a considerable impact on diagnostics of all kinds. Whereas traditional methods are often multi-stage, time consuming, labour intensive and subjective, this new technology offers the potential for rapid, same-day analysis, specific and sensitive tests.

5. MOLECULAR DNA DIAGNOSTICS

A DNA probe is a short length of single-stranded DNA labelled with a radioactive or, increasingly, chemical moiety by which its presence can be detected. This probe is designed to hybridise to a specific sequence in the DNA of the target organism and will only hybridise to that particular sequence giving specific detection of the target. The presence of the probe/target hybrid can be revealed using the label attached to the probe.

During the mid-1980s the use of DNA probes began to appear in plant pathology for general diagnostic purposes and a number of reports concerned the use of such probes for detection of seed-borne bacteria. These reports have been reviewed elsewhere [5]. Probes were developed by various means, mainly by selection of randomly cloned DNA fragments about which little was known but a number of bacterial probes were developed from cloned genes. The use of probes was successful in the identification of the target organism where a pure isolate of the pathogen was available; they were less effective in the detection of the pathogen against a background of contaminating DNA which interfered with hybridisation and reduced sensitivity. This, coupled with the need for pure isolates, restricted the utility of DNA probes in routine seed testing but they were, nevertheless, useful where the identification phase of a test was lengthy and complicated (for an example see Ref. 6).

The use of DNA probes for diagnostic purposes was overtaken by the development of the Polymerase Chain Reaction (PCR) [7]. This technique (Fig. 1) involves the enzymic amplification of a specific DNA sequence of the target organism using a thermostable DNA polymerase directed by oligonucleotide primers. Specificity is conferred by choice of primers and implies knowledge of the DNA sequences to be detected. Moreover, because the reaction is cyclic with the products of earlier cycles being used as the template for subsequent cycles, analytical sensitivity can be extremely high.

Since its origination PCR has had an immense impact both as a research tool, because of its versatility, and on diagnostics because of the combination of specificity and sensitivity it offers. These characteristics potentially allow both detection and identification without a stage which isolates the pathogen as a pure culture and for seed health testing this implies direct detection in liquors obtained from seed soaks of short duration. Direct detection *in planta* is also possible, which is of significance for the screening of tissue cultures. Added to these attributes is the rapidity of the PCR process; a typical PCR amplification taking between 2 and 3 h, to which has to be incorporated sample preparation and detection of the PCR product, usually by electrophoresis. It is clear that the technique offers the accurate, sensitive and rapid diagnosis that is required by seed testing and in the screening of plantlets during micropropagation and for these reasons much effort has been focused on its development for use in seed health testing.

POLYMERASE CHAIN REACTION

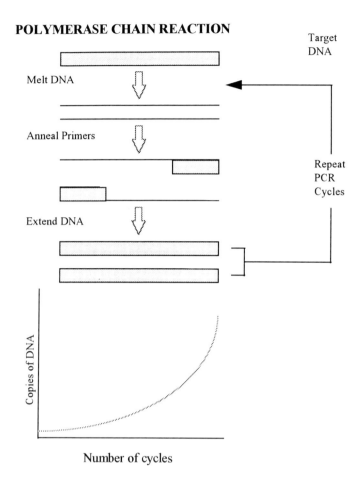

Figure 1. Schematic representation of the PCR reaction.

6. DEVELOPMENT OF PCR-BASED TESTS

For a PCR reaction to be specific it requires that there is prior DNA sequence knowledge which allows the design of pairs of oligonucleotide primers to confer the necessary degree of specificity. There are a number of general approaches to the acquisition of this basic information.

Where, for example, a DNA probe has been developed for the identification of a particular pathogen and shown, therefore, to have an element of specificity, it is possible partially or completely to sequence this probe and effect the requisite primer design based on the sequence data obtained. The pairs of primers thus designed can then be tested for their utility by confirming specificity and determining optimum reaction conditions for maximum sensitivity [6]. This approach could also be adopted where *ex libris* sequence data on, for example, a particular toxin or other genes are available.

In many cases, however, there will be no prior sequence data on which to base primer design and development will have to proceed *ex nihilo*. Two approaches are commonly adopted and the first of these involves the use of PCR based on arbitrary, and not necessarily specific, primers. This technique is known as RAPDs (random amplified polymorphic DNA) and is one of a number of DNA amplification techniques using single, short arbitrary primers some of which have been collectively termed MAAP (Multiple Arbitrary Amplicon Profiling) techniques [8]. RAPD differs from conventional specific PCR through the use of a single, short primer which is arbitrarily chosen and generates multiple amplification products by directing DNA amplification from a number of sites on the genome of the chosen target. Amplification products will be produced from many DNA templates by the same primer and a different primer will produce different products from the same DNA template. This comparison with specific PCR is illustrated in Figs. 2 and 3.

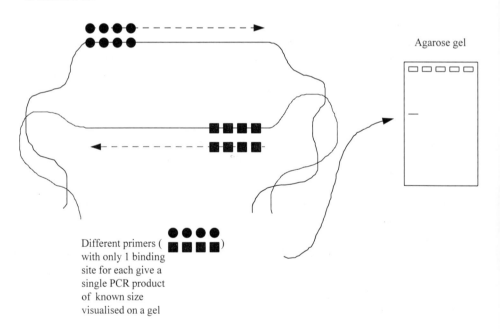

Figure 2. Schematic representation of specific PCR.

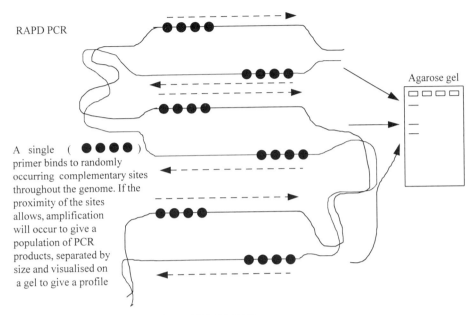

RAPD PCR

A single (●●●●)
primer binds to randomly
occurring complementary sites
throughout the genome. If the
proximity of the sites
allows, amplification
will occur to give a
population of PCR
products, separated by
size and visualised on
a gel to give a profile

Agarose gel

Figure 3. Schematic representation of RAPD PCR.

The RAPD amplification products are separated on an electrophoresis gel and, when visualised, represent a DNA profile of the target organisms. Components of this profile can be excised from the gel and cloned, thereby providing a source of DNA fragments for sequencing and subsequent design of pairs of primers for specific PCR. This procedure has been adopted with success both for bacterial pathogens [9] and for fungal pathogens [10].

A second approach to obtaining useful DNA sequences on which to base primer design for specific PCR involves the use of the internal transcribed spacer region (ITS) of the nuclear ribosomal unit. For fungi this lies between the 17S and 25S ribosomal RNA genes and contains two variable non-coding spacer regions and the 5.8S gene [11]. Sequences within these genes are highly conserved and standard PCR primers based on these conserved sequences have been designed to effect the amplification of the variable spacer regions (Fig. 4). These amplification products can be sequenced and the variability exploited to design primer pairs specific for the particular target organism. This approach has successfully produced primers for a number of pathogens [12–14] and, less successfully, for others [15].

7. PROBLEMS AND SOLUTIONS FOR PCR APPLICATIONS IN SEED HEALTH TESTING

Irrespective of the approach adopted, the primer pairs require to be tested for specificity and reaction conditions optimised for maximum sensitivity. It is often at this stage that

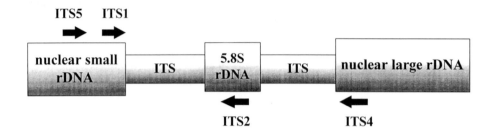

Figure 4. Schematic representation of ribosomal DNA showing location of standard PCR primers.

problems begin to materialise as the development of the test proceeds to determine sensitivity when used in the presence of plant material rather than purified DNA. Many plant and seed matrices contain chemicals which inhibit the Taq polymerase enzyme used in PCR to a degree which can reduce sensitivity to a level where no amplification occurs. However, there are methods by which this inhibition can be overcome. At worst, this involves extraction of DNA from the test material and subjecting this to PCR but there are other more sophisticated means of separating the target pathogen from inhibitors, for example by using immunocapture PCR (for a short review see Ref. 3). This technique uses immobilised polyclonal antisera or monoclonal antibodies to capture the target organism, which can then be washed free of contaminants before PCR proceeds. The easiest and most cost-effective means of avoiding inhibition in seed health testing is to reduce the concentration of inhibitors by limiting the length of time the seeds are soaked to release the pathogen before PCR. The success of this approach depends on the species of seed involved and the location of the pathogen in or on the seed.

Most, but not all, seed health tests require a quantitative result which gives some indication of the level of infection carried by a seed sample. This is usually given by the percentage of seeds infected rather than the amount of inoculum carried on each seed and presents problems for basic PCR which, unmodified, gives only a presence or absence result. In certain circumstances this is perfectly satisfactory but in the majority of seed health applications there is a limitation. It is possible, however, to use a PCR-based rapid test to screen seed samples to identify those which are uninfected, allowing them to pass quickly into commerce, and concentrate effort on those which are infected. Since many samples in routine testing may not be infected, this approach has some attraction but its value does rely on a pathogen not being ubiquitous, infecting the majority of seed samples to a greater or lesser degree. Nevertheless, there are methods of quantitative PCR which allow the determination of the concentration of the initial template in the reaction and it may be possible to use these in seed health testing. Other methods using a statistical approach could also be used [3].

A further limitation of PCR for seed health testing relates to the format of the tests and the determination of the presence of the diagnostic PCR product. Usually this is done by electrophoresis and the product of the expected molecular weight revealed by staining the gel in some way. This is not especially suited to high throughput rapid testing. However, other formats using different PCR product detection systems are possible. For example, Fig. 5 shows schematically how colorimetric determination of PCR product could be achieved.

Recently, there have been developments which could have significant implications for rapid, high throughput applications such as seed health testing and the testing of plants during micropropagation. These developments are based on the use of fluorescent probes which detect the PCR product and which potentially allow high throughput rapid and automated sample testing. In addition, the concentration of the PCR product can be monitored in real time.

The first of these is a proprietary development by the Perkin-Elmer Corporation known as TaqMan^{TM}. This uses probes fluorescently labelled with a reporter dye at one end of the probe and a quencher molecule at the other. The physical proximity of these molecules prevents fluorescence. During the formation of PCR product the probe hybridises to an internal sequence between the primers and is subsequently cleaved by the exonuclease activity of the Taq polymerase. This separates the reporter dye from the quencher molecule giving rise to an incremental increase in fluorescence at each PCR cycle. The reactions are

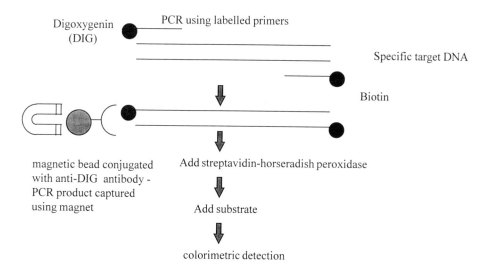

Figure 5. A possible approach to colorimetric detection of PCR products.

performed in microwells or, more recently, in closed tubes and avoids the need for gel detection of the PCR product.

Based on a similar principle, a very recent report [16] describes probes termed molecular beacons. These probes fluoresce as they undergo a conformational change during hybridisation to the target DNA sequence. This is achieved by designing the probe in a single-stranded nucleic acid stem and loop structure. The loop comprises the sequence specific for hybridisation to the target and the stem is formed by annealing two complementary arm sequences which flank, but are unrelated to, the loop sequence. A fluorescent reporter molecule is attached to the end of one of the arms with its complementary quencher molecule attached to the other. This proximity prevents fluorescence. When this probe and the target meet and hybridisation proceeds, the arm sequences are separated by the conformational change that the probe undergoes allowing the fluorophore to fluoresce under UV light. Real time quantification of the PCR product is, therefore, made possible.

8. CONCLUSIONS

The potential benefits these developments offer, particularly as prices for the hardware reduce, make them extremely attractive where high numbers of samples require rapid analysis and economies of scale can be generated. Coupled with the ability to multiplex PCR reactions using a number of primer pairs for the detection of multiple pathogens in a seed lot or plant sample, these benefits make PCR-based diagnostics a genuine option to meet disease monitoring demands.

Nevertheless, much development work remains to be done to realise this potential and this will concentrate on issues such as quantification and automation. However, work is also needed on the interpretation of results of PCR tests with regard to how these relate to the development of disease in the field crop and what thresholds to apply before treatment is recommended.

REFERENCES

1. Cassells, A.C. (1992) in Techniques for the Rapid Detection of Plant Pathogens (Duncan J.M. and Torrance L., eds) pp. 179–192, Blackwell Scientific Publications, Oxford, UK.

2. Paveley, N.D., Rennie, W.J., Reeves, J.C., Wray, M.W., Slawson, D.D., Clark, W.S., Cockerell, V. and Mitchell, A.G. (1996) HGCA Research Review No. 34, Home-Grown Cereals Authority, London, UK.

3. Reeves, J.C. (1995) in New Diagnostics in Crop Sciences (Skerritt J.H. and Appels, R., eds) pp. 127–149, CAB International, Wallingford, Oxon, UK.

4. Sheppard, J.W. (1993) Proceedings of the 1st ISTA Plant Disease Committee Symposium on Seed Health Testing, pp. 132–142, Ottawa, Canada.

5. Rasmussen, O.F. and Reeves, J.C. (1992) J. Biotechnol. 25, 203–220.

6. Reeves, J.C., Rasmussen, O.F. and Simpkins, S.A. (1994) Proceedings of the 8th ICPPB. Les Colloques de INRA, pp. 383–390.

7. Saiki, R.K., Gelfand, D.H., Stoffel, S., Scharf, S.J., Higuchi, RG., Mullis, K.B. and Erlich, H.A. (1998) Science 239, 487–491.

8. Caetano-Anolles, G., Bassam, B.J. and Gresshoff, P.M. (1992) Bio/Technology 10, 937.

9. Blakemore, E.J.A. and Reeves, J.C. (1993) Proceedings of the 1st ISTA Plant Disease Committee Symposium on Seed Health Testing, pp. 19–22, Ottawa, Canada.

10. Jaccoud, D. (1996) Ph.D. Thesis, University of Cambridge, Cambridge, UK.

11. Gardes, M. and Bruns, T.D. (1996) in Species Diagnostics Protocols; PCR and other Nucleic Acid Methods (Clapp J.P., ed.) pp. 177–186, Humana Press Inc., USA.

12. Xue, B., Goodwin, P.H. and Annis, S.L. (1992) Physiol. Mol. Plant Pathol. 41, 179–188.

13. Lee, S.B., White, T.J. and Taylor, J.W. (1993) Phytopathology 83, 177–181.

14. Morton, A., Carder, J.H. and Barbara, D.J. (1995) Plant Pathol. 44, 183–190.

15. Stevens, E.A., Blakemore, E.J.A. and Reeves, J.C. (1997) Seed Health Testing: Progress towards the 21st Century (Hutchins J.D. and Reeves J.C. eds) CAB International, Wallingford, Oxon., in press.

16. Tyagi, S. and Kramer, F.R. (1996) Nature Biotechnol. 14, 303–308.

ISOLATION AND ANALYSIS OF DOUBLE-STRANDED RNA IN PLANT TISSUES TO DETECT VIRUS AND VIRUS-LIKE AGENTS

A. TEIFION JONES

Scottish Crop Research Institute, Invergowrie, Dundee, DD2 5DA, Scotland, UK

1. INTRODUCTION

It is an accepted premise of cell molecular biology that most, if not all, plant RNA is transcribed directly from DNA. As such, normal healthy plants should contain no double-stranded RNA (dsRNA) molecules. In practice, small amounts of dsRNA of low M_r are sometimes detected in such plants but the reasons for this are not clear. However, the detection of dsRNA species of M_r greater than about 1×10^6 usually indicates some form of abnormality, the most likely being the presence of an agent with an RNA genome. Most of the many plant viruses that have been described have single-stranded RNA (ssRNA) genomes and, during their replication cycle in plants, dsRNA is produced as a transitory product, although the amount of this dsRNA that accumulates at any point in time may often be very small. Replication occurs by the production of an RNA strand complimentary to that of the virus genome parts. The RNA then strands with complementary sequences to form a base-paired double-stranded template for transcription of genomic viral RNA. Complimentary strands should therefore be equal in size to the individual genomic component strand(s) from which they were made, and the dsRNA of these species will be twice that size. However, at any point in time during the replication process in plant cells, there will be various extents of double-strandedness. On disruption of plant cells for nucleic acid extraction, the single-stranded arms of these 'intermediate' dsRNA forms will be destroyed by ribonucleases, leaving less than full length sized dsRNA. These 'incomplete' dsRNA species are termed *replicative intermediates* (RI) in contrast to the full length *replicative forms* (RF). Although such RIs may occur sometimes in sufficient quantities to be detected by dsRNA analysis, the bulk of the dsRNA obtained is of RF.

In addition to ssRNA viruses, a few plant viruses and many fungal viruses have dsRNA genomes and, because the end-product of their replication is dsRNA, dsRNA usually accumulates in infected tissues to a greater extent than the transitory product produced by infection with some ssRNA viruses. Recovery and detection of dsRNA from infection with these dsRNA-containing agents is therefore often much easier and more reliable than from infections with some ssRNA-containing agents.

A.C. Cassells (ed.), Pathogen and Microbial Contamination Management in Micropropagation, 97–105.
© *1997 Kluwer Academic Publishers. Printed in the Netherlands.*

In principle, therefore, the isolation and detection of dsRNA in plants offers a broad-based means of detecting the presence of these different RNA-containing agents without the need for agent-specific materials such as antibodies or nucleic acid probes. However, the successful detection of dsRNA in plants is dependent on its separation from the vastly larger amounts of host ssRNA, DNA and protein present in all cells, and on sensitive techniques to detect and analyse the small amounts of dsRNA recovered.

2. METHODOLOGY OF dsRNA ISOLATION, PURIFICATION AND DETECTION

Several different procedures have been used for the purification of dsRNA from plant tissues, each based on the different responses of nucleic acid species to various treatments. The most commonly used procedures utilise the specific response of dsRNA to enzymic treatment with DNase and/or RNase [1], solubility in salt solutions such as LiCl [2], sedimentation rate in sucrose density gradients [3], buoyant density in Cs_2SO_4 [4] or binding to cellulose powder at specific ethanol concentrations [5,6]. Usually, for high purity preparations of dsRNA, combinations of these treatments are used. For plant material, the use of cellulose chromatography to purify dsRNA has been the method most widely used because of its simplicity and the need for only minimal equipment. The advent of new stains for increased sensitivity of detection of nucleic acids has also assisted the widespread use of this approach. The protocol used by various workers differ in detail but is essentially that described by Morris and Dodds [6]. There are several important components in this protocol:

1. Adequate disruption of plant tissues and cells to release nucleic acids.
2. A suitable buffer system to denature proteins and minimise ribonuclease activity.
3. Separation of nucleic acids from proteins and other cell components.
4. Fractionation of nucleic acids to give an enriched component of dsRNA.
5. Removal of any residual contaminating ssRNA and DNA.
6. Detection of the dsRNA components present in the final extract.

Extraction of nucleic acid from tissues of some plants is made difficult because they contain high levels of tannins, carbohydrate or fibres, and some modifications of protocols are usually required to overcome these or other difficulties. The following two protocols are based on methods developed for (i) simple and rapid detection of cryptic viruses in *Vicia faba* leaves [7] and (ii) dsRNA detection from strawberry leaves which contain high levels of mucilaginous material on extraction [8].

2.1. Method I (rapid analysis used for alpha- and beta-cryptoviruses)

Using this method on *Vicia faba* tissue, no significant amounts of contaminating ssRNA or DNA were detected so that action to remove these contaminating nucleic acids was not necessary. However, this may not be so for other plant/virus combinations and tests should always be made to determine that those nucleic acid species detected are indeed dsRNA.

Procedure:

- Powder *c.* 0.1–0.2 g of fresh plant material (usually leaves) in liquid nitrogen in a pestle and mortar and extract the powdered tissue in 2 ml double-strength STE buffer ($1 \times$ STE = 100 mM NaCl, 50 mM Tris, 1 mM EDTA, pH 8) containing 1% (w/v) SDS and 0.1% (v/v) 2-mercaptoethanol.
- Add 2 ml water-saturated phenol and 1 ml chloroform–pentanol (25:1) and stir.
- Clarify the extract by centrifuging at $10,000 \times g$ for 10 min. Recover the aqueous phase and add ethanol to a final concentration of 16–18%.
- Pour the solution into a small glass column or disposable syringe barrel packed with a small amount of cellulose powder (Cellex N-1 or Whatman CF-11) that is equilibrated with 18% ethanol in $1 \times$ STE buffer.
- Wash the cellulose with 3-4 volumes of 18% ethanol in $1 \times$ STE buffer to elute contaminating ssRNA and DNA.
- After the final wash, elute the bound dsRNA fraction with *c.* 0.5–1 ml $1 \times$ STE buffer without ethanol.
- To precipitate the dsRNA, add to the eluate $2.5 \times$ volume of ethanol and store for a few hours at -70°C or, overnight at -15°C. Recover the dsRNA by centrifuging at $10,000 \times g$ for 10 min and discard the supernatant fluid. Dry the pelleted dsRNA in air for a few minutes and resuspend in 50–100 ml TPE buffer (35 mM Tris, 30 mM NaH_2PO_4, 1 mM EDTA, pH 7.6) containing 10% (w/v) RNase-free sucrose and a trace of bromophenol blue marker dye.
- Centrifuge briefly at $5,000 \times g$ to clarify the solution before electrophoresis of 25- to 40-ml samples per gel track at 50 V for 16–20 h in 7% (w/v) polyacrylamide or for 3–5 h for 1% (w/v) agarose gels.
- After electrophoresis, stain the gels with ethidium bromide or silver nitrate following standard protocols.

2.2. Method II (for more difficult plant tissues)

- Powder *c.* 10 g of fresh or frozen plant material (usually leaves) in liquid nitrogen in a pestle and mortar and extract the powdered tissue in 30 ml TMS buffer (100 mM Tris, 10 mM magnesium acetate, 500 mM NaCl, pH 8.5) containing 1% (w/v) SDS, 2% (w/v) PVP (mol. wt 44,000), 1% (w/v) bentonite, 0.2% (w/v) DIECA and 0.1% (v/v) 2-mercaptoethanol .
- Add 30 ml water-saturated phenol containing 10% (v/v) *m*-cresol, 0.3% (w/v) 8-hydroxyquinoline and 30 ml chloroform–pentanol (25:1) and stir.
- Incubate the mixture at 60°C for 15 min before centrifuging at $10,000 \times g$ for 10 min.
- Recover the aqueous phase and, to each 1 ml of this, add 0.02 g cellulose powder (Cellex N-1 or Whatman CF-11) and 0.22 ml ethanol. Stir the suspension for 30 min at room temperature.
- Centrifuge the suspension at $10,000 \times g$ for 10 min, discard the supernatant fluid and

resuspend the cellulose pellet containing the bound nucleic acid in 2 ml STE buffer (200 mM NaCl, 50 mM Tris, 1 mM EDTA, pH 7) containing 18% (v/v) ethanol. Repeat this washing operation to elute ssRNA and DNA twice more.

- After the final resuspension in 18% ethanol in STE buffer, load this cellulose suspension into a glass column or disposable syringe barrel and allow to drain.
- Elute the bound dsRNA and any remaining contaminating ssRNA and DNA with 1–2 ml STE buffer without ethanol.
- To precipitate the nucleic acid from the eluate, add 2.5 × volume of ethanol and store for a few hours at -70°C or, overnight at -15°C.
- To remove any contaminating DNA, recover the nucleic acid by centrifuging at 10,000 × g for 10 min, discard the supernatant fluid, dry the pelleted nucleic acid in air for a few min and resuspend in 1 ml STE buffer containing 30 mM $MgCl_2$.
- Add 10 µg/ml DNase and incubate at 30°C for 60 min. Then add 1 ml 20 mM Tris buffer containing 1 mM EDTA and 4% (w/v) SDS and incubate at 60°C for a further 15 min.
- To remove the enzyme, add an equal volume of water-saturated phenol and emulsify. Break the emulsion by centrifuging at 10,000 × g for 10 min and recover the aqueous phase. To recover the dsRNA, add 2.5 × volume of ethanol and store for a few hours at -70°C or, overnight at -15°C.
- Centrifuge at 10,000 × g for 10 min, discard the supernatant fluid and dry the pellet in air before resuspending in 100 ml TPE buffer containing 10% (w/v) RNase-free sucrose and a trace of bromophenol blue marker dye.
- Centrifuge briefly at 5,000 × g to clarify the solution before electrophoresis in agarose or polyacrylamide gels as given under Method I.

It is important to use suitable sized dsRNA markers to estimate the mol. wt of any dsRNA detected. Markers obtained from purified particles of characterised dsRNA-containing phytoreoviruses or mycoviruses, or of dsRNA purified from herbaceous plants infected with well characterised viruses, such as cucumber mosaic and/or tobacco mosaic, have been used. When staining gels with silver nitrate, it should be remembered that this will stain proteins and carbohydrates as well as nucleic acids.

3. APPLICATIONS

3.1. Detecting the presence of dsRNA-containing viruses of plants

Only five virus genera with members containing dsRNA genomes are known to infect plants: *Alphacryptovirus*, *Betacryptovirus*, *Fijivirus*, *Oryzavirus* and *Phytoreovirus* (Table 1) [9]. Because of their genome constitution, replication of these viruses in plants accumulates dsRNA. Consequently, even though the concentration of particles of some of these viruses in plants is often very small, the amount of dsRNA produced in plants is frequently greater than that produced by many ssRNA viruses. This is particularly apparent with alpha- and betacryptoviruses that typically reach very low concentrations in plants. Viruses belonging to these two genera appear to be common in a wide range of plant species (Table 2) and are transmitted to a high frequency through pollen and seed but are not known

Table 1. Virus genera containing dsRNA viruses infecting bacteria, fungi or plants

Family	Virus genus	Main hosts
Cystoviridae	*Cystovirus*	Bacteria
Totiviridae	*Totivirus*	Fungi
Hypoviridae	*Hypovirus*	Fungi
Partitiviridae	*Partitivirus, Chrysovirus*	Fungi
	Alphacryptovirus, Betacryptovirus	Plants
Reoviridae	*Phytoreovirus, Fijivirus, Oryzavirus*	Plants

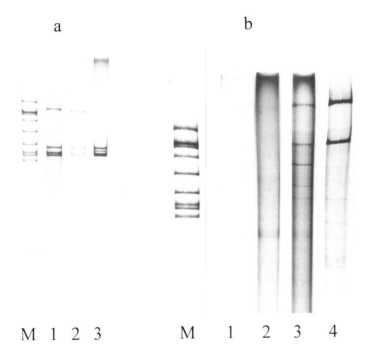

a   b

M 1 2 3   M 1 2 3 4

Figure 1. Polyacrylamide gel electrophoresis of dsRNA extracts from plants stained with silver nitrate. In each figure, track M contains dsRNA from purified particles of maize rough dwarf Fijivirus used as a mol. wt. marker. (a) Extracts from <1 g leaf from each of three different cultivars of *Vicia faba* infected with vicia cryptic virus (VCV; tracks 1, 2, 3) showing the dsRNA species characteristic of infection with this virus. (b) Extracts from *c.* 10 g leaf from each of four plants infected with raspberry ringspot nepovirus (RpRSV). Leaves are from raspberry (tracks 1, 2, 3) and *Nicotiana benthamiana* (track 4) showing the much greater recovery from herbaceous plants than from raspberry of the two main dsRNA species associated with infection with RpRSV.

Table 2. Some plant species reported to contain cryptoviruses

Alphacryptoviruses		Betacryptoviruses
Alfalfa	Pepper	Alfalfa
Beet	Poinsetta	Carrot
Carnation	Radish	Hop trefoil
Carrot	Ryegrass	Red clover
Cocksfoot	Spinach	White clover
Cucumber	*Vicia* spp.	
Fescue	White clover	
Hop trefoil		

to be transmitted by any other means, including grafting. Furthermore, their effects, if any, on plant growth and production appear to be mostly benign. Because of their cryptic nature in plants and the occurrence of their particles in only very low concentration, they have been difficult to study and detect in plants [10]. Abou-Elnasr *et al.* [7] were the first to use dsRNA analysis to reliably detect cryptic viruses in *Vicia faba* plants (Fig. 1a). They showed that as little as 0.5 g of leaf was sufficient for detection when silver staining was used to detect the dsRNA in gels.

Moreover, they were also able to detect these viruses in extracts from protoplasts of *Vicia faba* providing strong evidence that they were from plant cells and not contaminating fungi or other organisms [7]. Several other viruses that are probably species of the genera, *Alphacryptovirus* or *Betacryptovirus*, have been detected subsequently by dsRNA analysis in a range of plants including species of *Beta* [11], *Brassica* [12], *Cucumis* [13, 14] and *Dactylis* [15] (see also Table 2).

3.2. Detecting the presence of dsRNA-containing viruses of bacteria and fungi

Although comparatively few plant viruses contain dsRNA, many viruses of bacteria, fungi and related organisms known to parasitise plants can themselves be infected with dsRNA-containing viruses (Table 1). The presence of these virus-infected organisms in, or on, plant tissue can sometimes cause confusion in dsRNA analysis of plants. For example, recent studies at SCRI, Dundee, found that a single dsRNA species that was readily detected both in diseased and apparently healthy plants of several blackcurrant cultivars, had a nucleotide sequence of part of its genome that had about 50% homology with that of the RNA-dependent RNA polymerase of *Saccharomyces cerevisiae* L-A (S. Cox, A.T. Jones and M.A. Mayo, unpublished data), the type species of the genus *Totivirus*. This virus genus contains several viruses of plant-infecting fungi [16], suggesting strongly that the dsRNA species detected in the *Ribes* plants is not a virus of *Ribes* but of a possible parasite in, or on, such plants. Whilst this serves to illustrate the caution required in the interpretation of results from dsRNA analysis, it may also provide the opportunity to use dsRNA analysis to detect such viruses and hence their primary hosts in/on plant tissues. However, this approach would not provide a reliable test for such parasitic organisms because not all of them may be infected with such viruses.

3.3. Detecting the presence of ssRNA viruses of plants

As might be expected, because dsRNA is produced only as a transitory product in replication of ssRNA viruses, plants infected with such viruses usually contain considerably less dsRNA than when infected with viruses having dsRNA genomes.

However, in many instances this is compensated by the fact that the particles of some of these ssRNA viruses reach very high titres and are usually fully systemic in plants. Indeed, dsRNA is very readily detected in some plants infected with ssRNA viruses belonging to those virus genera that are known to reach such high virus particle concentrations in plants [17,18,19]. Nevertheless, dsRNA detection even for some of these viruses is greatly influenced by the plant host infected. For example, the nepoviruses, arabis mosaic and raspberry ringspot, reach high concentrations in herbaceous test plants and dsRNA is correspondingly detected readily in these hosts but, in *Rubus*, one of their natural woody hosts, detection can be much more difficult (Fig. 1b) [18,20].

For those ssRNA viruses reaching high concentrations in plants, dsRNA would not usually be the method of choice for their detection because other methods, such as serology, are an easier and more rapid means of assay but such tests do require the necessary virus-specific components. By contrast, for some ssRNA-containing viruses that are less amenable to study and detection using conventional virological techniques, dsRNA analysis has been of particular benefit, especially for the agents of carrot mottle, groundnut rosette and lettuce speckles mottle, for which no particles have been identified with confidence [21,22]. In other situations, dsRNA analysis has been useful to identify virus sub-genomic components and virus satellites as reviewed by Jones [23].

4. LIMITATIONS AND POSSIBLE PROBLEMS

The reliable detection of some RNA-containing agents, especially some ssRNA viruses, is limited by the amount of plant material available for assay. For example, dsRNA detection in plants infected by luteoviruses, that reach very low titres in plants and are confined to vascular tissue, usually requires relatively large amounts of plant tissue (>25 g). In many instances, such amounts are not likely to be available for tissue-cultured material. Consequently, dsRNA analysis of small amounts of plant material is likely to detect reliably only dsRNA-containing agents, especially viruses in the genera *Alphacryptovirus*, *Betacryptovirus*, *Fijivirus*, *Oryzavirus* and *Phytoreovirus* and viruses of fungi, and those ssRNA viruses that reach high concentrations in plants or contain satellite RNAs that reach high concentrations. The analysis is unable to identify the specific agent detected although, if it is a virus, the number and sizes of the dsRNA species detected may give some clues to the likely virus genus to which it belongs. However, the number of dsRNA species detected may be difficult to interpret if plants are infected with more than one agent or contain other virus-related RNAs. With a few exceptions, most reports of dsRNA analysis have been from plants infected with single viruses, but in nature plants are often infected with several different viruses which can cause difficulties in the interpretation of the data. Further difficulties in interpretation may arise from the presence in some virus isolates of satellite RNA molecules and/or the occurrence of sub-genomic RNAs, or defective interfering RNAs.

Additionally, the dsRNA obtained from plant extracts can be derived from sources other than the plant itself. The occurrence of organisms, such as fungi, that themselves may contain dsRNA viruses have been noted earlier. Another source is virus infection of arthropods that infest plants. A few such individuals, or even their carcasses on plants can be a potent source of contamination. For example, Watkins *et al.* [8] found that large quantities of about ten species of dsRNA were present in extracts of strawberry leaves containing the two-spotted spider mite, *Tetranychus urticae*, and in a single carcass of a leafhopper present on strawberry leaf material from the field.

In a few instances, such as *Vicia faba* [24], some cultivars of *Phaseolus vulgaris* [25] and *Rubus* species [26], some large molecular weight dsRNA species have been found to have sequences derived from the host genome and are now believed to occur in these plants by being transcribed from host DNA. The reason(s) for and significance of this are not clear.

Finally, the absence of dsRNA in assays of plant samples indicates only the absence of detectable dsRNA and does not necessarily indicate freedom from infection with RNA-containing agents.

5. CONCLUSIONS

Despite the limitations of the assay and the problems of interpreting the data outlined above, dsRNA analysis can provide a useful broad-based assay for screening plants for infection with RNA-containing pathogenic agents. This is because it does not require any of the usual pathogen-specific reagents, such as antiserum or nucleic acid-based probes, to the agents concerned. Although the analysis is unlikely to identify the precise agent(s) involved in infection, in many instances, it may be necessary only to determine the presence or absence of pathogens in source material. It is probably the method of choice for most alphacryptovirus and betacryptovirus that are often difficult to detect in plants by other means.

REFERENCES

1.   Breyel, E., Maiss, E., Ansa, O.A., Kuhn, C.W., Misari, S.M. and Demski, J.W. (1988) J. Phytopathol. 121, 118–124.

2. Diaz-Ruiz, J.R. and Kaper, J.M. (1978) Preparative Biochemistry 8, 1–17.

3. Condit, C. and Fraenkel-Conrat, H. (1979) Virology 97, 122–130.

4. Bozarth, R.F. (1976) Biochim. Biophys. Acta 442, 32–36.

5. Franklin, R.M. (1966) Proc. Natl. Acad. Sci. USA 55, 1504–1511.

6. Morris, J.T. and Dodds, J.A. (1979) Phytopathology 69, 854–858.

7. Abou-Elnasr, M.A., Jones, A.T. and Mayo, M.A. (1985) J. Gen. Virol. 66, 2453–2460.

8. Watkins, C.A., Jones, A.T., Mayo, M.A. and Mitchell, M.J. (1990) Ann. Appl. Biol. 117, 3–83.

9. Murphy, F.A., Fauquet, C.M., Bishop, D.H.L., Ghabrial, S.A., Jarvis, A.W., Martelli, G.P., Mayo, M.A. and Summers, M.D. (eds) (1995) Virus Taxonomy, Classification and Nomenclature of Viruses. Sixth Report of the International Committee on Taxonomy of Viruses. Arch. Virol. Suppl. 10, 586 pp.

10. Boccardo, G., Lisa, V., Luisoni, E. and Milne, R.G. (1987) Adv. Virus Res. 32, 171–213.

11. Xie, W.S., Antoniw, J.F., White, R.F. and Woods, R.D. (1993) Plant Pathol. 42, 465–470.

12. Jones, A.T., Abou-Elnasr, M.A., Mayo, M.A. and Hodgkin, J.R.T. (1988) Report of Scottish Crop Research Institute (SCRI) for 1985, SCRI, Scotland, UK, pp. 169–170.

13. Nameth, S.T. and Dodds, J.A. (1985) Phytopathology 75, 1293.

14. Jelkmann, W., Maiss, E., Casper, R. and Lesemann, D.E. (1988) J. Phytopathol. 121, 233–238.

15. Torrance, L., Jones, A.T. and Duncan, G.H. (1994) Ann. Appl. Biol. 124, 267–281.

16. Ghabrial, S.A., Bruenn, J.A., Buck, K.W., Wickner, R.B., Patterson, J.L., Stuart, K.D., Wang, A.L. and Wang, C.C. (1995) Arch. Virol. Suppl. 10, 245–252.

17. Dodds, J.A. (1986) in Developments and Applications in Virus Testing. (Jones R.A.C. and Torrance L. eds) pp. 71–86, Association of Applied Biologists, Wellesbourne, UK.

18. Jones, A.T., Abou-Elnasr, M.A., Mayo, M.A. and Mitchell, M.J. (1986) Acta Hortic. 186, 63–70.

19. Valvedere, R.A., Dodds, J.A. and Heick, J.A. (1986) Phytopathology 76, 459–465.

20. Jones, A.T. and Mitchell, M.J (1988) Ann. Appl. Biol. 113, 431–436.

21. Reddy, D.V.R., Murant, A.F., Raschke, J.H., Mayo, M.A. and Ansa, O.A. (1985) Ann. Appl. Biol. 107, 65–78.

22. Falk, B.W., Morris, T.J. and Duffus, J.E. (1979) Virology 96, 239–248.

23. Jones, A.T. (1992) in Techniques for the Rapid Detection of Plant Pathogens (Duncan J.M. and Torrance L., eds) pp. 115–128, Blackwell Scientific Publications, Oxford.

24. Grill, L.K. and Garger, S.J. (1981) Proc. Natl. Acad. Sci. USA 78, 7043–7046.

25. Wakarchuk, D.A. and Hamilton, R.I. (1985) Plant Mol. Biol. 5, 55–63.

26. Stace-Smith. R. and Martin. R.R. (1989) Acta Hortic. 236, 13–20.

A REVIEW OF THE *lux*-MARKER SYSTEM: POTENTIAL FOR APPLICATION IN PLANT TISSUE CULTURE

DUNCAN WHITE, KEN KILLHAM and CARLO LEIFERT

Department of Plant and Soil Science, University of Aberdeen, Cruickshank Building, Aberdeen, AB24 2UD, UK

1. INTRODUCTION

Bacterial contamination of plant tissue culture represents a large economic loss to plant propagators and it has been estimated that between 3 and 15% of plant culture is lost due to contamination [1]. Successful micropropagation relies on good aseptic technique in the initial and subsequent stages of tissue culture preparation [1]. Identification of the source of contamination and predicting its occurrence is, therefore, very important in establishing clean tissue culture lines.

The detection and subsequent monitoring of bacterial contaminants requires the ability to predict potential contamination. In many cases, the assessment of contamination is achieved through culturing methods which requires detailed taxonomic knowledge of likely contaminants. In many cases, however, contaminants may be present at very low levels which may escape detection during any of the subculturing procedures. This, therefore, makes predicting sources of contamination difficult. However, if likely contaminants are known then it is possible to examine routes of infection and plant–microbe interactions and contaminant behaviour by deliberately introducing marked bacteria either into the plant before or after preparing tissue cultures.

The aim of this paper is to outline the potential for using bioluminescence-based markers to determine the extent of specific bacterial infections in plant tissue culture and to discuss the potential for using this marker system to develop novel methods for the elimination of bacterial contamination from tissue cultures.

2. MONITORING OF SPECIFIC PLANT PATHOGENS: SPECIFIC PROBLEMS ASSOCIATED WITH PLANT TISSUE CULTURE

The monitoring of bacteria associated with plant tissue culture has concentrated on the growth and cultivation of the bacteria to obtain data relating to the location of the source(s)

A.C. Cassells (ed.), Pathogen and Microbial Contamination Management in Micropropagation, 107–113.
© *1997 Kluwer Academic Publishers. Printed in the Netherlands.*

of the contamination. Bacterial contaminants are often present without visible signs in the plant tissue culture medium [2]. The choice of medium used to culture bacterial contaminants limits the extent of determination of the contamination. This is further complicated by certain species having very specific nutritional requirements, particularly if they have been growing in plant tissue for a period. Common bacterial contaminants of plant cell culture include *Bacillus*, *Lactobacillus*, Staphylococci and Coryneforms amongst the Gram-positive types and *Acinetobacter*, *Agrobacterium*, *Erwinia*, Enterobacteracea and *Pseudomonas* amongst the Gram-negatives [1]. The use of indicator species such as *Bacillus* and *Staphylococcus* [3] allows the monitoring of the effectiveness of a procedure at a particular production stage without the need to identify all bacterial isolates, thereby removing some of the inherent problems in isolating contaminants.

The extraction of bacteria from the plant tissue/sap is important in determining the level of infection (i.e. cell numbers present) and also the types of bacteria present. Many bacteria, including those responsible for the infection, may be firmly lodged within the plant tissue, for example firmly bound to the cell wall. Therefore, either the plant tissue has to be treated to extract the bacteria or a non-extractive method has to be used. The majority of these common methods rely on either placing the sample source in direct contact with the growth media or through macerating the sample prior to making serial dilutions to obtain direct cfu counts. In certain cases, the bacterial contamination may not be present at sufficient cell densities at a specific location (in relation to disease loci) on the plant to enable detection even using plating methods (sensitivity of detection). In this case, if a simple contamination or non-contamination score is required, then an enrichment procedure could be used. With this procedure, the sensitivity of detection is increased [4].

3. USE OF MARKER TECHNOLOGY IN PLANT TISSUE CULTURE

Marker technology was developed to enable the tracking of specific bacteria released into the environment. The ideal marker system should provide information about cell densities, activity (and viability) and spatial location. Several bacterial marker systems can be used to monitor bacterial behaviour and interaction in the environment and include luciferase (*lux*) [5], *lac*ZY (β-galactosidase) [4], *xyl* E [6] and GUS (β-glucuronidase) [7]. These markers rely on a specific signal, not present in the environment in which they are released, and have been shown to be successful in soil and at the soil–plant interface [8,9].

The plant tissue culture environment represents a unique environment with which to study plant–microbe interactions since the relationship and spatial organisation between microbe and plant are very well defined. These interactions are likely to be of a different nature to plant–soil–microbe inter-relationships due to the relative complexity of the soil environment. The advantages to the plant tissue culture environment is that specific contaminants are known and can easily be marked, particularly if indicator species are used; tissue cultures are a 'contained system', routes of contamination and contaminants are relatively well known, and background signals or quenching of the marker signal is kept to a minimum. Specific problems associated with the plant tissue culture environment include relatively large sample areas (i.e. whole or part of plant) and degradation or quenching of the marker signal due to plant tissue, although with sufficiently sensitive equipment this can be largely overcome.

3.1. The *lux* marker system

The luciferase (*lux*) marker system was developed to enable the tracking of an introduced marked bacteria in the environment. *Lux* genes used in marking bacteria are derived from the marine bacteria *Vibrio fischeri* or *Vibrio harveyi*. No free-living terrestrial bacteria have, so far, been found to possess *lux*. A description of the *lux* genes is reviewed by Meighen [10] and the regulation enzymology by Hastings and Nealson [11]. A large number of Gram-positive and Gram-negative bacteria have now been marked and include both plant symbionts and plant pathogens [5,9]. Strains have been marked with either the entire complement of *lux* genes (*lux* ABCDE) or those genes responsible for the production of light (*lux* AB). In the latter case, the substrate for the reaction (aldehyde), normally synthesised by *lux* CDE, has to be supplied exogenously. So far, no difference in behaviour between the *lux*-marked strains and the parent (non-*lux*) has been found in terms of growth, activity (as determined by dehydrogenase activity) and ecological function, for example pathogenicity [5] and biocontrol [12]. The cloning of *lux* genes involves routine procedures including both chromosomal integration and the use of multicopy. Current legislation, however, requires that the use of *lux*-marked GMOs be within a contained environment.

Detection and enumeration of *lux*-marked bacteria can be through the use of plate counts and/or luminometry [9] and through the use of light-sensitive (Charge-Coupled Device) cameras to image metabolically active single cells [13], microcolonies [14] or bacteria associated with plant tissues [15]. The use of light-sensitive cameras can also be used to augment plate count information particularly if non-*lux* labelled colonies are present. Other imaging techniques used include the use of photographic film and a fibre-optic 'light-pipe'[16]. Bioluminescence can generally be monitored non-extractively and the amount of light emission (at 490 nm with *lux*) is related to the microbial biomass and metabolic activity (via the cellular pool of NAD(H)) of the cell [6]. Measurements of bioluminescence can be related to either the actual or potential activity, depending on whether substrates are present or have been added prior to measuring the luminescence output [17]. Measurements of potential luminescence, by adding nutrients such as yeast extract [18], indicate how responsive cells are to influx of nutrients from, for example, sites of root exudate leakage [8]. These particular areas have been shown to be susceptible to pathogen colonisation [19].

4. THE POTENTIAL FOR USING *lux* MARKER TECHNOLOGY IN TISSUE CULTURE

There are two main areas in plant tissue culture where bioluminescent (*lux*) marker technology could potentially be used. In both these areas, plants are deliberately inoculated with *lux*-marked strains to (i) study the interactions between host (plant) and pathogens, and (ii) track the source of contaminants

4.1. Interactions between host and pathogen

4.1.1. Use of bioluminescence to study location of pathogen within plant tissue

It is likely that different contaminants will have different loci within plant tissue and it is possible that a bacterial contaminant may be present at a different location to its point of infection. Obtaining information on the spread of bacteria within plant tissue is, therefore, important. Information on the location of bacteria using *lux*-marked strains associated with plant tissues has been obtained on the rhizoplane of wheat [8] and canola roots (*Brassica campestris*) [20], the phyloplane of French bean (*Phaseolus vulgaris*) [14] and in potato tuber [15]. It is likely, therefore, that images can be obtained from tissue culture preparations.

4.1.2. Use of bioluminescence to study survival of plant pathogens

Bacteria introduced into plant tissue show a gradual decline in culturable numbers (H. Dodd, personal communication). This is similar to the decline of culturable numbers observed when bacteria are introduced into soil [21]. For those bacteria unable to proliferate, the plant (tissue) environment could be viewed as oligotrophic with the decline in numbers attributed to these cells undergoing starvation to form microcells (<0.1 μm diameter). These cells may enter into a viable but non-culturable form (VBNC) in which they cannot be cultured using current methods. VBNC bacteria studied in pure culture and from soil have been shown to be metabolically active [18] and, therefore, cells in the plant tissue could be potentially capable of producing disease (perhaps by coming out of their VBNC state under the right conditions). Duncan *et al.* [18] used bioluminescence to study starvation and metabolic activity in *V. harveyi* and *lux*-marked *E. coli* and *Pseudomonas fluorescens*. Cell starvation and formation of VBNC forms were dependent upon the type of bacterium and temperature. Microbial activity was determined by both actual luminescence and potential luminescence measurements. Actual activity correlated with the formation of VBNC forms, whilst potential luminescence correlated with the ability of the cells to divide.

4.1.3. Use of bioluminescence to study latent infections

Latent infections occur when apparently non-pathogenic bacteria become pathogenic after prolonged incubation with the plant tissue cultures or when these plants are introduced into the soil [22]. The consequences of these infections are often serious because the pathogens can be hard to detect during tissue culture and can rapidly result in reduced plant vigour and necrotic infection [23]. Problems of detection could be attributed to the difficulty of culturing, perhaps due to the unsuitability of media used and/or the relatively low numbers of latent bacteria which may be present although in many cases there is the potential for these plants to become re-inoculated with contaminants [1]. The source of the bacteria responsible for latent infections may also not be readily known due to the time-lag between initial infection and the onset of disease symptoms [24]. Some bacteria responsible for latent infections have been shown to have very specific nutritional requirements such as growth factors. For example, *Bacillus cereus* and *Bacillus circulans* were shown to have a requirement for thiamine and biotin [25]. The monitoring of bacteria responsible for latent infections is currently hindered by lack of direct knowledge relating to the constraints for latency. The use of potential bioluminescence using specific growth substrates may enable monitoring of the bacteria as they enter into the latent phase and if latency is related to a 'reversible-starvation state', perhaps related to the VBNC phase and whether latency is related to the formation of L-forms in plants [26].

4.1.4. The use of bioluminescence to study plant–microbe interactions

Lux-marking of bacteria deliberately introduced into plant tissue culture may enable direct monitoring of the metabolic status of the infecting pathogen in response to changes in nutrient composition (e.g. nutrients released by the plant tissue) and will also enable study of possible defence mechanisms formed by the plant in response to the pathogens. For example, one can determine specific responses in terms of potential bioluminescence.

The direct linking of *lux*-genes with those responsible for environmental activity, for example using *lux* as a reporter gene for naphthalene degradation [27] has great potential. White *et al.* [9] suggested a possible role of *lux* in monitoring biocontrol and it is possible that *lux* could be linked to specific genes involved in the pathogenesis or virulence factors.

4.2. The use of *lux* in monitoring tissue culture contamination: detecting the source of contamination

Contamination of tissue culture stock can be due to either contamination of the original explant and/or infection during transfers. Specific bacteria are known to originate from contamination during subculture whilst others are from the original explant [1]. Potential specific sources of bacterial contamination include airborne and vector-associated transfer by mites and thrips [28] and incomplete sterilisation and subsequent maintenance of sterility procedures. Control of infections in plant tissue culture can, therefore, only be brought about by the careful monitoring of likely pathogens and good housekeeping of tissue culture stocks.

Several schemes have been devised to reduce or eliminate microbial contamination; for example, Leifert and Waites [3] applied the Hazard Analysis Critical Control 3-Point system to prevent or reduce contamination in commercial plant tissue culture systems. This system relates the identification of hazards associated with a particular plant growth stage to possible mechanisms of controlling the disease with reference to particular indicator organisms known to be associated with that growth stage. The re-introduction of specific *lux*-marked indicator organisms in combination with such schemes will allow monitoring of the transmission of contaminants from one explant to another. This should lead to better understanding of the epidemiology of tissue culture infections and better monitoring of production stages.

ACKNOWLEDGEMENTS

We would like to thank the Ministry for Agriculture, Food and Fisheries (MAFF, UK) for the support for this work (Open Contract Grant CSA 2767).

REFERENCES

1. Leifert, C., Morris, C.E. and Waites, W.M. (1994) CRC Crit. Rev. Plant Sci. 13, 139–183.

2. Debergh, F.C. and Vanderschaeghe, A.M. (1988) Acta Hortic. 225, 77–81.

3. Leifert, C. and Waites, W.M. (1994) in Physiology, Growth and Development of Plants in Culture (Lumsden, P.J., Nicholas J.R. and Davies W.J., eds) pp. 363–378, Kluwer Academic Publishers, Dordrecht, The Netherlands.

4. Ryder, M.H. (1994) FEMS Microbiol. Ecol. 15, 139–146.

5. Shaw, J.J., and Kado, C.I. (1986) Bio/Technology 4, 560–564.

6. Prosser, J.I. (1994) Microbiology 140, 5–17.

7. Wilson, K.J. (1995) Soil Biol. Biochem. 27, 501–514.

8. Rattray, E.A.S., Prosser, J.I., Glover. L.A. and Killham, K. (1995) Appl. Environ. Microbiol. 61, 2950–2957.

9. White, D., Leifert, C., Ryder, M.H. and Killham, K. (1996) New Phytol. 133, 173–181.

10. Meighen, E.A. (1991) Microbiol. Rev. 55, 123–142.

11. Hastings, J.W. and Nealson, K.H. (1977) Annu. Rev. Microbiol. 31, 549–595.

12. Fravel, D. (1988) Annu. Rev. Phytopathol. 26, 75–91.

13. Silcock, D.J., Waterhouse, R.N., Glover, L.A., Prosser, J.I. and Killham, K. (1992) Appl. Environ. Microbiol. 58, 2444–2448.

14. Waterhouse, R.N., Silcock, D.J., White, H.L., Buhariwalla, H.K. and Glover, L.A. (1993) Mol. Ecol. 2, 285–294.

15. McLennan, K., Glover, L.A., Killham, K. and Prosser, J. I. (1992) Lett. Appl. Microbiol. 15, 121–124.

16. de Weger L.A., Dunbar, P., Mahafee, W.F., Lugtenberg B.J.J. and Sayler, G.S. (1991) Appl. Environ. Microbiol. 57, 3641–3644.

17. Meikle, A., Glover, L.A., Killham, K., and Prosser, J.I. (1993) Soil Biol. Biochem. 26, 747–755.

18. Duncan, S., Glover, L.A., Killham, K. and Prosser, J.I. (1994) Appl. Environ. Microbiol. 60, 1308–1316.

19. Jang, S.S., Jeong, M.J., Park, C.S. and Kim, H.K. (1993) Korean J. Plant Pathol. 9, 7–11.

20. Boelens, J., Zoutman, D., Campbell, J., Verstraete, W. and Paranchych, W. (1993) Can. J. Microbiol. 39, 329–334.

21. Postma, J., van Veen, J.A., and Walter, S. (1989) Soil Biol. Biochem. 21, 437–442.

22. Herman, E.B. (1992) Agricell Rep. 18, 41–42.

23. Long, R.D., Curtin T.F. and Cassells, A.C. (1988) Acta Hortic. 225, 83–91.

24. Gunson, H.E. and Spencer-Phillips, P.T.N. (1994) in Physiology, Growth and Development of Plants in Culture (Lumsden, P.J., Nicholas J.R. and Davies W.J., eds) pp. 379–396, Kluwer Academic Publishers, Dordrecht, The Netherlands.

25. Trick, I. and Lingens, F. (1985) Appl. Microbiol. Biotechnol. 21, 245–249.

26. Aloysius, S.K.D. and Paton, A.M. (1983) J. Appl. Bacteriol. 56, 465–477.

27. Burlage, R.S., Sayler, G.S. and Larimer, F. (1990) J. Bacteriol. 172, 4749–4757.

28. Blake, J. (1988) Acta Hortic. 225, 163–166.

DNA PROFILING AND STRATEGIES FOR IN PLANTA LOCALISATION OF BACTERIA INTIMATELY ASSOCIATED WITH *BILLBERGIA MAGNIFICA* SSP. *ACUTISEPALIA*

GEORGIOS TSOKTOURIDIS[1,2], SINCLAIR MANTELL[1], ELENI BANTINAKI[2] and MADAN THANGAVELU[2]

[1] *Unit for Advanced Propagation Systems and* [2]*Molecular Diagnostics Laboratory, Horticulture Section, Wye College, University of London, Wye, Kent, TN25 5AH, UK*

1. INTRODUCTION

Billbergia, Tillandsia, Guzmania, Aechmea are examples of bromeliads with colourful foliage and floral spathes. Bromeliads are often difficult to propagate. An efficient propagation system is an essential requirement for high multiplication rates especially if it is also necessary to preserve the characteristics of hybrids and sports. Our attempts to obtain aseptic *in vitro* cultures of explants of *Tillandsia* and *Billbergia* from glasshouse-grown plants and to optimise a micropropagation procedure have been hindered by persistent bacterial or fungal contamination. Bacterial contamination in tissue culture is well documented [1]. Various experimental procedures including chemical sterilisation and antibiotics have been used to varying levels of success to minimise or eliminate such contamination. This study reports the effect of different chemical sterilants on the recovery by enrichment culture of bacterial species intimately associated with glasshouse-grown stock plants of the ornamental bromeliad *Billbergia magnifica* ssp. *acutisepalia*. RAPD analyses of the DNAs of 365 bacterial isolates revealed the existence of at least 35 different RAPD profile groups. Some of these bacteria also contain plasmids. The significance of these bacteria in the biology of the plant is unclear. Experiments using aseptically cultured 'sterile' seedlings of the mother plant are in progress to monitor the reinfestation of these bacteria, the nature of the intimate association and their localisation in plants.

2. PROCEDURE

2.1. Tissue preparation and treatment with chemical sterilants

Mature leaves were harvested from 5-year-old mother plants maintained in the glasshouse in Wye College. Leaf sections (6–7 cm) were washed in running tap water for 5 h, transferred to a conical flask containing 70% alcohol, shaken for 3 min and rinsed with sterile distilled water. All subsequent treatments were done in a sterile laminar flow hood. The leaf sections were soaked in sodium hypochlorite, chloroform (Analar BDH) or

A.C. Cassells (ed.), Pathogen and Microbial Contamination Management in Micropropagation, 115–122.
© 1997 Kluwer Academic Publishers. Printed in the Netherlands.

formaldehyde (37–41%, Analar BDH) for varying duration (listed below in section 3.1) with shaking. The sterilants were removed by washing in 100 ml of sterile distilled water, 4 × 2 min each. The leaf sections were cut and macerated with a flamed-sterile tweezers and scalpel, then transferred into autoclaved flasks containing 30 ml enrichment culture. The flasks were shaken (180 rpm) at room temperature.

2.2. Enrichment culture of bacteria

Bacteria were recovered by enrichment culture in Trypto Soya Broth (TSB, Oxoid, Unipath Ltd.) and Dextrose Broth (DB, Oxoid, Unipath Ltd.). On days 1, 3 and 5 post-inoculation, aliquots of the liquid media were streaked out on four different solid media namely Trypto Soya Agar (TSA, Oxoid, Unipath Ltd.), Dextrose Tryptone Agar (DTA), Columbia CNA Medium (CNA, Difco Laboratories Ltd.), and 523 bacterial medium (Sigma; used earlier for characterising bacterial contamination in woody plant tissue; [2]).

2.3. Isolation of bacterial DNA and RAPD analysis

Aliquots (10 ml) of TSB were inoculated with single bacterial colonies and incubated overnight with shaking at room temperature. Bacterial DNA was isolated according to Lawson [3]. The integrity and quality of the DNA was verified by electrophoresis through a 0.5% agarose gel in TAE buffer and quantified by comparing with bacteriophage lambda DNA concentration standards. RAPD analysis was performed in 30-µl reactions. RAPD products were separated by electrophoresis through 1.5% agarose gels (20 × 20 cm^2) in TAE buffer at 100 V, stained in 0.5 µg/ml ethidium bromide for 20 min, destained in water for 5 min. The DNA profile was visualised by UV transillumination and photographed using a Polaroid MP4 camera and Polaroid 665 film. DNA profiles were scored visually on 10" × 8" prints of the negatives.

3. RESULTS AND DISCUSSION

3.1. Sterilisation protocol

The 13 different chemical sterilant treatments were obtained by varying the duration of exposure to sodium hypochlorite (0.5% and 1%, w/v), chloroform and formaldehyde. The different treatments and the codes are listed below.

Chemical	Duration (treatment code)
Bleach 10%	15 min (B15), 60 min (B60)
Bleach 20%	10 min (L10), 30 min (L30), 60 min (L60)
Chloroform	15 min (C15), 20 min (C20), 40 min (C40)
Formaldehyde	8 min (F8), 12 min (F12), 18 min (F18), 25 min (F25) and 35 min (F35)

3.2. Isolation and enumeration of bacterial types

A total in excess of 500 individual single colonies, selected on the basis of contrasting phenotypic criteria (colony shape, colour and texture), were obtained from the two liquid media and the four subsequent selection processes. The distribution of the different bacterial accessions based on the analysis of 365 representative single colonies and the different RAPD profile groupings are presented in Tables 1 and 2. The two relatively mild chemical treatments (sodium hypochlorite) yielded more isolates with different RAPD profiles (Table 1). The formaldehyde treatments were more effective for minimising or eliminating growth of bacteria in the enrichment culture assays. The 25 min formaldehyde treatment yielded three RAPD profile types on the fifth day after inoculation of enrichment media while the 35 min treatment either killed all bacteria or inhibited the growth of all culturable bacteria. Isolates were subsequently grown in TSB and total bacterial DNA was isolated and fingerprinted on the basis of their RAPD profiles.

3.3. Bacterial DNA isolation and RAPD fingerprinting

The rapid mini-preparation protocol enabled the isolation of DNA from as many as 60 isolates in 8–10 h. The DNA obtained was sufficiently pure to be used as templates for PCR analysis. A representative RAPD gel for the Operon 10-mer primer OPF16 is presented in Fig. 1. The total number of bacterial accessions obtained from the different treatments and the different RAPD profile types are presented in Table 2. There was a significant decrease in the different RAPD types with increasing "harshness" of the sterilisation procedure. The seemingly mild treatment of 15 min in 0.1% sodium hypochlorite yielded ten different RAPD profile types. By superimposing the RAPD profile grouping with the resolution achieved using the two liquid media (TSB and DB) and the four subsequent solid selection media (TSA, DTA, CNA and 523) some striking trends were noticeable. RAPD profile type 1 was common to all the selection treatments. Types 7, 11 and 14 were selected only from dextrose broth and type 13 from TSB.

4. CONCLUSIONS

The control of bacteria and bacteria-like contaminants is essential for successful micropropagation. This study was designed to test the efficacy of chemical sterilants to eliminate bacteria in leaf tissue obtained from glasshouse-grown plants. By using a combination of selective media and the DNA-based RAPD profiling technique as many as 35 DNA-profile groups representing both Gram-positive and Gram-negative types were recognised. The speed, ease of use, cost-effectiveness and reproducibility has proved that RAPD profiling can be an extremely powerful and efficient procedure for characterising the large numbers of bacterial isolates which will be recovered in such an experiment. Some of these bacteria also harbour plasmids (e.g. isolates 99, 103, 291, 292, 293, 296, 305). The disappearance or elimination of specific RAPD types by the different chemical treatments suggests varying levels or degrees of association with the plant tissue. Some of these bacteria are potentially intimately associated with the explants and possibly also contribute

Table 1. Effects of different surface sterilisation treatments on bacteria profiles isolated on selective media from glasshouse-raised mother stock plants of *Billbergia magnifica* ssp. *acutisepalia*

Sterilising agent	Treatment (min)	Treatment code	Total bacteria accessions obtained	RAPD profile groupings															Total groups
				1	2	3	4	5	6	7	8	9	10	11	12	13	14	Others	
Formaldehyde	8	F8	33								+	+						1	3
	12	F12	32				+				+							6	8
	18	F18	13															3	3
	25	F25	6															2	2
	35	F35	None																
Chloroform	15	C15	50	+							+	+	+					6	10
	20	C20	20			+						+	+					3	6
	40	C40	37			+		+		+								9	12
10% Bleach	15	B15	44			+	+	+								+	+	6	11
	60	B60	42			+		+	+	+				+				7	11
20% Bleach	10	L10	25	+										+	+			1	4
	30	L30	38	+	+													4	6
	60	L60	33	+	+	+												2	5

Table 2. Effects of different Selection Plating treatments on the recovery of different bacteria group from glasshouse-raised mother stock plants of *Billbergia magnifica* ssp. *acutisepalia*

Liquid medium	Treatment	Control accessions obtained	Total bacteria accessions obtained	RAPD profile groupings															Total groups
				1	2	3	4	5	6	7	8	9	10	11	12	13	14	Others	
Trypto-Soya broth	Trypto-Soya agar	1	55	+	+	+	+		+	+	+	+	+		+	+		15	25
	Dextrose Tryptone agar	2	62	+	+	+	+	+	+		+		+		+	+		15	25
	Columbia CNA medium	2	29	+		+	+		+		+							7	12
	523	none	48	+	+	+	+	+	+		+		+		+	+		10	20
Dextrose broth	Trypto-Soya agar	1	49	+	+	+	+	+	+	+	+	+		+	+		+	10	21
	Dextrose Tryptone agar	2	54	+	+	+				+	+	+			+		+	8	16
	Columbia CNA medium	none	25	+	+	+	+				+			+			+	2	9
	523	1	50	+	+	+	+	+		+	+	+	+	+	+		+	11	22

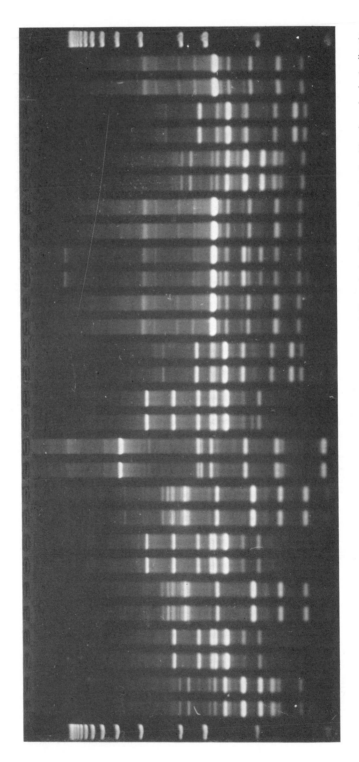

Figure 1. RAPD profiling of bacteria obtained by enrichment culture of *Billbergia* leaf tissue treated with chemical sterilants. Aliquots (10 ng) of genomic DNA isolated using the procedure described by Lawson [3] were amplified in a RAPD reaction using the Operon primer OPF16. The RAPD reaction products were separated on a 1.5% (w/v) agarose gel in 1 × TAE buffer. Gels were stained for 30 min in 0.5 µg/ml ethidium bromide and visualised by UV transillumination. Individual templates were amplified in duplicate reactions to estimate experimental error.

to the covert forms of contamination in tissue culture. Some of the chemical sterilant treatments used for complete elimination of contaminants are far from appropriate for obtaining viable plant tissue. The harsh chemical treatments used, and the subsequent recovery of culturable bacterial species suggests that some of the *Billbergia* isolates may be sheltered well within the plant tissue. The prolonged incubation which was sometimes necessary to encourage the growth of some isolates also suggests that bacteria have been dormant — possibly as endospores. It is not known whether some of the chemical sterilants stimulate or suppress the growth of bacteria.

Seedling-derived "sterile plants" of *Billbergia* — which showed no signs of culturable bacteria in repeated enrichment culture assays are being used to address questions about the localisation of the different bacteria within plants. These *in-vitro* micropropagated "clean plants" have been inoculated with some of the bacterial isolates obtained during this study. Electron microscopic examination of plant tissue derived from such *in vitro* infected plants may indicate which microenvironments, if any, within plants are preferred by these bacteria. The reinfestation of "sterile" plants with single or different isolates in combination may also verify the intimacy of the interactions seen in the mother plant. Molecular tagging using gene markers (e.g. *luc* gene in *Rhizobium*; [4]) is another method of probing such interactions.

The significance of such intimately associated bacteria and their roles, if any, in the normal biology of these plants is far from clear. Many of these interactions might be coincidental with no special significance. However, some may have significant or beneficial roles. For example, bromeliads are known to harbour nitrogen-fixing bacteria (e.g. *Pseudomonas stutzeri* in *Tillandsia recurvata*; [5]) or complex nitrogen-fixing microbial communities on the leaf [6]. The molecular classification of the different bacterial isolates based on the nucleotide sequence of the 16S rDNA genes is being pursued to classify these bacteria at the genus, species and possibly at the sub-species levels. This information may provide new directions to appreciate the significance and possible exploitation of these bacteria in the future.

ACKNOWLEDGEMENTS

Georgios Tsoktouridis was supported by a Postgraduate Research Studentship from the Greek State Scholarships Foundation.

REFERENCES

1. Leifert, C., Morris, C.E. and Waites, W.M. (1991) Crit. Rev. Plant Sci. 13, 139–183.
2. Viss, P.R., Brooks, E.M. and Driver, J.A. (1991) In Vitro Cell Dev. Biol. 27P, 42.

3. Lawson, P.A., Gharbia, S.E., Shah, H.N. and Clark, D.R. (1989) FEMS Microbiol. Lett. 113, 87–92.

4. Cresswell, A., Skot. L. and Cookson, A.R. (1994) J. Appl. Bacteriol. 77, 656–665.

5. Puente, M.E. and Bashan, Y. (1994) Can. J. Bot. 72, 406–408.

6. Brighigna, L., Montaini, P., Favilli, F. and Trejo, A.C. (1992) Am. J. Bot. 79, 723–727.

COST ANALYSIS OF DETECTION OF BACTERIA AND PHYTOPLASMAS IN PLANT TISSUE CULTURES BY PCR

SUSAN SEAL

Natural Resources Institute, University of Greenwich, Central Avenue, Chatham Maritime, Kent, ME4 4TB, UK

1. INTRODUCTION

Rapid, simple and highly sensitive detection methods are required to ensure that mother stocks are free from low-level latent bacterial infections. Such infections are a particular risk as they can be passed on to progeny plants efficiently through vegetative propagation and they may also provide a means by which pathogens can be distributed internationally.

In the past decade numerous diagnostic tests have been developed based on polymerase chain reaction (PCR) amplification of specific bacterial sequences. In many instances these tests have been shown to have advantages over detection tests based on serology or culturing, particularly in terms of sensitivity and specificity. PCR-based tests have the greatest advantage for detection of organisms that are non-culturable or difficult to culture, such as plant pathogenic phytoplasmas. They are very valuable where tests of the highest sensitivity are required, e.g. to certify that no bacteria can be detected within exported tissue cultured plantlets.

However, disadvantages of PCR-based tests, such as the high cost of labour and consumables and the presence of PCR-inhibitory compounds in plant tissues, have hindered the uptake of PCR as a routine diagnostic tool. This paper aims to highlight ways to minimise these disadvantages.

2. REDUCING THE COST OF PCR CONSUMABLES

The use of PCR requires a supply of high quality water, a scientific freezer and the availability of three work areas in separate rooms to minimise cross-contamination of samples with amplified PCR product; one room for carrying out sample preparation, another for setting up PCR reactions, and a third for gel electrophoresis. Depending on the DNA extraction method used, various centrifuges and liquid nitrogen storage facilities may also

A.C. Cassells (ed.), Pathogen and Microbial Contamination Management in Micropropagation, 123–130.
© *1997 Kluwer Academic Publishers. Printed in the Netherlands.*

be needed. For a small throughput of samples the initial capital expenditure for purchase of a thermal PCR cycler, electrophoresis equipment, UV transilluminator, camera and micropipettes will generally lie in the range of £6,000–10,000 pounds sterling.

Despite a large amount of competition between molecular biology reagent suppliers, PCR for many bacteria remains a more expensive diagnostic technique than culturing or serology. Table 1 shows that, in the UK, the approximate cost of PCR-associated consumables from DNA preparation to agarose gel electrophoresis lies in the range of £0.43 to £1.85 per sample. The figures show that the greatest variation in cost is represented by reagents required for DNA sample preparation and for gel electrophoresis with only a small proportion of the overall cost range being accounted for by the PCR consumables (£0.31–0.69). The cost of DNA extraction will depend greatly on the type of tissue being analysed, and it may not be possible to use the lower cost methods for some particularly recalcitrant plant species. Significant savings can be made in the agarose used for gel electrophoresis, and in the source of DNA polymerase, reaction tubes, deoxynucleotide triphosphates (dNTPs) and oligonucleotide primers. The cost of a single PCR test can also be reduced easily by decreasing the reaction volume from 50 µl down to 25 or even 10 µl. However, as it is good practise to run two replicate PCRs for each DNA sample, the cost range of £0.43–£1.85 per sample provides a reasonably accurate indication of cost.

It should, however, be noted that care should be taken in the purchase of the reagents and that the least costly reagents may not represent the best value. The source of DNA polymerase can significantly affect the nature of resulting PCR products and units of polymerase activity supplied are in some instances not directly comparable between suppliers. It should also be noted that PCR is a process covered by patents owned by Hoffman La Roche, Inc., and hence use of the technique at a commercial level may preclude the use of the less expensive non-licensed products available on the market.

At the lower end, the consumable cost (£0.43/sample) of PCR compares favourably to selective plating or an enzyme-linked immunosorbent assay (ELISA). However, PCR is a more time-consuming process, particularly for sample preparation and gel electrophoresis of PCR products, and in countries with high labour costs, it is the hands-on time of staff that accounts for the majority of the high costs associated with PCR.

3. REDUCING THE LABOUR COSTS ASSOCIATED WITH PCR

3.1. PCR sample preparation

PCR will only be an appropriate tool if rapid and reliable methods of sample preparation are available, and these methods should ideally be amenable to automation. There are many methods for DNA extraction and the method of choice will depend on the target bacterium

Table 1. Approximate cost (excl. VAT) of PCR consumables in the UK per sample with use of a 50 µl PCR volume

	Cost (£sterling)
DNA sample preparation	0.100–1.000
DNA polymerase (1U) and buffer	0.170–0.370
dNTPs	0.031–0.052
Primers (RP1 cartridge purified)	0.020–0.040
Thin-walled PCR tube	0.019–0.125
Mineral oil	0.004–0.020
Filtered pipette tip	0.066–0.083
Total PCR reagents	0.310–0.690
Gel electrophoresis (30 sample gel)	0.017–0.158
Total Cost of PCR Analysis	0.427–1.848

and plant tissue. Table 2 outlines a selection of methods that have been used in our laboratory, and we have found that the reliability of the more rapid methods appears to be strongly related to the plant tissue tested. Methods published in the literature often require minor modifications to make them suitable for another plant family.

PCR inhibitors are a common problem when extracting DNA from plant tissue samples for PCR analysis as many inhibitors, such as polysaccharides and phenolics, co-purify with nucleic acids. There are a large range of matrices commercially available to clean up DNA samples, e.g. Wizard (Promega, USA), Elutip-d columns (Schleicher & Schuell, Germany) but these are very expensive and hence generally unsuitable for routine diagnostic screening.

The production or co-extraction of polyphenols in DNA preparations can be minimised by inclusion of a variety of anti-oxidants (e.g. ß-mercaptoethanol, dithiothreitol, sodium bisulphite) and/or compounds that remove polyphenols such as polyvinyl polypyrrolidone. Co-purification of complex carbohydrates is often a more severe problem, but can be minimised through use of tissue extraction buffers containing substances such as sodium chloride and cetyltrimethylammonium bromide (CTAB), or cation-exchange resin like Chelex-100 (Bio-Rad labs).

Table 2. Comparison of the steps and cost involved in five DNA extraction methods. A, leaf area used for DNA extraction; B, approximate number of samples that can be processed by one person in an 8 h working day; C, cost of consumables, £££ = >£0.35, ££ = £0.10–0.35 and £ = £0.10 cost per sample; D, literature reference.

	DNA extraction method				
Step	CTAB/PVP/NaCl buffer	Clean-up resins	Single precipitation	Alkaline lysis	Tris/KCl/ EDTA lysis
1	Grind in N_2	Grind in N_2	Grind	Add NaOH	Add Tris/KCl/ EDTA buffer
2	Add CTAB/PVP NaCl buffer	Add SDW/ PVP buffer	Add Tris/NaCl/ SDS/EDTA buffer	Boil for 30 s	Heat at 95°C, 10 min
3	Heat at 60°C, 25 min	Mix and low spin (2000×g)	Vortex and centrifuge	Add HCl/Tris and Nonidet P40	Dilute 10-fold
4	Mix with chloroform– octanol	Centrifuge 13,000×g	Mix with iso-propanol	Boil 2 min	
5	Centrifuge	Add DNA purification matrix	Centrifuge		
6	Add NaCl and ethanol	56°C, 15 min vortex	Dry pellet		
7	4°C, 30 min	Boil 8 min vortex	Dissolve DNA		
8	Centrifuge	Centrifuge			
9	Wash pellet 70% EtOH				
10	Dry DNA pellet				
11	Dissolve DNA				
A	>1000 mm²	500 mm²	50 mm²	2–5 mm²	2 mm²
B	50	100	200	400	>500
C	£££	££ (£££)	££	£	£
D	[1]	[2]	[3]	[4]	[5]

Alternatively, if target sequences are abundant, simpler extraction methods can be used and inhibitors diluted out to a level where PCR can occur. The inhibitory effect of some compounds on *Taq* polymerase can also be minimised by additions of detergent (Tween 20), dimethyl sulfoxide or polyethylene glycol 400 in the PCR buffer [6].

3.2. Immunocapture PCR

When testing for low numbers of bacteria, there is a choice of either testing a large number of samples, or testing fewer samples of bulked material in which any bacteria have been selectively concentrated. One method of concentrating bacteria or phytoplasmas is to centrifuge the sample as outlined in the "clean-up resins" method in Table 2. Alternatively, if there is an antibody that binds to the organism of interest, it is possible to separate it away from the PCR-inhibitory plant compounds, release the DNA by cell lysis (achieved, e.g., by boiling, alkali or microwave treatment) and carry out PCR directly. This technique is referred to as "immunocapture PCR" (IC–PCR). Advantages of the system are that lengthy nucleic acid extraction steps are not required, and that increased sensitivities can be achieved through the use of greater sample volumes and DNA preparations not needing to be diluted prior to use [7].

IC–PCR has been used successfully for the detection of a number of pathogenic bacteria found in food [8,9] and plant samples [7]. An immunocapture PCR test for *Ralstonia ("Pseudomonas") solanacearum* has been developed (S. Seal, unpublished) and results have shown that PCR-inhibitory compounds found in soil are removed by a few phosphate buffer washes. The test was found to be more sensitive and reproducible once it was modified to a magnetic bead format (MIC–PCR), presumably due to the beads presenting a larger surface to which bacteria can bind, and also providing better access to bacteria in the sample.

Magnetic beads can add considerable cost per sample depending on the volume and source of beads used. The source of the beads can also affect the effectiveness of the method; van der Wolf and colleagues [7] reported that "Advanced Magnetics" Protein A particles were over 2-fold better than their anti rabbit IgG particles, which in turn were over 2-fold better than those of a different manufacturer.

3.3. PCR product visualisation

PCR products have traditionally been most commonly visualised through agarose gel electrophoresis, which is one of most labour intensive, and hence costly steps associated with PCR. There are a number of alternative analysis methods which, depending on throughput, may save time and money. A simple method is to include a DNA concentration indicator, such as ethidium bromide, in the reaction and observe fluorescence under UV light. However, this method has limited applications as non-specific PCR products are relatively common, and positive results will be generated regardless of the nature of the DNA product.

To confirm the identity of products and increase the sensitivity, PCR reactions can be immobilised on nylon or nitrocellulose membranes and probed with internal PCR product sequences to confirm the product's identity. A modification of this is to coat microtitre plates (e.g. CovaLink NH plates supplied by Nunc) with the probe sequence, and then detect specific PCR product in the reactions by hybridisation. PCR product that has hybridised to the well can be detected either by use of another labelled probe, or by including a labelled primer in the PCR reaction. A number of PCR product detection kits are available commercially based on similar principles, and suitable ones combine confirmation of presence of one of the primers with an internal PCR sequence to avoid detection of non-specific products. The cost of some commercially available kits (Boehringer Mannheim "PCR–ELISA" system currently costs £181 for 192 detection reactions) may make them economically unjustifiable for many diagnostic laboratories. Moreover, for laboratories with small throughput the methods described above will not result in a saving in labour costs over that required for gel electrophoresis.

4. CHOICE OF PRIMERS OF DESIRED SPECIFICITY

A vast range of specific PCR primer sets for plant pathogens have been published [10–13] and further diagnostic PCR tests for such micro-organisms appear in the scientific literature almost weekly. However, bacterial contamination in plant tissue cultures is often caused by microbes of unknown identity, and hence a general highly sensitive PCR test for any bacterium or phytoplasma might be the most useful.

Amplification of DNA encoding ribosomal RNA (e.g. 16S rRNA) has been found to be desirable, as such DNA is often present in multiple copies, and there is a great wealth of sequence data available from genebanks accessible through the World Wide Web. Comparison of 16S rDNA sequences has allowed the identification of highly conserved regions in all bacteria [14] or subgroups such as alpha proteobacteria [15], *Staphylococcus* spp. [16] or particular groups such as phytoplasmas [17], and species thereof. Oligonucleotide primers can be designed to such regions and assessed for their suitability as PCR primers.

5. INTERNAL CONTROLS

The amplification efficiency of PCR is readily affected by small differences in reaction composition. As a result false-negative results can easily occur through operator errors or the presence of inhibitory substances in DNA samples. The presence of primer dimers can act as crude indicators of inhibitors, but will not reveal low level inhibition. The inclusion of an internal control is recommended to allow identification of such false-negative results. Internal controls can be based on amplification of a plant sequence, or of a DNA added to the reaction. Considerable care should be taken to find an internal control whose amplification does not hinder exponential amplification of the target sequence. For example, if the internal control has considerable sequence homology with the target sequence, this can interfere with the amplification of the target by heteroduplex formation of the amplified products. Likewise, all primers should be checked both singly and in combination with each other, on healthy and contaminated samples, to ensure that there is no interference between the primer pairs.

Interference between primer pairs can easily be avoided by adding a control template which is amplified using the same primer pair. Such a control can be constructed by amplification of a template (of similar G/C content to the target) with primers for that template which have been modified by addition of the target template primer sequences at their 5′ ends. Providing the amplification efficiency of the control template equals that of the target template, such a control can also be used to quantify the level of bacterial contamination.

REFERENCES

1. Lodhi, M.A., Ye, G-N., Weeden, N.F. and Reisch, B.I. (1994) Plant Mol. Biol. Rep. 12, 6–13.

2. Firrao, G. and Locci, R. (1993) Lett. Appl. Microbiol. 17, 280–281.

3. Edwards, K., Johnstone, C. and Thompson, C. (1991) Nucleic Acids Res. 19, 1349.

4. Klimyuk, V.I., Carroll, B.J., Thomas, C.M. and Jones, J.D.G. (1993) Plant J. 3, 493–494.

5. Thomson, D. and Dietzgen, R.G. (1995) J. Virol. Methods 54, 85–95.

6. Demeke, T. and Adams, R.P. (1992) BioTechniques 12, 332–334.

7. Van der Wolf, J.M., Hyman, L.J., Jones, D.A.C., Grevesse, C., van Beckhoven, J.R.C.M., van Vuurde, J.W.L. and Pérombelon, M.C.M. (1996) J. Appl. Bacteriol. 80, 487–495.

8. Widjojoatmodjo, M.N., Fluit, A.C., Torensma, R., Keller, B.H.I. and Verhoef, J. (1991) Eur. J. Clin. Microbiol. Inf. Dis. 10, 935–938.

9. Fluit, A.C., Torensma, R., Visser, M.J.C., Aarsman, C.J.M., Poppelier, M.J.J.G., Keller, B.H.I. and Verhoef, J. (1993) Appl. Env. Microbiol. 59, 1289–1293.

10. Bereswill S., Pahl, A., Bellemann, P., Zeller, W. and Geider K. (1992). Appl. Environ. Microbiol. 58, 3522–3526.

11. Minsavage, G.N., Thompson, C.M., Hopkins, D.L., Leite, R.M. and Stall, R.E. (1994) Phytopathology 84, 456–461.

12. McManus, P.S. and Jones, A.L. (1995) Phytopathology 85, 618–623.

13. Seal, S.E., Jackson, L.A., Young, J.P.W. and Daniels, M.J. (1993) J. Gen. Microbiol. 139, 1587–1594.

14. Lane, D.J., Pace, B., Olsen, G.J., Stahl, D.A., Sogin, M.L. and Pace, N.R. (1985) Proc. Natl. Acad. Sci. USA 82, 6955–6959.

15. Young, J.P.W., Downer, H.L. and Eardly, B.D. (1991) J. Bacteriol. 173, 2271–2277.

16. Bentley, R.W., Harland, N.M., Leigh, J.A. and Collins, M.D. (1993) Lett. Appl. Microbiol. 16, 203–206.

17. Lee, I.-M., Hammond, R.W., Davis, R.E. and Gundersen, D.E. (1993) Phytopathology 83, 834–842.

MICROBES INTIMATELY ASSOCIATED WITH TISSUE AND CELL CULTURES OF TROPICAL *DIOSCOREA* YAMS

SINCLAIR H. MANTELL

Unit for Advanced Propagation Systems, Horticulture Section, Wye College, University of London, Wye, Ashford, Kent, TN25 5AH, UK

1. INTRODUCTION

The true yams belong to the monocotyledonous genus *Dioscorea* containing 600 species, many of which are of economic significance in the tropical and sub-tropical regions of the world as food or as sources of the steroid precursor diosgenin. Selected cultivars have been vegetatively propagated over centuries and, as a consequence, systemic pathogens such as viruses, bacteria and fungi have accumulated. When cultured *in vitro* from nodal segment and shoot tip explants of vigorously growing vines, most *Dioscorea* produce clean healthy shoots which can be multiplied satisfactorily using standard microcutting techniques [1]. Plantlets derived from yam apical meristems can also yield planting materials free of yam viruses [2]. It has been our experience, however, that apparently healthy yam shoot cultures can support a surprisingly high level of exopolysaccharide-embedded surface microflora in the absence of obvious signs of microbial contamination. Surface contaminants were first recognised during a series of SEM studies on the development of conidiospores of *Colletotrichum gloeosporioides* (the causal agent of anthracnose disease) on leaf surfaces of different yam species [3]. It was only when cultures were shipped by air mail (i.e. in the dark under fluctuating temperature conditions) or when shoot cultures were allowed to age for several months before subculturing, did conspicuous signs of bacterial growth on tissue culture media begin to appear at the bases of shoot cultures. Also when yam shoot tissues were used as sources of explants for cell [4] or protoplast [5] culture studies the presence of covert bacteria contaminants in two key food yams, *Dioscorea alata* and *Dioscorea cayenensis* [6,7], were discovered. Reported here are the results of our studies over a 10-year period on the natures of microbial contamination in tissue, cell and protoplast cultures of *Dioscorea* spp. using electron microscopy, tissue print immuno-blotting and PCR techniques.

A.C. Cassells (ed.), Pathogen and Microbial Contamination Management in Micropropagation, 131–138.
© 1997 *Kluwer Academic Publishers. Printed in the Netherlands.*

2. PROCEDURES

2.1. Yam mother stock plants

Field-grown plants were sampled in Jamaica during July/August 1991 and at Guapiles, Costa Rica in January 1992. At Wye, yam plants were raised from 100 g tuber setts planted in a 4:1 peat/grit mixture in 30-cm diameter clay pots in a glasshouse under 16-h photoperiods, $25 \pm 5°C$.

2.2. Yam tissue, cell and protoplast culture

Yam nodal segment culture, yam cell suspension culture and yam protoplast culture techniques employed throughout these studies were the same as those already described [1,5,6]. For all culture work, Murashige and Skoog salts-based media [7] were employed with minor modifications as regards sugar levels, growth regulator types and concentrations and the osmotica used. Immobilisation of yam protoplasts in alginate was carried out by dropping 5×10^6 protoplasts/ml and osmotically adjusted 3% sodium alginate mixtures into a 12 g/l $CaCl_2$ solution (pH 5.8).

2.3. Bacteria isolation, enrichment culture and identification

The standard surface sterilisation technique was brief immersion of explants in 70% aqueous ethanol prior to 15–20 min in a 10% Jeyes "Brobat" bleach solution containing 0.01% Tween 20 wetting agent, followed by four separate washes in sterilised distilled water. This was applied to yam nodal segments of *in vitro* shoot cultures, healthy vines at between 4 and 6 months after tuber sett germination (in the case of field-grown plants) and nodal and internal segments of developing tubers on glasshouse-raised yam plants. Sterilised samples were placed onto a plain MS medium supplemented with 0.3% phytagel and 2% sucrose for at least 4 weeks and only those samples which did not show signs of obvious microbial contamination were selected for further bacterial isolations. Samples of apparently clean nodal and tuber segment materials were macerated in Luria Broth (LB) and incubated on a shaker (230 rpm) for 14 days. Samples of the broth were then streaked onto LB plates and recovered colonies maintained on Trypto-Soya Broth Agar plates for long-term storage at 4°C and, in the case of a few isolates, for identification by means of fatty acid profiling and comparison with profile databases (MAFF, Harpenden: Plant Pathogenic Bacteria ID Service).

2.4. Yam genetic transformation experiments

The indirect and direct methods of transformation used for *Dioscorea* spp. are described by Tor [6,7].

2.5. Electron microscopy

For scanning electron microscopy studies, leaf and stem pieces were fixed in 2.5% glutaraldehyde in 0.1 M phosphate buffer (pH 7.0) overnight at 4°C, then washed twice in phosphate buffer. After dehydration in an acetone series, the samples were either mounted intact on stubs or the internal surfaces of samples were first exposed by freeze-fracturing in liquid nitrogen. The specimens were dried to critical point in liquid carbon dioxide, gold-coated and examined under a Hitachi S430 microscope at 15 kV accelerating voltage. For transmission electron microscopy, samples of leaves (ca 0.5×1.0 mm^2 in size) were taken from either glasshouse-grown plants (4–6 months after germination) or *in vitro* yam nodal segment cultures and fixed in 3% glutaraldehyde in 0.1 M cacodylate buffer at 4°C overnight. The following day glutaraldehyde was washed out of the samples using two washes of 0.1 M cacodylate buffer, fixed in 1% osmium tetroxide in the same buffer for 1 h and then dehydrated in a 50–100% acetone series. After embedding in epoxy resin (Araldite/Agar/Dodecenyl succinic anhydride and benzyldime methyldime), blocks were trimmed and 2-μm sections cut using a glass knife. Contrast in preliminary light microscope observations was obtained by staining in either 1% methylene blue or 1% toluidine blue in 1% disodium tetraborate. Ruthenium red and silver proteinate staining were used to determine whether β1–3 (prokaryotic and eukaryotic) or β1–4 (eukaryotic) glucans surrounded bacterial cells.

2.6. Tissue immunoblotting

The methods used for raising polyclonal antibodies in rabbits against whole cells of covert *Sphingomonas paucimobilis* and the tissue immunoblotting studies on the localisation of this bacterium in yam nodal segment cultures were carried out according to procedures used for tracing persistent *Agrobacterium tumefaciens* in transformed tobacco plants [9].

2.7. REP-PCR analyses

Repetitive Extragenic Palindromic (REP)–polymerase chain reactions (PCR) were carried out to confirm the presence of *Sphingomonas paucimobilis* in yam nodal segment cultures using the REP1R-I / REP2-1 primer, i.e. upper 5′-IIIICGICCGICATCIGGC-3′/lower 5′-ICGICTTATCIGGCCTAC-3′, respectively, under defined PCR conditions, i.e. delayed denaturation for 6 min at 95°C followed by 35 cycles of 1 min at 94°C, 1 min annealing at 40°C and 8 min polymerase activity at 65°C followed by a delay of 16 min at 65°C and a cooled soak to 4°C [3].

3. RESULTS AND DISCUSSION

A key target of the current studies was to establish the frequency with which microbes could be isolated from surface sterilised yam tissues of mother stock plants either grown *in vivo* under field conditions in the tropics or under protected glasshouse conditions at Wye, UK. This was of interest because of the known associations of symbiotic bacteria which colonise the acuminate (leaf-tip) glands of some forest yams, e.g. *Dioscorea sansibarensis* [10] and because of the increasing numbers of observations that some food yams were difficult to decontaminate during establishment of *in vitro* cultures. With this in mind, a survey of yam shoot and tuber tissues for the presence of intimate microbial associations was carried out on Caribbean yams during 1992.

3.1. Microflora in leaf tips, specialised leaf glands and tubers of yams in the field

Isolation of bacteria from surface sterilised nodal segments and young tuber pieces field- and glasshouse-raised plants in Jamaica and Costa Rica showed that at least 95% of the yam tissues cultured on bacteriological media following standard surface sterilisation treatments produced bacterial contaminants belonging mainly to the genera *Serratia*, *Enterobacter, Xanthomonas, Curtobacterium* and *Pseudomonas*. Not only bacteria were found as contaminants of food yams. In one Costa Rican *D. alata* cultivar 'Antillano', an unidentified microyeast was consistently present in nodal segment cultures to such an extent that it has proved impossible to obtain aseptic cultures of this yam. Yet, despite the presence of what appears to be a truly endophytic microbe, this yam is extremely vigorous and produces tubers of good quality.

3.2. Bacteria and microyeasts intimately associated with *in vitro* shoot cultures of yam

Having established that yams carry microflora intimately associated with their shoot and tuber tissues (the traditional source of planting material) the locations of bacteria co-existing in contaminated nodal segment cultures was studied. Detailed SEM studies revealed that, in most tissue culture accessions, small numbers of bacteria appeared to survive standard surface sterilisation procedures and this was attributed to the presence of a copious matrix covering bacteria around and under glandular hairs (Fig. 1a). The matrix was probably exopolysaccharide in nature of both host origin, as indicated by some degree of silver proteinate staining (Fig. 1a), and of bacterial origin, as indicated by strong ruthenium red staining (Fig. 1b). Large numbers of glandular hairs as well as sunken internal glands cover the surfaces of leaves and stems of many yam species [11]. The exopolysaccharide matrices on the surfaces of yam leaves probably prevented the adequate penetration of normal surface sterilising agents to underlying bacterial cells. Following the immersion of yam explants in solvents such as acetone, chloroform or hexane for a few seconds just prior to standard surface sterilisation treatment, we have been able to produce yam nodal segment cultures, up to 80% of which appeared to contain no bacterial contaminants even after several repeated enrichment culture cycles. However, as some cultures remained contaminated, we investigated whether or not microbes could exist in internal positions within yam tissue culture plants. Results of these studies showed that an unidentified microyeast could coexist in intercellular positions between cells of yam roots below the surface of culture media (Fig. 1c) and that xylem vessels could harbour bacilliform structures resembling bacterial cells (Fig. 1d). The truly internal positions of some bacteria contaminating yam shoot cultures was also dramatically demonstrated during a series of genetic transformation experiments which we carried out in 1990–1991, in which many false-positives were detected following standard histochemical β-glucuronidase (GUS) assays [6]. By making serendipitous use of the histological GUS staining procedure we found that bacteria causing

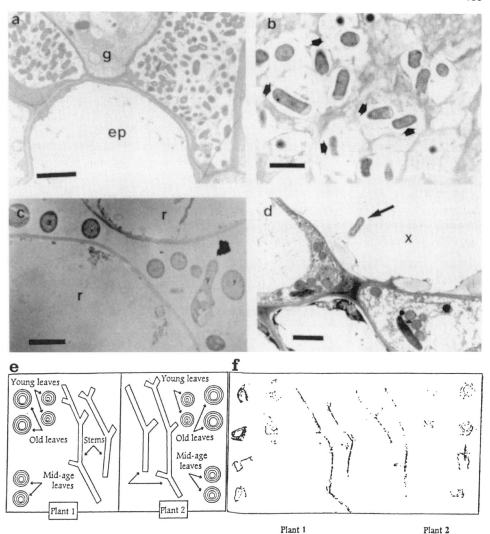

Figure 1. Bacteria and microyeasts intimately associated with shoots and roots of *in vitro* cultures of *Dioscorea* spp. (a) *Xanthomonas* spp. surrounding glandular hair on the upper epidermis of a leaf of *D. bulbifera* stained with silver proteinate. g, glandular hair; ep, epidermal cell. Bar represents 10 μm. (b) A close-up of the same area shown in Fig. 1a but stained with ruthenium red. Bacterial cells arrowed. Bar represents 5 μm. (c) An unidentified microyeast located between root cells of *D. alata*. y, microyeast cell; r, root cell. Bar represents 5 μm. (d) Bacilliform bacterium (arrow) in xylem vessel of *D. sansibarensis*. x, xylem cell. Bar represents 10 μm. (e) Layout of stems and leaves of two *D. alata* microplants printed onto nitrocellulose membrane shown in Fig. 1f. Bar represents 5 mm. (f) Tissue print immunoblots of microplants described in Fig. 1e. Bar represents 6 mm.

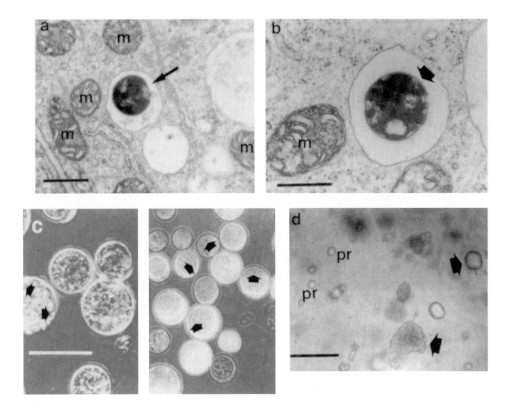

Figure 2. Endophytic bacteria in intracellular positions in embryogenic cells and protoplasts of *D. alata* cultivar 'Oriental Lisbon'. (a) A vesicle-bound bacteria-like structure (arrow) of *Citrobacter diversus* in ultra-thin section of an embryogenic cell of *D. alata*. m, mitochondrion of yam host. Bar represents 0.5 μm. (b) L-form (arrow) of *Sphingomonas paucimobilis* in *D. alata* callus cell. m, mitochondrion. Bar represents 0.1 μm. (c) Protoplasts of *D. alata* 'Oriental Lisbon' containing bacteria-like bodies (arrows). Bar represents 50 μm. (d) *D. alata* protoplasts embedded in alginate beads showing endogenously originating bacterial growth (arrow). Non-contaminated protoplasts are indicated by "pr". Bar represents 150 μm.

false-positives were located in some cases well within the vascular tissues of roots and stems and that the main species responsible was *Curtobacterium flaccumfaciens*. We went on to show that this covert bacterium in yam contained a gene (1.4 kb) homologous to the *uid*A gene (0.9 kb) of *Escherichia coli*. Our work confirmed that yam covert microbes were capable of taking up intercellular positions within yam shoot cultures *in vitro*.

3.3. Bacteria intimately associated with yam cell cultures

In light of the above results, it was not surprising that when cell suspension cultures were derived from shoot culture explants, microbial contamination proved to be problematic. Other workers have reported having problems culturing cells and protoplast culture of *Dioscorea* spp. because of persistent contaminating bacteria, e.g. in *Dioscorea opposita* [13] and mesophyll protoplasts of *Dioscorea batatas* [14]. One contaminant which we found persistently in a *D. alata* cultivar, 'Oriental Lisbon', embryogenic cell suspension line was *Citrobacter diversus*. This bacterium proved to be sensitive to kanamycin so it was not a persistent problem during genetic transformation experiments [7] in which this antibiotic was used as a selective agent. Results of a TEM study on the ultrastructure of cells in apparently "clean" embryogenic cell suspensions of *D. alata* revealed bacteria-like cells within double unit membrane-bound vesicles (Figs. 2a and 2b) in a proportion of the cells. These structures (20–30 per cell) are remarkably similar to the recently described symbiotic L-forms of endophytic nitrogen-fixing *Acetobacter* found in sugar cane [14]. Isolation of the bacteria from yam cell suspensions proved difficult although not impossible. Yam embryogenic cell protoplasts, when freshly isolated, contain small rapidly moving granules of the size and appearance of bacteria released from neighbouring broken protoplasts (Fig. 2c). These bacteria when released from yam cells did not grow in protoplast culture media. However, when protoplasts were immobilised in alginate beads, endophytic bacteria were able to grow within degenerating yam cells to produce small colonies (Fig. 2d). One of the bacteria which has been isolated from a morphogenically competent embryogenic cell suspension line "C26" of *D. alata* 'Oriental Lisbon'[4] is a slow growing strain of *Sphingomonas paucimobilis* (referred to as "HG1"), the *in vitro* locations of which are being studied by us in detail using a combination of tissue immunoblotting and PCR techniques. Polyclonal antibodies raised in rabbits against whole cells of this bacterium and antibodies conjugated to alkaline phosphatase have been used to visualise the distributions of cells of this bacterium in yam shoot cultures imprinted onto nitrocellulose membranes (Figs. 1e and 1f). In this way it has been possible to show that HG1 exists predominantly on the surfaces of the aerial parts (stems and leaves) of *in vitro* yam shoots and it is possible that it could also contaminate cell suspensions derived from tissue explants of these. The recovered bacteria were shown to be the same as the original isolated using REP–PCR analyses [3]. We are currently investigating the microecology of HG1 using gold-labelled antibodies on ultra-thin sections of yam cells and deploying a genetic tagging approach. Transformed *Sphingomonas* lines have been recovered following electroporation in the presence of plasmids containing *lux* and *npt*II genes. It is planned to use stably transformed bacteria strains to deliberately infect yam cells and tissue cultures to follow the development of bacteria on the surfaces of *in vitro* shoots and to determine whether or not they are capable of taking up intercellular and intracellular positions in cells and protoplast materials while these are being cultured under artificial conditions in the laboratory.

Contamination of *in vitro* cultures by phylloplanic microbes, because of ineffective surface sterilisation procedures and the absence of normal *in planta* ranges of antimicrobial biochemicals in both tissue-cultured plants and non-differentiated cells *in vitro,* may present potentially novel scenarios for microbes to develop opportunistic partnerships with plants. These could lead either to new pathogenesis situations (a negative feature) or possibly new beneficial symbioses (a positive feature) via interactions not normally possible in the highly selective *in vivo* environment where natural barriers to commensalism, such as heavily suberised epidermal cells, lignified cell walls and the production of a wide spectrum of antimicrobial secondary compounds, prevent rather than encourage intimate interactions. The application of sensitive detection and diagnostics methods like PCR and artificial genetic tagging will assist us to determine if these situations can occur *in vitro* through "forced" forms of intimacy.

ACKNOWLEDGEMENTS

The contribution of B.Sc., M.Sc. and Ph.D. students Donna Hartley, Pauline Jordan, Helen Sheperd, Mahmut Tor, Kodi Kandasamy and Cedric Twyford to the above studies is gratefully acknowledged, as are the valuable contributions of postdoctoral scientist Dr Helen Gunson and associates Dr Joachim Schiemann, Anja Matzk (Germany), Professor Henrietta Zhiznevskaya (Russia) and Professor Jacques Boccon-Gibod (France).

REFERENCES

1. Mantell, S.H., Haque S.Q. and Whitehall A.P. (1978) J. Hortic. Sci. 53, 95–98.

2. Mantell, S.H., Haque S.Q. and Whitehall A.P. (1980) Trop. Pest Manage. 26, 170–179.

3. Kandasamy, K.I. (1996) Ph.D. Thesis, Wye College, University of London.

4. Twyford, C.T. and Mantell, S.H. (1996) Plant Cell, Tissue Org. Cult. (in press).

5. Tor, M., Twyford, C.T., Funes, I., Boccon-Gibod, J. and Mantell, S.H. (1996) Plant Cell Tissue Org. Cult. (submitted).

6. Tor, M., Mantell, S.H. and Ainsworth C.C. (1992) Plant Cell Rep. 11, 452–456.

7. Tor, M., Ainsworth, C.C. and Mantell, S.H. (1993) Plant Cell Rep. 12, 468–473.

8. Murashige, T. and Skoog, F. (1962) Physiol. Plant. 15, 473–497.

9. Matzk, A., Mantell, S.H. and Schiemann, J. (1995) Mol. Plant Microbe Int. 9, 373–381.

10. Miller, I.M. and Reporter, M. (1987) Plant Cell Environ. 10, 413–424.

11. Orr, M.Y. (1926) Notes R. Bot. Gard. Edinburgh 73, 133–147.

12. Jordan, P. (1996) B.Sc. Integrated Study, Wye College, University of London.

13. Nagasawa, A. and Finer, J.J. (1989) Plant Sci. 60, 263–271.

14. Sauer, A. and Walther, F. (1987) in Progress in Protoplast Research (Pinte K.J. *et al.*, eds) pp. 43–44, Kluwer Academic Publishers, Dordrecht, The Netherlands.

15. James, E.K., Reis, V.M., Oliveares, F.L., Baldani, J.I. and Dobereiner, J. (1994) J. Exp. Bot. 45, 757–766.

DETECTION AND IDENTIFICATION OF BACTERIAL CONTAMINANTS OF STRAWBERRY RUNNER EXPLANTS

PIYARAK TANPRASERT and BARBARA M. REED

USDA/ARS National Clonal Germplasm Repository, 33447 Peoria Road, Corvallis, OR 97333-2521, USA

1. INTRODUCTION

Bacterial contamination is a continuing problem for research and commercial plant tissue culture labs [1,2]. Bacterial contamination is often difficult to detect [3,4]. Plants which appear healthy can contain bacteria [3,5], and some plant exudates may look similar to bacterial growth [6,7]. Contaminated plants may lack symptoms, have reduced multiplication rates, reduced rooting rates, or may die [8,9].

Bacteria in plant tissues can be detected by several types of indexing media: 523 medium [4,10]; Murashige and Skoog [11] medium alone or with additions such as yeast extract, peptone, glucose [8,12]; or coconut water [13]. Liquid MS medium alone detects most contaminants from mint explants with a few additional contaminants detected later [10]. Additional experiments with contaminated mint explants found that medium with peptone and yeast extract produced more obvious and faster bacterial growth without causing toxicity to the plants (P. Tanprasert, unpublished).

Many types of bacteria have been detected in plant cultures: *Acinetobacter, Agrobacterium, Bacillus, Corynebacterium, Curtobacterium, Enterobacter, Erwinia, Flavobacterium, Lactobacillus, Micrococcus, Pseudomonas, Staphylococcus, Xanthomonas*, and yeasts such as *Torulopsis* [5,8,12,14]. Bacterial contaminants found in plants cultured for 4 weeks or less are often motile Gram-negative bacteria, while contaminants from plants cultured for at least 12 months are more likely to be Gram-positive [8]. Kneifel and Leonhardt [15] found that some plants contain mixtures of Gram-negative rods, Gram-positive rods, and cocci. Mint cultures are contaminated mostly with Gram-negative bacteria, such as *Agrobacterium* and *Xanthomonas*, that are usually associated with plants and soils [14]. Different types of bacteria are found on various plant species and at specific stages of plant cultures.

The goals of this study were to develop good detection methods for bacterial contaminants of strawberry explants, isolate and identify bacteria from explants, characterize the antibiotic response of the isolated bacteria and successfully treat contaminated plants.

A.C. Cassells (ed.), Pathogen and Microbial Contamination Management in Micropropagation, 139–143.
© *1997 Kluwer Academic Publishers. Printed in the Netherlands.*

2. PROCEDURE

2.1. Plant material

In vitro cultures were initiated from runners taken from over 100 cultivars of pot-grown strawberries in a screenhouse at the USDA–ARS National Clonal Germplasm Repository, Corvallis, OR. Runners were disinfested by immersion in 10% household bleach (5.25% sodium hypochlorite; Clorox, Oakland, CA) solution containing 1% Tween-20 (Sigma Chemical Co., St. Louis, MO) for 10 min, rinsed twice in sterile deionized water, and grown in individual 16 × 100 mm tubes containing half-strength liquid Murashige and Skoog (MS) medium [11], pH 6.9, at 25°C for 10 days with 16 h of light (25 $\mu Em^2/s$).

2.2. Detection of contaminants

Runners (2–12 from each genotype) from 70 genotypes of *Fragaria* spp. were collected from June to August 1994 and screened for contaminants by partially submerging them in liquid medium (above). Runners (2–12 from each genotype) from 72 additional genotypes were collected from June to August 1995, and screened as above but in medium with 265 mg/l peptone and 88 mg/l yeast extract.

2.3. Isolation and purification of bacteria

Contaminants were transferred with a sterile cotton swab to 0.8% nutrient broth (Difco; Sigma Chemical, St. Louis, MO) with 1% glucose and 0.1% yeast extract at pH 6.9 and incubated at 25°C. Bacteria were purified by repetitive streaking on nutrient agar (NA) plates.

2.4. Diagnostic tests and identification of bacteria

Gram stain, KOH, motility, oxidase, starch hydrolysis, and gelatinase tests were performed according to the methods described by Klement *et al.* [16]. The ability of bacterial strains to oxidize 95 substrates was tested on Biolog Gram-negative and Gram-positive Microplates (Biolog, Inc., Hayward, CA). Identification was made by comparison with standard cultures, Biolog computer analysis and Bergey's Manual [17].

2.6. Minimal bactericidal concentrations (MBCs)

MBCs were tested for sensitivity to single antibiotics (Timentin, gentamicin, streptomycin sulfate). All two-way combinations of Timentin (125 and 250 µg/ml) and gentamicin (6.25 and 12.5 µg/ml), Timentin (125 and 250 µg/ml) and streptomycin sulfate (250 and 500 µg/ml), and streptomycin sulfate (250 and 500 µg/ml) and gentamicin (6.25 and 12.5 µg/ml) were tested. All possible three-way combinations of Timentin (125 and 250 µg/ml), streptomycin sulfate (250 and 500 µg/ml), and gentamicin (6.25, 12.5, and 25 µg/ml) were tested.

2.7. Treatment of contaminated plants

Contaminated plants were treated (10 days) with combinations of three antibiotics (Timentin, streptomycin sulfate, and gentamicin) at (mg/ml) 500 (T) + 250 (S) + 25 (G), 1000 (T) + 250 (S) + 25 (G), and 1000 (T) + 500 (S) + 25 (G).

3. RESULTS AND DISCUSSION

3.1. Detection of contaminants

Contaminants were detected in 45 of 70 genotypes initiated in 1994 in basal medium (Table 1). Only in four genotypes were all explants contaminated. Similar results were seen in 1995 in enriched medium with 53 of 72 genotypes contaminated. Only in six genotypes were all explants contaminated. The majority of contaminants were bacteria and yeast. The 10% increase in contaminants detected in 1995 might be due to the use of peptone and yeast extract in the medium; however, this was only a screening process and these results need to be verified in a controlled experiment. This technique did not distinguish between endophytic and epiphytic contaminants.

Several bacterial and yeast contaminants cannot be detected on proliferation or elongation medium, but are apparent and often lethal on rooting medium [12]. Peptone and yeast extract added to the proliferation and elongation medium allowed detection and elimination of all bacterial contaminants within two subcultures [12].

Table 1. Detection of contaminants in strawberry runner explants partially submerged in liquid medium for 10 days

Year	Tested	Only bacteria & yeasts	Only fungi	Mixed[a]	Total
		Number of genotypes			
		Contaminated			
1994[b]	70	23	11	11	45
1995[c]	72	32	12	9	53

[a]Some explants of these genotypes were contaminated with bacteria and others with fungi.
[b]Half-strength MS liquid medium.
[c]Half-strength MS liquid medium with 265 mg/l peptone and 88 mg/l yeast extract.

3.2. Characterization and identification of bacteria

Bacterial contaminants isolated from 22 strawberry genotypes included approximately 16 bacterial strains as determined by the Biolog database, standard cultures, and biochemical tests. One to four bacterial strains were found per genotype. More than half of the genotypes tested contained more than one bacterial strain. Most of the contaminants were Gram-negative, rod-shaped, motile, non-spore forming bacteria, and most were *P. fluorescens* types. In this study, the most common bacteria found were *P. fluorescens* type F (13 isolates). Two *Xanthomonas* spp., *Enterobacter cloacae* A, and several *Pseudomonas* spp. were identified, but five other Gram-negative bacteria, two Gram-positive bacteria, and two yeasts were not identified in this study. They are likely to be soil-, water-, or plant-related microorganisms because runners of strawberry plants are closely associated with the soil and could easily be contaminated by soil-borne bacteria.

3.3. Minimal bactericidal concentrations (MBCs) and plant treatment

Single antibiotics were ineffective for most bacterial isolates. Combinations of Timentin, streptomycin, and gentamicin showed promising results against all the bacteria. All bacterial strains were killed even at the lowest concentrations: (mg/ml) 125 (T) + 250 (S) + 6.25 (G). Ten-day treatments of two strawberry genotypes infected with two different bacterial isolates produced 23–100% bacteria-free plants. The three antibiotics combined produced different results depending on the plant and bacterial genotypes. These antibiotics were not phytotoxic.

4. CONCLUSIONS

Microbial contaminants were successfully detected in strawberry runners partially submerged in half-strength liquid MS medium. Contaminants were detected in 75% of genotypes tested in 1994 and 73% of genotypes tested in 1995. Half of the contaminants were bacteria and yeasts, 25% filamentous fungi and 25% mixed. Most genotypes had some contaminated explants and in a few genotypes all explants were contaminated. This detection method made it possible to quickly identify contaminated explants.

Most of the bacteria isolated from strawberry runner explants were Gram-negative, rod-shaped, motile, non-spore forming *Pseudomonas* species. *P. fluorescens* types A, F, and G were the most common contaminants found. *P. corrugata*, *P. tolaasii*, and *P. paucimobilis*, one *X. campestris*, two *Xanthomonas* spp., *Enterobacter* spp. and *Enterobacter cloacae* were also identified. Five Gram-negative and two Gram-positive contaminants could not be identified by the Biolog test. Biochemical tests confirmed Biolog test results and were useful for bacterial characterization.

Bacterial isolates were evaluated for antibiotic sensitivity. Combinations of antibiotics were most effective for killing these bacterial isolates. Initial treatment of contaminated plants produced promising results and little phytotoxicity.

REFERENCES

1. Cassells, A.C. (1990) in Micropropagation: Technology and Application (Debergh P.C. and Zimmerman, R.H., eds.) pp. 31–44, Kluwer Academic Publishers, Dordrecht, The Netherlands.

2. Leifert, C. and Waites, W.M. (1992) J. Appl. Bacteriol. 72, 460–466.

3. Debergh, P.C. and Vanderschaeghe, A.M. (1988) Acta Hortic. 255, 77–81.

4. Viss, P.R., Brooks, E.M. and Driver, J.A. (1991) In Vitro Cell. Dev. Biol. 27P, 42.

5. Leggatt, I.V., Waites, W.M., Leifert, C. and Nicholas, J. (1988) Acta Hortic. 255, 93–102.

6. Finer, J.J., Saxton, R.W., Norris, B.L., Steele, J.A. and Rahnema, S. (1991) Plant Cell Rep. 10, 380–383.

7. Bastiaens, L. (1983) Meded. Fac. Landbouwwet. Rijksuniv. Gent 48, 1–11.

8. Leifert, C., Waites, W.M. and Nicholas, J.R. (1989) J. Appl. Bacteriol. 67, 353–361.

9. Leifert, C., Pryce, S., Lumsden, P.J. and Waites, W.M.(1992) Plant Cell Tissue Org. Cult. 30, 171–179.

10. Reed, B.M., Buckley, P.M. and DeWilde, T.N. (1995) In Vitro Cell. Dev. Biol. 31P, 53–57.

11. Murashige, T. and Skoog, F. (1962) Physiol. Plant. 15, 473–497.

12. Boxus, P. and Terzi, J.M. (1987) Acta Hortic. 212, 91–93.

13. Norman, D.J. and Alvarez, A.M. (1994) Plant Cell Tissue Org. Cult. 39, 55–61.

14. Buckley, P.M., DeWilde, T.N. and Reed, B.M. (1995) In Vitro Cell. Dev. Biol. 31P, 58–64.

15. Kneifel, W. and Leonhardt, W. (1992) Plant Cell Tissue Org. Cult. 29, 139–144.

16. Klement, Z., Rudolph, K. and Sands, D.C., eds. (1990) Methods in Phytobacteriology, Akademiai Kiado, Budapest.

17. Krieg, N.R. and Holt, J. G., eds (1994) Bergey's Manual of Systematic Bacteriology. Vol. l, Williams and Wilkins, Baltimore.

EVIDENCE FOR THE OCCURRENCE OF ENDOPHYTIC PROKARYOTIC CONTAMINANTS IN MICROPROPAGATED PLANTLETS OF *PRUNUS CERASUS* CV. 'MONTMORENCY'

R. KAMOUN[1], P. LEPOIVRE[1] and P. BOXUS[2]

[1]*Laboratoire de Phytopathologie, Faculté universitaire des Sciences agronomiques 5030 Gembloux, Belgium*
[2]*Station des Cultures fruitières et maraîchères, Centre de Recherches Agronomiques de Gembloux, 5030 Gembloux, Belgium*

1. INTRODUCTION

A significantly damaging problem encountered by the plant tissue culture industry is the presence of covert microbial contaminations [1]. In this respect, bacterial contamination is responsible for considerable losses at each step of the micropropagation process, and also for the final consumers of the products (e.g. the greenhouse and nursery industries). Moreover, the increasing applications of micropropagated material as sources of reputedly pathogen-free status is placed in jeopardy since phytopathogenic bacteria can be found among these contaminating micro-organisms [2].

This report presents evidence that aseptically 'micropropagated' *Prunus cerasus* plantlets contained populations of endophytic bacteria which persist in latent form.

2. PROCEDURE

2.1. Plant material

Prunus cerasus L. 'Montmorency' (mericlone 8665) was 'aseptically' multiplied by clonal propagation on medium 706 as previously described [3].

2.2. Isolation of bacteria

In vitro-grown plantlets of *P. cerasus* were examined for the presence of endogenous bacteria by grinding the tissues in water or in CPW (Cell and Protoplast Washing) solution containing 13% mannitol [4]. After different incubation times (at 25°C and in the dark) aliquots of the homogenate were either directly plated onto 868 medium (20 g/l glucose, 10 g/l peptone, 10 g/l yeast extract and 15 g/l Difco agar), or incubated in CPW solutions containing lower concentrations of mannitol before plating them onto 868 medium.

A.C. Cassells (ed.), Pathogen and Microbial Contamination Management in Micropropagation, 145–148.
© *1997 Kluwer Academic Publishers. Printed in the Netherlands.*

2.3. PCR detection of *Pseudomonas* sp.

DNA extracts were obtained from *P. cerasus* plantlets or axenic cultures of bacteria using the method described by Sambrook *et al.* [5]. Primers (PS1 and PS2) designed to amplify a region of the gene coding for the receptor siderophore in fluorescent *Pseudomonas* sp. were used according to the protocol described by de Vos [6]. The following thermal cycling scheme was used for 30 reaction cycles (Biometra cycler): template denaturation at 94°C for 1 min, primer annealing at 60°C for 1 min and DNA synthesis at 72°C for 2 min. Amplification products were analysed by electrophoresis in a 1% agarose gel, in Tris–acetate–EDTA buffer [5]. Bands were visualised by ethidium bromide staining.

3. RESULTS AND DISCUSSION

3.1. Description and identification of endophytic bacteria

Bacterial contaminations of protoplast suspensions prepared from *P. cerasus* (mericlone 8665) led us to investigate whether these contaminants originated from the material itself rather than from casual contaminations introduced during tissue culture procedures.

Plating the homogenate of *P. cerasus* (mericlone 8665) tissues onto 868 medium after grinding and incubating (up to 5 days) it in water did not reveal any bacterial contaminant. On the other hand, plating the homogenate onto 868 medium after grinding the tissues in solutions in which osmotic pressure decreased with incubation time (either the media used in the protocol of protoplasts isolation and incubation [4] or CPW solutions containing mannitol diluted by one-third every 3 days of incubation) allowed the recovery of bacteria (after 2 days of incubation in protoplast media or after 7 days of incubation in the successive CPW solutions). Different types of morphologically distinct colonies were observed and identified with biochemical tests as *Pseudomonas syringae* and *Agrobacterium rhizogenes* species.

Moreover, electron microscopical examinations of leaf tissues indicated that wall-less prokaryotic cells were present within the cytoplasm of mesophyll cells of the micropropagated plantlets as well as in the centrifugation pellet of the homogenate of the leaf tissues.

3.2. PCR detection of *Pseudomonas*-like sequence in micropropagated plantlets

The PCR protocol applied on total nucleic acids extracted from micropropagated plantlets and from the *Pseudomonas aeruginosa* strain (PAO1) used to design the PS1 and PS2 primers allowed the amplification of a 270 pb fragment, thus suggesting the presence of bacteria belonging to the genus *Pseudomonas* in the plant material (Fig. 1). Moreover, a similar amplified product was observed from the centrifugation pellet of the leaf

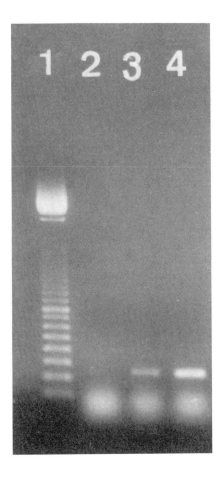

Figure 1. Analysis of PCR products amplified with PS1 and PS2 primers. Lane 1, DNA size markers. Lane 2, control without DNA. Lane 3, 270 pb amplified product from *Pseudomonas aeruginosa* (PAO1).

homogenate and from the *Pseudomonas syringae* isolate recovered from plant tissues (Kamoun, unpublished results).

Overall observations led us to conclude that *Prunus* shoot culture may contain associated bacteria which appear to be truly endophytic and which persist in latent form during the micropropagation. Similar observations of endophytic bacteria in shoot cultures have already been reported in *Dioscorea* species [7] and potato [2]. In the case of potato, evidence of wall-less prokaryotic (L-form) *Erwinia carotovora* var. *atroseptica* was reported both *in vitro* and *in vivo*. These closely associated bacteria are not directly detected by streaking explants onto classical nutrient media. Our results suggest that a preliminary incubation in solutions with decreasing osmotic pressure is a prerequisite for their recovery and growth on classical nutrient media. That preliminary step could be necessary to regenerate the cell wall of these hypothetical L-form of bacterial contaminants.

Due to its rapid execution and sensitivity threshold, the PCR technique described here can be useful for the detection of fluorescent *Pseudomonas* sp. contaminants. Compared to microbiological tests, its main limitation is its specificity. Primers targeting more conserved regions of bacterial genomes would be necessary to develop broader spectrum PCR tests to detect bacteria belonging to different species or even families.

REFERENCES

1. Reed, B.M. and Tanprasert, P. (1995) Plant Tissue Cult. Biotechnol. 1, 137–142.

2. Jones, S.M. and Paton, A.M. (1973) J. Appl. Bacteriol. 36, 729–737.

3. Druart, P. (1988) Acta Hortic. 227, 369–380.

4. Kamoun-Mehri, R., Lepoivre, P. and Boxus, P. (1994) in Quel avenir pour l'amelioration des plantes? (Aupelf-Uref, ed.) pp. 321-330. John Libbey, Paris, France.

5. Sambrook, J., Fritsch, E.F. and Maniatis, T. (1989) in Molecular Cloning: A Laboratory Manual. Cold Spring Harbor Laboratory Press, New York.

6. De Vos, D., Lim, A., De Vos, P., Sarniguet, A., Kersters, K. and Cornelis, P. (1993) J. Gen. Microbiol. 139, 2215–2223.

7. Tor, M., Mantell, S.H. and Ainsworth, C. (1992) Plant Cell Rep. 11, 452–456.

OCCURRENCE AND INFLUENCE OF ENDOGENOUS BACTERIA IN EMBRYOGENIC CULTURES OF NORWAY SPRUCE

D. EWALD[1], G. NAUJOKS[1], I. ZASPEL[1] and K. SZCZYGIEL[2]

[1]*Federal Research Centre for Forestry and Forest Products, Institute for Forest Tree Breeding, Eberswalder Chaussee 3, 15377 Waldsieversdorf, Germany*
[2]*Forest Research Institute, Section of Genetic and Physiology of Woody Plants, Bitwy Warszawskiej 1920 roku nr 3, 00-973 Warsaw, Poland*

1. INTRODUCTION

A regeneration system based on suspension cultures was established starting from embryogenic calli of Norway spruce. After 1 year with increasing times of cultivation in suspension, the regeneration capacity showed a decrease in the formation of somatic embryos ending with the failure of regeneration. Slow-growing and difficult-to-detect bacteria were found to be the cause. Several attempts were carried out to measure the influence of these bacteria, to detect them and to overcome the problem.

2. MATERIAL AND METHODS

2.1. Plant material

(1) Norway spruce seeds which were used from the German partner (G) originated from different controlled crosses. The donor trees derived from surviving plants in spruce populations in the Saxon Ore Mountains which are heavily influenced by air pollution. Embryos were harvested from ripe seeds (January 1992) and from seeds during ripening (August 1993).

(2) Seeds which were used from the Polish partner (P) were collected from the best Polish provenances. Immature as well as mature embryos were used for establishing embryogenic lines.

2.2. Culture conditions

2.2.1. Formation of embryogenic lines

(1) Embryogenic lines (G) were established on SPE-nutrient medium, according to Gupta and Durzan [1], supplemented with 1 mg/l kinetin, 2 mg/l NAA, 450 mg/l glutamic acid, 500 mg/l casamino acids and 30 g/l sucrose. Cultures were transferred to fresh medium

A.C. Cassells (ed.), Pathogen and Microbial Contamination Management in Micropropagation, 149–154.
© *1997 Kluwer Academic Publishers. Printed in the Netherlands.*

every fourth week and kept in the dark at 22°C as described by Süss *et al.* [2]. Suspension cultures were established only from these lines.

(2) Embryogenic lines (P) were initiated on a basal medium containing 2,4-D (2 mg/l) and BAP (1 mg/l) according to Szczygiel [3]. The multiplication medium contained BAP in a reduced concentration (0.5 mg/l). The cultures were transferred to fresh medium every second week and kept at 25°C in the dark.

2.2.2. Liquid cultures

Callus parts, at least 1 g fresh weight (FW), were transferred into 50 ml of nutrient medium in 100-ml Erlenmeyer flasks on a rotary shaker at 75 rpm in the dark. When continuous growth started, portions of 2 g fresh weight of the embryo-suspensor masses (ESM) were used as inoculum for precultures. Precultures were transferred to fresh medium every week. The culture step from which the ESM was harvested for regeneration (main culture) was inoculated with the same density but cultured for 2 weeks under identical conditions. The carbon source of the main culture medium was either sucrose (30 g/l, variant Z1), sucrose/starch (15:15 g/l, variant Z2) or sucrose/starch (7.5:22.5 g/l, variant Z3).

2.2.3. Regeneration of somatic embryos

The ESM from the main culture variants were harvested and resuspended in nutrient media with the basal medium (SPE) containing either 8 mg/l (30.26 µM) abscisic acid (ABA) alone or in combination with 7.5% polyethylene glycol 4000 (PEG) and 30 g/l sucrose to induce embryo formation. Portions (0.25 g FW) were transferred onto viscose tissue sheets. The sheets with the attached ESM were placed onto the same, but solidified medium, in petri dishes. After 4 weeks the viscose sheets with cells and proembryos were transferred onto petri dishes with the SPE basal medium without phytohormones but with 0.5% activated charcoal (AC) sealed with cellophane to gradually reduce air humidity. Four weeks later the number of early stage embryos, ripe embryos with cotyledons and precocious germinated embryos of these two stages was counted. Twelve to 24 viscose sheets were used per variant. The cultures were kept in the dark at 20°C.

2.2.4. Detection of bacteria

The extremely slow-growing bacteria were detected under the viscose tissue support after the ABA-treatment period as a very slight clouding on the surface of the solidified medium.

The medium which was found to be useful for a reliable detection of the bacteria was a Tryptone–Yeast-medium (TY). Five grams of Bacto-tryptone (DIFCO), 3 g yeast extract (SERVA 24540) and 10 g agar–agar were dissolved in 1 litre distilled water. The pH was adjusted to 6.8. An amount of 1.029 g (7 mM) $CaCl_2$ was dissolved in 5 ml distilled water and added as a filter-sterile solution after autoclaving. The medium was used in plastic petri dishes for detection of bacteria.

Callus parts or drops of nutrient medium from the suspension culture were placed in the middle of a cross (10 mm diameter) which was cut with a sterile scalpel into the surface of the TY medium. After 2–4 days the bacteria were visible as white clouding in the cross-cut.

2.2.5. Attempts to select axenic cultures

Suspension cultures in Erlenmeyer flasks were examined. Clouded cultures were discarded. From clean flasks six drops of nutrient medium were tested on petri dishes with TY medium. After 4 days, contaminated cultures were discarded. Only cultures without detectable bacteria were subcultured. This procedure was continued over 6 months. After this period a regeneration experiment was started with at least 200 replications each of 0.25 mg FW per variant.

2.2.6. Treatment with antibiotics

To eliminate bacteria, suspension cultures were subcultured for 2 weeks in the medium Z2 containing different amounts of two antibiotics. Rifampicin was tested at 50, 75 and 100 mg/l and carbenicillin was used at a concentration of 500 mg/l. Afterwards, 0.25 g of the ESM was placed on a viscose support and subcultured for 4 weeks on the same variant but on a solidified medium. From the antibiotics-containing media the calli were transferred to the medium Z1 without antibiotics. Every fourth week they were subcultured and the agar surface was examined for clouding.

3. RESULTS

3.1. Influence on somatic embryo formation

The undetected occurrence of the slow-growing bacteria after 1 year of subculturing in liquid media prevented embryo formation in the experiments (Table 1). Thus, the first sign was the failure of regeneration. The normal growth of the ESM was also influenced later on. The growth of ESM after establishment is characterised as a white callus showing suspensor cells and dense cell clusters. Calli which contain bacteria are of a greyish colour.

Although the variant Z3 in experiment 2 produced the same amount of somatic embryos as in the first experiment it has to be considered that 20.8% of all explants produced no embryos (and were not calculated). In the uninfluenced experiment 1 all explants regenerated embryos.

3.2. Attempts to select axenic cultures

A critical evaluation of cultures showed that not all vessels of suspension cultures were cloudy. This led to an attempt to test the possibility of selecting clean cultures over longer time periods. Although only tested (TY) cultures were propagated over 6 months and although only clean and well-grown cultures were used for the regeneration experiment,

Table 1. Influence of slow-growing and difficult-to-detect bacteria on the average number of somatic embryos formed in the same experiment at different times. Twenty-four explants of clone 159 were used per variant

		Number of somatic embryos formed/0.25 g FW			
Medium	Experiment[a]	Without cotyledons	With cotyledons	Precocious germination without cotyledons	Precocious germination with cotyledons
		Mean±SD	Mean±SD	Mean±SD	Mean±SD
Experiment 1					
Z1		5.1 ± 3.8	10.1 ± 5.3	3.0 ± 2.8	1.0 ± 1.0
Z2		7.2 ± 4.9	10.1 ± 8.3	2.2 ± 1.9	1.5 ± 1.3
Z3		6.5 ± 3.8	12.8 ± 6.0	2.6 ± 2.4	2.0 ± 1.6
Experiment 2					
Z1	+	0	0	0	0
Z2	+	0	0	0	0
Z3	+	9.7 ± 7.1	12 ± 10.8	1.5 ± 1.3	4.1 ± 2.8

[a]+: influenced by bacteria.
SAS Proc GLM, (Tukey test) Model Type III (unbalanced design).

no embryo formation was obtained. Explants with visible bacteria were found in all variants (Table 2). This led to the conclusion that a visual selection of axenic cultures over longer periods is impossible.

3.3. Treatment with antibiotics

After a treatment in suspension culture, and later on solid medium, with antibiotics a trend was detected in the number of sterile explants compared with the control (Table 3).

The survival of the ESM was negatively influenced. Only one explant treated with rifampicin survived whereas carbenicillin showed a better mode of action as regards the number of sterile as well as surviving explants. The sterile and growing calli were visually examined for bacteria every fourth week when they were subcultured anew. After 6 months it became evident that all surviving calli showed evidence of bacterial contamination. This led to the conclusion that the chosen treatments only suppressed bacterial growth. This was one piece of evidence for the supposition that also in newly established callus lines a constant propagation of those bacteria over longer time periods takes place.

Table 2. Influence of slow-growing and difficult-to-detect bacteria on the average number of somatic embryos formed after selection of apparent axenic cultures

Medium	No. of explants	Without cotyledons	With cotyledons	Precocious germination without cotyledons	Precocious germination with cotyledons	No. of explants with visible bacteria
Z1	214	0	0	0	0	36
Z2	208	0	0	0	0	14
Z3	205	0	0	0	0	11

3.4. Occurrence of the slow-growing bacteria in ESM

From 12 established ESM lines it was possible to obtain five lines as suspension cultures. After the visible influence of bacteria on the regeneration capacity of suspension cultures all ESM were tested with TY medium. Bacteria were detected in all lines.

In two ESM lines which were established from Polish provenances and not expected to be contaminated these slow-growing bacteria also occurred. The influence on growth and regeneration on solid media was not as extreme as in suspension cultures. Although the bacteria in the suspension culture were detectable by clouding, in some cultures no clouding was observed. White globules were visible which at first appeared to be proembryos, but which were later on detected with TY medium to be the slow-growing pleiomorphic bacteria.

Table 3. Occurrence of slow-growing bacteria after treatment with antibiotics and a subculture period of 4 weeks without antibiotics (24 viscose tissue sheets each with 0.25 g FW were used per variant)

Variant	No. of sheets with growing callus	Sterile explants %
Control	0	29.2
50 mg/l rifampicin	1	79.2
75 mg/l rifampicin	0	79.2
100 mg/l rifampicin	0	87.5
500 mg/l carbenicillin	15	87.5

154

3.5. Description of bacteria

Attempts to isolate these bacteria revealed three different types and gave some evidence for a very slow-growing coccoid-like strain and a rod-shaped strain. It is assumed that the third strain, characterised by its distinctive pigmentation, is *Micrococcus luteus*. It is presumed that this strain of *M. luteus* is not an endogenous spruce-specific strain but a contaminant caused by the range of cultivation steps.

4. DISCUSSION

The occurrence of slow-growing and difficult-to-detect bacteria influenced the regeneration potential of embryogenic cultures. Visual selection of contaminated suspension cultures over even longer periods was not successful. Using a special medium for the detection of these bacteria showed that at least all embryogenic lines in suspension, but also on solid media, were contaminated. A selection by the help of this medium over a longer time also failed. Experiments with different antibiotics (rifampicin, carbenicillin) either led to the death of cultures or showed only a short-lasting influence (several months) over the occurrence of these bacteria. The negative influence on callus growth or embryo development observed on solid media was not as severe as in liquid media. The bacteria became visible especially after induction of maturation by abscisic acid and the forced inhibition of callus growth. Testing of embryogenic lines from other laboratories subcultured on solid media also detected bacteria characterised by the same hidden growth. Following the results, it is presumed that, in plants, bacteria exist which were introduced by the process of establishing plant tissue *in vitro*. These organisms normally live in balance with plant tissue and do not produce visible symptoms. As a result of ageing processes and changing culture conditions (e.g. transfer to a hormone-containing medium) stress seems to influence the plant tissue cultures and the bacteria appear. They are able to interrupt the growth of the cultures and, in consequence, to destroy them.

Two strategies are now striven for to overcome the problem: first, to find combinations of other antibiotics and, second, to change the culture medium in such a way that the growth and influence of the bacteria are reduced, e.g. by replacing organic nitrogen.

REFERENCES

1. Gupta, P.K. and Durzan, D.J. (1987) Biotechnology 5, 147–151.

2. Süss, R., Ewald, D. and Matschke, J. (1990) Beitr. Forstwirtschaft. 24, 126–130.0

3. Szczygiel, K. (1995) Recent Advances in Plant Biotechnology, Book of Abstracts, p. 34, Nitra, Slovak Republic.

ANTIBIOTICS IN PLANT TISSUE CULTURE AND MICROPROPAGATION – WHAT ARE WE AIMING AT?

FREDERICK R. FALKINER

Department of Clinical Microbiology, Trinity College, Dublin, CPL St. James's Hospital, Dublin 8, Ireland

1. INTRODUCTION

Antibiotics have been used and abused since the 1940s. Their original use was the treatment of infection in man, so a single host was involved and a relatively small number of different bacterial species were targeted. These agents were then put to veterinary use, involving many different species of host and a wider range of pathogens. At this point there was effectively a loss of control of the use of antibiotics and large quantities were used without prescription or even sensitivity testing for the treatment of animals and huge amounts of crude antibiotics were included in animal feedstuffs as growth promoters. In the early 1960s resistance became a problem amongst organisms isolated from animals and residues of antibiotics were found in a range of human foodstuffs. In 1969 the 'Swann Report' [1] recommended, amongst other things, that the antibiotics used for veterinary purposes should be distinct from those used in clinical practice. The aim was clearly to remove the risk of resistance to antibiotics amongst human pathogens arising from exposure to antibiotic residues or to resistant animal bacteria. This recommendation has failed, not least because the limitation to veterinary use by prescription has been imposed only relatively recently.

At the outset it must be clearly stated that the use of antibiotics in plant tissue culture must be carefully controlled so as to avoid these problems. This has not been the case and it may now be too late, despite the protestations that antibiotics should not replace asepsis, because antimicrobials have been used for these purposes since the early 1950s [2–4]. In addition it has probably escaped the attention of many that two antibiotics, tetracycline and streptomycin, have been registered for horticultural use in the USA for more than 40 years [5]. While these are long established agents not widely used in clinical practice, their use in great quantity for spraying fruit trees and various vegetables is a matter of concern for many reasons including toxicity for humans and the risk of encouraging resistance amongst environmental organisms.

A.C. Cassells (ed.), Pathogen and Microbial Contamination Management in Micropropagation, 155–160.
© *1997 Kluwer Academic Publishers. Printed in the Netherlands.*

There are two expectations related to the use of antibiotics. The first is that they should kill the pathogen/contaminant at which they are aimed. The second is that, like hospitals in the words of Florence Nightingale, they "should do ..... no harm". These expectations should clearly apply to the prophylactic or therapeutic use of antibiotics in plant tissue culture primarily because of the financial costs involved but, for a variety of reasons, this may not be so. There are many possible consequences of antibiotic usage beyond the scope of the immediate application which are not considered. These include the effects on the personnel involved, the effects on any possible beneficial organisms or environmental strains and the emergence of resistance generally.

2. CONTAMINATING BACTERIA – PATHOGEN OR COMMENSAL?

Very few bacteria are pathogenic to humans or animals and many may be simply contaminants or merely harmless commensals. The normal bacterial flora are beneficial in many ways to the animal or human and it seems reasonable to expect that the same applies to the plant kingdom – the most obvious example being *Rhizobium* spp. in root nodulation in legumes. Most bacteria produce bacteriocins which are antibiotics generally active only against organisms of the same species and which are thought to enable the producers to establish themselves in the environment. Many bacteria produce extracellular enzymes such as proteinases and lipases and some produce specialised enzymes like chitinases which may be inhibitory to fungi. So natural production of these substances may be beneficial. However, it has been demonstrated at last, that not all organisms are harmful and may even have a beneficial or protective role (Wilhelm, this volume). Clearly much work is required to determine the roles of the bacteria in the ecosystem of micropropagation and to determine whether damage to plant tissue is due to a single organism or indeed to a combination of different bacteria. It would appear too that some organisms only become significant at a later phase of propagation when there is significant overgrowth of the organism. Many plant pathogens are highly species specific; for example the many hundreds of different pseudomonads are pathogenic only for the species or sometimes genus after which they are named. This problem requires urgent attention so that the harmful organisms can be recognised when they are isolated and appropriate susceptibility tests performed to facilitate a more focused employment of antibiotics.

3. SOURCES OF BACTERIAL CONTAMINANTS

There are many possible sources of contaminating bacteria. These include the personnel, the laboratory environment and the plant tissue to be propagated. According to Leifert and Waites [6], most problems of contamination arise from the use of inefficient methods for: sterilising explants taken from *in vivo* plants; detecting contaminants in *in vitro* plant cultures; handling aseptic plant material and for the sterilisation of culture vessels, instruments and media. It would seem that the latter should not be an issue and aseptic technique is as good as the operator. This leaves the problems of identifying contaminated raw material and decontaminating it prior to propagation which have been recently reviewed [7].

4. PROPHYLACTIC USE OF ANTIBIOTICS IN PLANT TISSUE CULTURE

The incorporation of antibiotics in plant tissue culture media for the duration of tissue growth can be described as prophylaxis. For this to be effective the pathogen should be recognised and remain sensitive to the antibiotic. For an antibiotic to function, the target organisms should be metabolising actively so that the antibiotic can be taken up. The tissue culture medium is often hostile to the organisms through high salt or sugar content or unsuitable pH. The organisms will therefore not multiply and instead remain dormant and therefore resistant to, or at least tolerant of, any antimicrobial agent present. In addition the conditions may not favour the activity or stability of the agent. Micropropagation/plant tissue culture could reasonably be construed to be a laboratory technique in which a level of control and precision in the use of antibiotics may be expected. Prophylaxis should be of short duration and carefully aimed. In fact such use of antibiotics when the pathogen is unknown, is little more than a shot in the dark, indeed in greater darkness than in what one would normally consider to be prophylaxis. This is because the pathogen/contaminant is covert or unrecognised and unidentified. Further difficulties arise from the infinite range of hosts (i.e. the many species of plant tissue) likely to be cultured and the wide range of bacteria which might contaminate the tissue culture.

5. ANTIBIOTIC THERAPY

All papers discussing the use of antibiotics in plant tissue culture indicate that they should not provide an alternative to asepsis. However, there seems to be a new approach in the topical use of antibiotics. This is unlikely to be beneficial partly due to the short period of exposure but also due to the lack of precision in the dose applied. As an approach to management of contamination it should be discouraged. On the other hand, allowing the plant material to take up antibiotic systemically prior to propagation is a tactic to be preferred. All the same, treatments should not be blind but should follow culture and sensitivity testing of the contaminating organism(s). It is not possible to predict what organisms might arise from the personnel and the environment and their antibiotic susceptibility but it is feasible to culture plant tissue prior to propagation to determine whether, and with what, it is contaminated and subsequently determine the antibiotic susceptibility of the contaminants (Reed *et al.*, this volume).

6. ANTIBIOTIC RESISTANCE

From wherever the contamination arises there is a good chance that the contaminants are resistant to more than one antibiotic in common use. If the plant material is contaminated at the outset by, for example, compost or particularly farmyard manure then the organism is very likely to be resistant to a range of agents. Given that the basic rule of prophylaxis and therapy is 'Know your pathogen' it is important to identify the organisms contaminating the product if not the production line. To provide adequate blind or empiric cover for the wide range of environmental or human organisms which might contaminate the plant tissue is a tall order. As in medical microbiology, modern technology for the detection of bacteria is of little use. It is often over-sensitive or imprecise and, at present, such technology does not allow the susceptibility to antimicrobials to be determined. As successful therapy depends primarily on the susceptibility of target organisms, the infecting organism must therefore be isolated by culture, identified, and susceptibility assessed by disc testing and by determination of minimum inhibitory concentration (MIC) and minimum bactericidal concentration (MBC) (normally the MBC is 2× or 4× MIC). Whatever the approach, these tests should be carried out in conditions that reflect the conditions under which the antibiotic will be used [8,9]. In summary, the diagnosis of the infective organisms and treatment of plant material with antibiotics prior to culture (Reed *et al.*, this volume) seems to be a logical approach to the problem.

Resistance to antibiotics is defined by the levels of the agent achievable. Clearly there is no particular limit to the concentration that may be added to the culture medium except that phytotoxicity will result. So, in addition to determining the susceptibility of the bacteria, both qualitatively and quantitatively, an assessment of the toxicity to the plant tissue is important. Some bacteria may become tolerant to antimicrobials especially following prolonged exposure. (The best way of defining tolerance is that the MBC greatly exceeds the MIC, perhaps by 32-fold.) Such bacteria are therefore able to 'persist' until the antibiotic is removed.

Some organisms are inherently resistant to certain antibiotics while others acquire resistance. There are a number of ways that the bacterium may approach the problem: firstly, loss of permeability in cell membrane; secondly, modification of site of action; thirdly, destruction of drug; fourthly, alteration of metabolic pathway. These are usually genetically encoded by plasmids. As we have seen, some organisms may become insensitive while dormant or not actively metabolising or growing. Those organisms which can produce L-forms in the presence of β-lactam agents are therefore resistant to these agents because of the loss of the target and attachment sites. This may turn out to be a real rather than a theoretical, problem.

Genetically controlled resistance can be acquired by various means which results in endless possibilities of transfer. These have been outlined elsewhere [8]. There is clearly little that can be done about resistance other than to formulate a policy for the rational use of antibiotics and to focus prophylaxis and treatment on the harmful organisms. However, one possible approach to the problem of resistance and indeed to that of polymicrobial contamination, is the use of combinations of antibiotics. The first requirement of a combination of antibiotics is that they will act synergistically which may enable them to be used in lower, less toxic concentrations. It is often more difficult for an organism to become resistant to more than one antibiotic and this is the rationale behind the multiple therapy for tuberculosis. (*Mycobacterium tuberculosis* is a slow-growing organism which because of this, seems to be able to mutate to resistance quite readily. It is not certain whether other organisms which may be slow growing due to culture conditions might mutate in a similar way.) As always there is a catch. One is that the antibiotics should not be antagonistic with each other or for that matter, with any constituents of the growth medium. Synergy and antagonism are easily visualised on a sensitivity plate and the results will vary from one organism to another. It is essential that even if the antibiotics have only an additive effect they should be compatible. There is a long list of agents which are synergistic in combination but chemically incompatible and when they are administered by one route, i.e. in this case, by incorporation in the culture medium – then the agents will neutralise each other.

7. CHOICE OF ANTIBIOTICS

There is a wide range of agents from which to choose and although the pharmaceutical industry is striving to produce more 'powerful' antibiotics, new antimicrobials are all related to or derived from existing compounds. For this reason it is not difficult for the organisms to become resistant to them and indeed resistance amongst supposed target organisms is often reported long before the agent reaches the market. Given the variety of micro-organisms that might contaminate plant tissue culture and the uncertainty of the significance of some of them the choice of antibiotics is fraught. The criteria for choosing an antibiotic have been discussed elsewhere [9] but in the intervening years there seems to be little to add. Central to successful therapy is the concept of selective toxicity, i.e. antibacterial agents should be toxic to bacteria and not to the host. However, despite the fact that bacteria are procaryotic organisms and the hosts are eucaryotic, many antibacterial agents have some toxic side-effects. A more extreme problem arises with antiviral therapy. Because viruses multiply intracellularly any attempt to inhibit viral replication with an antiviral agent will interfere with host cell metabolism and in most cases will have noticeable toxic side-effects. By extrapolation it is therefore logical to choose an agent where the site of action does not exist in the host; thus, the β-lactam agents, including the various penicillins and cephalosporins, and newer agents, such as the carbapenems and monobactams which act on bacterial cell walls, remain the most likely to be beneficial and the least phytotoxic; however, the newer agents, like all recently introduced antibiotics, are expensive. For many reasons it is encouraging to see agents such as novobiocin used because it is an 'old' antibiotic with no clinical application (Benjama and Charkaoui, this volume). However, it acts on DNA replication as do the new fluoroquinolones and may in some cases inhibit the host tissue.

There are a number of characteristics of an antibiotic which are desirable when choosing such an agent for use in the control of bacterial contamination of plant tissue culture and these are reiterated here with the treatment approach in mind. This may not be an exhaustive list but some features are essential. Clearly the agent must be soluble in water and be unaffected by pH, by the constituents of any media in which the tissue is grown or maintained. It should be stable in the growth conditions, e.g. light, temperature, etc. It is useful if it can be used in combination with another agent. It should be bactericidal. Bacteriostatic agents will not achieve the desired effect of killing the contaminants and many of these agents are likely to be phytotoxic [9]. A minimal chance of emergence of resistance is to be desired. The agent should be suitable for systemic use and should be taken up by the plant tissue. It is always an advantage if the agents are inexpensive which usually means that these are the long established and perhaps generic preparations.

8. CONCLUSION

There are certain elements which should be considered in the formulation of a policy for antibiotic usage in plant tissue culture or micropropagation. These have been discussed above but they should be re-emphasised. No antibiotic can or should be considered as a replacement for the aseptic routines of surface decontamination and handling. It is essential to identify the organisms and undertake sensitivity tests and aim the therapy at specific organisms. A more focused narrow spectrum agent is preferable so as not to eliminate possibly beneficial bacteria but broad-spectrum agents and combinations of agents may be required for mixed infections and where the susceptibility has not been determined or perhaps to reduce toxicity by reducing concentrations. A long-established bactericidal agent is required and one that is not used in clinical applications.

REFERENCES

1. Joint Committee on the Use of Antibiotics in Animal Husbandry and Veterinary Medicine Report (1969) HMSO, London.

2. Katznelson, H. and Sutton, M.D. (1951) Can. J. Bot. 29, 270–278.

3. Morgan, B.S. and Goodman, R.N. (1955) Plant Dis. Rep. 39, 487–490.

4. Watts, J.W. and King, J.M. (1973) Planta 113, 271–277.

5. Levy, S.B. (1992) The Antibiotic Paradox. Plenum Publishing Corporation, New York.

6. Leifert, C. and Waites, W.M. (1990) Newsl. Int. Assoc. Plant Tiss. Cult. 60, 2–13.

7. Reed, B.M. and Tanprasert P. (1995) Plant Tiss. Cult. Biotechnol. 1 (3), 137–141.

8. Falkiner, F.R. (1988) Acta Hortic. 225, 53–56.

9. Falkiner, F.R. (1990) Newsl. Int. Assoc. Plant Tiss. Cult. 60, 13–23.

ELIMINATION OF EXTERNAL AND INTERNAL CONTAMINANTS IN RHIZOMES OF *ZANTEDESCHIA AETHIOPICA* WITH COMMERCIAL FUNGICIDES AND ANTIBIOTICS

E.M. KRITZINGER[1], R. JANSEN VAN VUUREN[1], B. WOODWARD[1], I.H. RONG[2], M.H. SPREETH[1] and M.M. SLABBERT[1]

[1]*Agricultural Research Council, Vegetable and Ornamental Plant Institute, P/Bag X293, Pretoria, 0001, Republic of South Africa*
[2]*Agricultural Research Council, Plant Protection Research Institute, P/Bag X134, Pretoria, 0001, Republic of South Africa*

1. INTRODUCTION

The genus *Zantedeschia*, commonly known as the arum or calla lily, is native to Africa. Apart from the more commonly known white *Zantedeschia aethiopica* there are six other *Zantedeschia* species in South Africa with flower colours ranging from pink, cream and yellow to maroon. As cut flowers they are very popular all over the world and a multitude of cultivars and hybrids are available (Plate 1A).

In nature, *Zantedeschia* plants grow in marshy habitats that provide a good climate for the growth of rotting bacteria and fungi. To free roots, bulbs, rhizomes and other soil-grown organs from contamination is problematic and more difficult than from non-soil-grown organs [1]. Cohen also found this to be the case, with internal contamination being the biggest problem [2]. The availability of rhizomes as a source of explant material makes it difficult to establish contamination-free material *in vitro*. Cohen [2] initiated *Zantedeschia* in tissue culture by washing the rhizomes thoroughly followed by removal of a section of the rhizome containing a bud. This was then dipped in 95% ethanol and flamed twice.

The conventional disinfestation methods [2,3] appeared to be unsuccessful, since plant material becomes damaged during the long exposure to NaOCl which is necessary for the efficient removal of external contamination. The above methods were therefore found to be unsatisfactory for the removal of contaminants. A gentler method where rhizomes are pretreated with heat and a fungicide, followed by a mild ethanol/NaOCl treatment [4] was also not sufficient to free explant material from contaminants. This also resulted in low survival rates of rhizome buds. Internal contaminants such as bacteria can be very difficult to eliminate but apart from discarding such infected explant material, treatment with antibiotics is an alternative. Care has to been taken with the use of antibiotics, as several cases of toxicity of antibiotics to plants have been reported [5–7], although many authors include antibiotics in the culture medium [8,9]. Geier [10] managed to free a satisfactory

A.C. Cassells (ed.), Pathogen and Microbial Contamination Management in Micropropagation, 161–167.
© 1997 Kluwer Academic Publishers. Printed in the Netherlands.

162

Plate 1. Establishment of *Zantedeschia aethiopica* rhizomes in tissue culture. (A) The white flower of *Zantedeschia aethiopica*. (B) Removal of the outermost layers of the rhizomes. (C) Plantlets established in culture, following fungicide and antibiotic treatment.

percentage of *Cyclamen persicum* tubers from internal contamination by treatment of the explants for 1–4 days in ashromycin before transfer to a culture media.

In this study a pretreatment of *Zantedeschia* rhizomes with a combination of two commercially available fungicides Captab 500 WP (active ingredient: dicarboximide, 500 g/kg) and Dithane M-45 (active ingredient: dithiocarbamate, 800 g/kg) was tested over a period of several days. We also investigated a cocktail of the antibiotics, ABM1, used by Gilbert *et al.* [11] and Imipenem, one of the antibiotics shown to be promising in the study by Kneifel and Leonhardt [12]. Combinations of ABM1 and Imipenem were tested by treating the explants in these solutions of antibiotics over different periods before transferring them to culture media.

2. MATERIALS AND METHODS

2.1. Plant material

Zantedeschia aethiopica rhizomes were used as explant material. The rhizomes were obtained from Hadeco, a large plant and bulb nursery near Johannesburg, where they were removed from heavy clay soil and therefore needed to be cleaned thoroughly. They were then allowed to air dry.

2.2. Disinfestation procedure

Any remaining soil or dead plant parts were removed from the rhizomes, according to the recommendation of Pierik [13]. The outermost layer of the rhizomes was removed (Plate 1B), followed by rinsing under running tap water for 1 h. Rhizomes were pretreated with a mixture of broad-spectrum commercial fungicides, Captab 500 WP (5 g/l) and Dithane M-45 (5 g/l), by shaking them in the fungicides on an orbital shaker in the dark. Rhizomes were treated for 12, 24, 48 and 52 h. Rhizomes were removed after each treatment and rinsed three times with sterile distilled water. An adaptation of the conventional sterilization method [4] was applied. Rhizomes were sterilized with 70% ethanol (v/v) for 5 min after which they were soaked for 5 min in 2% NaOCl (m/v) solution, to which 0.25 ml Tween 20 had been added. The rhizomes were again rinsed three times with sterile distilled water. Sections of the rhizomes with a bud were dissected and damaged tissue surrounding the buds was discarded. These explants were dipped in 0.5% NaOCl (m/v) solution for 1 min after which they were rinsed three times in sterile distilled water.

2.3. Antibiotics

Following a 48 h fungicide pretreatment and sterilization as described above, some explants were further used for the antibiotic treatments, while the rest were placed on culture media as controls. ABM1 (SIGMA catalogue number A9909), containing 10,000 units of penicillin, 10 mg/ml streptomycin, 25 µg/ml amphotericin B and 0.9% sodium chloride was used at concentrations of 0, 50 and 100 mg/l in combination with the antibiotic,

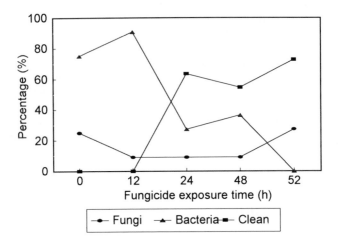

Figure 1. Percentage of fungal and bacterial contamination as well as clean cultures found after exposure to a fungicide pretreatment at various exposure times.

Imipenem at concentrations of 0, 5, 7.5, 10 and 12.5 mg/l. Antibiotics were diluted in sterile distilled water. Surface sterilised explants were exposed to the different combinations in 10 ml antibiotic mixture by shaking each explant in a 25 ml Erlenmeyer flask on an orbital shaker, in the dark at $25 \pm 1°C$. This treatment was applied for 5 days. Each treatment comprised ten replicates in two separate experiments.

2.4. Tissue culture

After each treatment, explants were transferred to Murashige and Skoog [14] growth medium supplemented with 3 mg/l BA (benzyladenine) [2]. Explants were cultured for 10 days at $25 \pm 1°C$ under low light intensity after which fungal and bacterial contamination were scored.

3. RESULTS AND DISCUSSION

The results of this study indicated that it was possible to achieve satisfactory disinfestation of *Zantedeschia* rhizomes using a fungicide pretreatment and an antibiotic or combinations of antibiotics. Figure 1 shows that a fungicide pretreatment of at least 24 h was necessary in the disinfestation procedure to obtain maximum contamination control. A large number of different fungi were detected from rhizomes before cleaning (Table 1, column A). After the outermost layer of the rhizomes had been removed and the rhizomes rinsed under running tap water for an hour, only five different fungi were detected (Table 1, column B). Following the 48 h fungicide pretreatment, cultures were contaminated by only three fungi, *Rhizoctonia, Alternaria* and *Stilbella* (Table 1, column C).

Table 1. Fungi isolated from *Zantedeschia* rhizome before wash (A), after surface cleaning (B) and from culture after 48 h of fungicide treatment (C)

Genera of fungi isolated		A	B	C
Alternaria		+	+	+
Artrobotrys		+	+	-
Aureobasidium		+	-	-
Cladosporium		+	+	-
Coemansia		+	-	-
Epicoccum		+	-	-
Fusarium		+	+	-
Gliocladium		+	-	-
Gliomastix		+	-	-
Mortierella		+	-	-
Mucor		+	-	-
Penicillifer		+	-	-
Penicillium		+	-	-
Phoma		+	-	-
cf. *Pleurocatena*	Hyphomycetes	+	-	-
cf. *Pythium*	Oomycetes	+	-	-
Rhizoctonia		+	+	+
Stilbella		+	+	+
Volutella		+	+	-

Table 2. Percentage of bacterial contamination after treatment of explants with ABM1 and Imipenem in different combinations, over 5 days

[Imipenem] (g/l)	[ABMl] (μl/ml)		
	0.00	5	10
0.00	90%	0%	0%
5	60%	0%	0%
7.5	20%	0%	0%
10	10%	0%	0%
12.5	0%	0%	0%

These are endogenous fungi in plants and would not be eliminated by surface sterilization of plant material alone. The fungicide Dithane, with dithiocarbamate as active ingredient belongs to the same group of fungicides as Benomyl (Methyl [1-[(butylamino) carbonyl]-lH-benzimidazol-2-yl] carbamate), that is also a carbamate. These repress mycorrhizal development of several fungi and have no inhibitory effect on tissue cultures [7,15,16]. Although Captab was found to be phytotoxic for the orchid *Cattleya aurantiaca* and *Stanhopea occulata* seedlings [7], no inhibitory effects were observed on *Zantedeschia* rhizomes in this study.

Bacterial contaminants isolated from rhizome material as well as from contaminated cultures were identified as *Bacillus* and *Pseudomonas* species. These two bacterial genera are most often associated with bacterial contamination detected in tissue cultures [11,17]. Due to the resistant endospore phase of *Bacillus* species, they are highly resistant to the lethal effects of heat, drying and many disinfectants [18].

According to our results (Fig. 1) the fungicides showed an increasingly antibacterial effect over time. ABM1 proved to be very effective against contaminants even at the lowest concentration of 5 µl/ml (Table 2). Imipenem was shown to be effective only at a concentration of 12.5 mg/l. The optimal treatment appeared to be ABM1 on its own at 5 µl/ml, or a combination of 5 mg/l Imipenem and 5.0 µl/ml ABM1 (Table 2). The latter treatment was most effective after 3 days exposure (Table 3). This method of using antibiotics as a treatment before transfer to culture media appears to be very successful [10]. The inclusion of the antibiotics in sterile distilled water instead of culture medium limited the bacterial growth as no nutrients occur that could increase bacterial growth. This resulted in faster and more effective antibiotic action. ABM1 consisting of penicillin, streptomycin and amphotericin B has been shown to be very effective in plant tissue culture [6,7,11,19]. Although Imipenem is not a well-known antibiotic that is used with plant tissue cultures, it appears to be effective when used in combination with other antibiotics [12]. After

Table 3. Percentage of bacterial contamination after treatment of explant material with 5 µl/ml ABM1 and 5 mg/ml Imipenem over time

Time (days)	Contamination (%)
0	100
1	70
2	20
3	10
4	10
5	10

2 months in culture the *Zantedeschia* explants appeared free of contamination and the daughter plants appeared normal and had started to produce new buds (Plate 1C).

ACKNOWLEDGEMENTS

We would like to thank Dudley Wilson from Hadeco (Pty) Ltd. for supplying the *Zantedeschia* rhizomes, Nico Mienie of VOPI for identification of the bacterial contaminants and colleagues at VOPI for review of the manuscript.

REFERENCES

1. Hussey, G. (1975) Sci. Hortic. 3, 21–28.

2. Cohen, D. (1983) Proc. Int. Plant Prop. Soc. 31, 312–316.

3. Bapat, V.A. and Narayanaswamy, S. (1976) Torrey Bot. Club 103, 53–56.

4. De Bruyn, M.H., Ferreira, D.I., Slabbert, M.M. and Pretorius, J. (1992) Plant Cell Tissue Org. Cult. 31, 179–184.

5. Maliga, P., Sz-Breznovits, A. and Marton, L. (1975) Nature 2S5, 401–402.

6. Pollock, K., Barfield, D.G. and Shields, R. (1983) Plant Cell Rep. 2, 36–39.

7. Thurston, K.C., Spencer, S.J. and Arditti, J. (1979) Am. J. Bot. 66, 825–835.

8. Barrett, C. and Cassells, A.C. (1994) Plant Cell Tissue Org. Cult. 36, 169–175.

9. Mathias, P.J., Alderson, P.G. and Leaky, R.R.B. (1987) Acta Hortic. 212, 43–48.

10. Geier, T. (1977) Acta Hortic. 78, 167–174.

11. Gilbert, J.E., Shohet, S. and Caligari, P.D.S. (1991) Ann. Appl. Biol. 119, 113–120.

12. Kneifel, W. and Leonhardt, W. (1992) Plant Cell Tissue Org. Cult. 29, 139–144.

13. Pierik, R.L.M. (1987) In Vitro Culture of Higher Plants. MTP Press, Dordrecht.

14. Murashige, T. and Skoog, F. (1962) Physiol. Plant. 15, 473–497.

15. Hauptman, R.M., Widholm, J.M. and Paxton, J.D. (1985) Plant Cell Rep. 4, 129–132.

16. Shields, R., Robinson, S.J. and Anslow, P.A. (1984) Plant Cell Rep. 3, 33–36.

17. Long, R.D., Curtin, T.F. and Cassells, A.C. (1988) Acta Hortic. 225, 83–90.

18. Sneath, P.H.A. (1989) in Sneath P.H.A., Staley J.T., Bryant M.P., Pfennig N. and Holt J.H. (eds), Bergey's Manual of Systematic Bacteriology, Vol. 2, pp. 1104–1105, Williams and Wilkins Press, Baltimore.

19. Fisse, J., Battle, A. and Pera, J. (1987) Acta Hortic. 212, 87–89.

INTERNAL BACTERIAL CONTAMINATION OF MICROPROPAGATED HAZELNUT: IDENTIFICATION AND ANTIBIOTIC TREATMENT

BARBARA M. REED, JESSICA MENTZER, PIYARAK TANPRASERT and XIAOLING YU

USDA/ARS National Clonal Germplasm Repository, 33447 Peoria Road, Corvallis, OR 97333-2521, USA

1. INTRODUCTION

Commercial micropropagation laboratories very often report that persistent bacterial and fungal contamination is a serious problem [1-3]. Failure of surface sterilization procedures to produce aseptic cultures is a problem especially with woody plants. Isolated meristems [4] and explants from stock plants grown under controlled conditions [5] have been used to obtain aseptic cultures for some plants. Contamination is not always seen at the culture establishment stage; some internal contaminants become evident at later subcultures and are difficult to eliminate [6]. Detection at an early stage can aid in selecting bacteria-free cultures [7]. Antibiotic or other treatments may be needed to eliminate persistent microbial contamination [8-11], but the type and level of antibiotics and the duration of treatment useful for different plant tissue cultures vary and therefore need to be determined before use [11,12].

Internal bacterial contamination was observed in hazelnut shoot cultures in our laboratory. Contaminants were evident at culture establishment or became apparent after several subcultures. Loss of plants resulted when bacteria overgrew plant material but some explants survived and continued to grow with bacteria present. In this study we isolated bacteria from hazelnut shoot cultures, characterized and identified them, and determined the effects of antibiotic treatments on bacteria and plant materials.

2. PROCEDURE

2.1. Plant material

Shoot cultures from infected hazelnut cultures (*Corylus avellana* L., *C. contorta*, C.) were from the USDA-ARS National Clonal Germplasm Repository (NCGR) collections: 'Tonda Gentile Romana' (Cor 5), 'Hall's Giant' (Cor 16), 'Cutleaf' (Cor 18), *C. contorta* (Cor 50), 'OSU 20-58' (Cor 79), 'Giresun 54–60' (Cor 96), 'Fitzgerald' (Cor 105), *C. avellana* (Cor 187), 'Bergeri' (Cor 262), 'Badem' (Cor 415), and 'Cosford Sel 3L' (Cor 494).

A.C. Cassells (ed.), Pathogen and Microbial Contamination Management in Micropropagation, 169–174.
© *1997 Kluwer Academic Publishers. Printed in the Netherlands.*

2.2. Detection and isolation of bacteria

Segments of tissue cultured plantlets were inoculated into a liquid nutrient broth containing 0.8% nutrient broth (Difco; Sigma Chemical Co., St. Louis, MO) with 1% glucose and 0.5% yeast extract at pH 6.9 and incubated at room temperature until visibly turbid. The bacteria were then streaked onto nutrient agar plates and purified by repeated streaking.

2.3. Identification of bacterial isolates

Standard bacteriological tests were performed on the cultures and on known standard organisms. Results of these tests and colony morphology were used to classify the bacteria. Gram-stain, oxidase, oxidative/fermentive (O/F), starch hydrolysis (SH), motility and gelatinase tests [13] and colony description were performed on isolates. Three-day-old bacterial cultures (7-day for two slow-growing ones) were used for these tests. Comparisons were made with descriptions in Bergey's Manual of Determinative Bacteriology [14]. Some cultures were tested for carbon metabolism using the Biolog system and the following procedure. Cultures were grown for 24 h on tryptic soy agar (Sigma) without a carbon source. Cultures were inoculated into physiological saline solution to a predetermined turbidity and pipetted onto Biolog plates containing 96 different carbon sources. Plates were read manually at 4 and 24 h and at 48 h for some slow growing cultures. Results were compared to the Biolog database for identification.

2.4. Effect of antibiotics on bacterial isolates

Minimal inhibitory concentrations of the antibiotics were found by using a tube dilution method for standard bacterial cultures and for bacterial isolates from the infected hazelnut plants. Tubes were inoculated with one drop of bacteria from 3- to 4-day-old cultures into liquid MS medium [15] containing 2 ml of the following concentrations. For single antibiotics, streptomycin sulfate or Timentin at 62.5, 125, 250, 500, or 1000 µg/ml or gentamicin at 6.25, 12.5, 25, or 50 µg/ml. For combinations of antibiotics, Timentin + streptomycin at 125+250, 125+500, 250+250, or 250+500 µg/ml; Timentin + gentamicin at 125+6.25, 125+12.5, 250+6.25, or 250+12.5 µg/ml; gentamicin + streptomycin at 6.25+250, 6.25+500, 12.5+250, 12.5+500 µg/ml. Effectiveness was determined by putting a drop of each culture onto sections of nutrient agar plates and checking for growth after 4 to 6 days.

2.5. Antibiotic treatment of plant material

Timentin (500 µg/ml) + streptomycin sulfate (1000 µg/ml) or gentamicin (12.5 µg/ml) + streptomycin (1000 µg/ml) were used in liquid MS to treat C. avellana cv. 'OSU 20-58'. Shoot tips (1 cm) and first node cuttings (0.5 cm) from six plantlets were submerged into individual tubes containing 3 ml of the two treatments for 6 days. After 6 days, the condition of the plant material was noted, and liquid was removed from the tubes so that only the bases were submerged for the remaining 4 days. Controls (plant tissues grown in liquid MS without antibiotics) were also included with each experiment. After antibiotic treatment

(10 days), the plant condition was again noted and the plants were placed in individual tubes of semi-solid *Corylus* multiplication medium. The bases were placed into nutrient broth for detection of bacterial growth. At the next transfer (3 weeks), the plant bases were streaked onto a bacterial detection medium, 523 agar plates [16], before transfer to new medium. The bases were streaked on 523 medium on all subsequent transfers.

The second experiment treated three genotypes with two antibiotic combinations. 'Hall's Giant', 'OSU 20-58', and 'Giresun 54-60' were treated as above with Timentin (500 μg/ml or 1000 μg/ml) + streptomycin sulfate (1000 μg/ml).

Phytotoxicity of antibiotics was determined visually by checking for browning, chlorosis, and morphological changes. All shoot cultures were kept at 25°C and 16-h photoperiod (25 μmol/m^2/s). Tests for additional treatments followed this same procedure.

3. RESULTS AND DISCUSSION

3.1. Identification of bacterial isolates

Colonies were visible on nutrient agar plates in 3 days for most bacteria but some were slow-growing and required 7 days for colonies to be visible. Colony pigmentation varied from white and beige to yellow and pink to pink–red. The results of oxidase, starch hydrolysis, oxidative/fermentive, motility, and gelatinase tests varied with the isolates. Isolates identified included *Agrobacterium radiobacter* B, *Pseudomonas fluorescens*, *Xanthomonas* spp., *Enterobacter asburiae*, *Flavobacterium* spp., and *Alcaligenes* spp. Many isolates were not identified and were not similar to bacteria in the Biolog database.

3.2. Effect of antibiotics on bacterial isolates

Initial experiments showed that streptomycin and Timentin were ineffective on most of the bacterial isolates. Gentamicin was the most effective, controlling approximately half of the isolates including those from Cor 96 and Cor 415b (*Enterobacter* spp.) at concentrations as low as 6.25 μg/ml. No single antibiotic was effective for all bacterial isolates from hazelnut shoot cultures. Alcaligenes and eight others were not inhibited by any single treatment. When gentamicin and Timentin were effective, it was usually at a very low concentration. Streptomycin often required the highest concentration for effective treatment. Young *et al.* [17] also reported that among rifampicin, tetracycline, cefotaxime, and polymyxin B, no single antibiotic was bactericidal against all of the bacterial isolates from shoot cultures of several woody plants, but each of the isolates was killed by at least one of the antibiotics.

Table 1. Percentage of bacteria-free hazelnut plantlets produced from contaminated shoot tips treated with two concentrations of combinations of Timentin (T) and streptomycin (S) for 10 days

Plants treated	Bacteria	Bacteria-free plants/treated plants	
		T(500)+S(1000)[a]	T(1000)+S(1000)
'Hall's Giant'	Gram-negative	5/10	7/10
'OSU 20-58'	Alcaligenes	6/10	9/10
'Giresun 54-60'	Enterobacter	6/10[b]	na

[a]Antibiotic treatment (μg/ml).
[b]Six were bacteria free, but three died after treatment was completed.
na, data not available for this treatment.

Broader testing with Timentin, gentamicin, and streptomycin showed that combinations of the antibiotics were more effective for killing bacteria than single antibiotics. Streptomycin combined with Timentin or gentamicin was effective in killing all of the bacteria tested. Timentin and gentamicin combined were effective with most isolates. Combinations of all three antibiotics were also effective.

3.3. Antibiotic treatment of plant materials

Initial tests showed that a 10 day treatment with a combination of streptomycin and Timentin (250 μg each) was ineffective for eliminating unidentified Gram-negative bacteria from 'Tonda Gentile Romana', and 'Hall's Giant', or *Alcaligenes* from 'OSU 20-58'. Initially some cultures appeared bacteria free, but later indexing showed bacterial growth. 'OSU 20-58' shoot tips treated for 10 days with Timentin (500 μg/ml) + streptomycin (1000 μg/ml); or gentamicin (12.5 μg/ml) + streptomycin (1000 μg/ml) produced 12–25% bacteria-free cultures.

'Hall's Giant', 'OSU 20-58', and 'Giresun 54-60' treated with Timentin (500 μg/ml or 1000 μg/ml) + streptomycin (1000 μg/ml) resulted in more bacteria-free cultures than the earlier tests (Table 1). Treatments of 'Giresun 54-60' infected with *Enterobacter* produced 60% bacteria-free shoots, but half of them died due to the phytotoxicity of the treatment. Repeat tests are in progress.

Combinations of antibiotics are used against bacteria from plant tissue cultures [10,17]. Young *et al.* [17] used a combination of 25 μg cefotaxime, 25 μg tetracycline, 6 μg rifampicin, and 6 μg/ml polymyxin B to treat bacteria in tissue cultures of apple, *Rhododendron*, and Douglas-fir but later found that the bacteria were still present. Leifert *et al.* [10] reported that a range of different bacteria were eliminated from contaminated plant tissues of *Hemerocallis*, *Choisya*, and *Delphinium* using combinations of gentamicin, streptomycin, carbenicillin, cephalothin, and rifampicin.

Hazelnut shoot tips showed some antibiotic phytotoxicity but severe damage was evident in nodal cuttings. The combination of gentamicin and streptomycin produced the greatest phytotoxicity.

Both gentamicin and streptomycin belong to the group of aminoglycoside antibiotics. They bind to 30S ribosomal subunits in bacterial cells and inhibit olein synthesis and may also inhibit protein synthesis in chloroplasts and mitochondria in plant tissues [17], therefore resulting in small and yellow leaves. Phytotoxicity to gentamicin was shown in *Mentha* [12] and to cell growth of *Helianthus tuberosus* by streptomycin [11] and to shoot cultures of *Clematis, Delphinium, Hosta, Iris,* and *Photinia* by streptomycin [18]. Gentamicin 50 μg/ml was added to pear culture medium without harm to the plants [8]. Streptomycin was effective in eliminating bacteria from infected mint cultures with little phytotoxicity [7].

4. CONCLUSIONS

Internal bacterial contaminants in tissue cultured hazelnuts were eliminated by antibiotic treatment. Single antibiotics were ineffective, but combinations of two or more eliminated most contaminants. Streptomycin combined with Timentin or gentamicin killed all of the isolated bacteria tested, as did a combination of all three. Timentin combined with gentamicin was effective for most isolates. In plant tissues, antibiotic concentrations 3-4 times higher than those effective on isolated bacteria were needed to eliminate internal bacteria. Combinations of streptomycin with gentamicin and streptomycin with Timentin were effective in eliminating persistent bacterial contamination in hazelnut plants. Phytotoxicity varied with antibiotic type and plant genotype.

REFERENCES

1. Boxus, Ph. and Terzi, J.M. (1988) Acta Hortic. 225, 189–191.

2. Cassells, A.C. (1990) in Micropropagation: Technology and Application (Debergh P.C. and Zimmerman R.H., eds), pp. 31-44, Kluwer Academic Publishers, Dordrecht, The Netherlands.

3. Kunneman, B.P.A.M. and Faaij-Groenen, G.P.M. (1988) Acta Hortic. 225,183–188.

4. Hakkaart, F.A. and Versluijs, J.M.A. (1983) Acta Hortic. 131, 299–301.

5. Messeguer, J. and Mele, E. (1983) Avellino 22–23, 293–295.

6. Thorpe, T.A. and Harry, I. S. (1990) in Plant Aging: Basic and Applied Approaches, (R. Rodriguez *et al.*, eds) pp. 67–74, Plenum Press, New York.

7. Reed, B.M., Buckley, P.M. and DeWilde, T.N. (1995) In Vitro Cell. Dev. Biol. 31P, 53–57.

8. Chevreau, E., Skirvin, R.M., Abu-Qaoud, H.A., Korban, S.S. and Sullivan, J.G. (1989) Plant Cell Rep. 7, 688–691.

9. Kneifel, W. and Leonhardt, W. (1992) Plant Cell Tissue Org. Cult. 29, 139–144.

10. Leifert, C., Camotta, H., Wright, S.M.,Waites, B., Cheyne, V.A. and Waites, W.M. (1991) J. Appl. Bacteriol. 71, 307–330.

11. Phillips, R., Arnott, S.M. and Kaplan, S.E. (1981) Plant Sci. Lett. 21, 235–240.

12. Buckley, P., DeWilde, T.N. and Reed, B.M. (1995) In Vitro Cell. Dev. Biol. 31P, 58–64.

13. Sands, D.C. (1990) in Methods in Phytobacteriology (Klements Z., Rudolph K. and Sands D.C., eds) pp. 133–143, Akademiai Kiado, Budapest.

14. Kreig, N.R. and Holt, J.G., eds. (1994) Bergey's Manual of Systematic Bacteriology. Vol. 1. Williams and Wilkins, Baltimore.

15. Murashige, T. and Skoog, F. (1962) Physiol. Plant. 15, 473–497.

16. Viss, P.R., Brooks, E.M. and Driver, J.A. (1991) In Vitro Cell. Dev. Biol. 27P, 42.

17. Young, P.M. , Hutchins, A.S. and Canfield, M.L. (1984) Plant Sci. Lett. 34, 203–209.

18. Leifert, C., Camotta, H. and Waites, W.M. (1992) Plant Cell Tissue Org. Cult. 29, 153–160.

ELIMINATION OF SEVERAL BACTERIAL ISOLATES FROM MERISTEM TIPS OF *HYDRANGEA* SPP.

V.I. TAHMATSIDOU and A.C. CASSELLS
Department of Plant Science, University College, Cork, Ireland

1. INTRODUCTION

Both surface and endophytic bacteria are responsible for bacterial contamination in micropropagation. The main sources of the above have been extensively discussed [1–5]. Surface inhabiting bacteria, i.e. those accessible to surface sterilants, can be effectively eliminated at explant preparation [6]. The problem of endophytic bacteria, i.e. non-accessible bacteria, should be approached with caution. Although meristem culture may, in general, eliminate phloem-restricted micro-organisms, this has proved to be difficult in practice in some species, e.g. in *Pelargonium* [7]. Problems with the establishment of small meristem explants and the risk of losing desirable starting material prompted the inclusion of antibiotics in the meristem culture medium [8]. The use of antibiotics, however, often yields unrepeatable results as they depend on the contaminant, the antibiotic dose, the plant cultivar and the antibiotic or combination of antibiotics used ([9–12]; Falkiner, this volume).

Hydrangea species are cultured *in vitro* to produce, mainly, clean stock plants for cutting production [13–19]. Meristem tips, axillary buds, shoot tips and unfolded young leaves are reported as explants used.

The objective of this work was to eliminate several bacterial contaminants isolated and characterised from meristem tips of various *Hydrangea* species [21]. Cefotaxime and carbenicillin were the antibiotics tested because of their bactericidal and broad-spectrum activity against both Gram-positive and Gram-negative bacteria [6].

2. PROCEDURE

2.1. Preparation and sterilization of plant culture and bacteriological media

All the media, glassware and instruments were sterilised at 105 kPa, 121°C for 15 min. Heat-labile substances were double filter-sterilised, using a sterile acrodisc with pore size

A.C. Cassells (ed.), Pathogen and Microbial Contamination Management in Micropropagation, 175–181.
© *1997 Kluwer Academic Publishers. Printed in the Netherlands.*

0.2 mm (Gelman Sciences Ltd., UK), into a sterile container and added to the culture medium which had been autoclaved and allowed to cool to approximately 35–45°C.

2.2. Plant tissue culture

Apical or axillary buds were excised from the stems of 71 donor plants of 19 species of *Hydrangea* from different locations in Ireland and the UK. The species, subspecies and cultivars are listed in Table 1. All the donor plants were grown outdoors and had no visible symptoms of plant pathogens.

The buds were dipped in 50% aq. (v/v) ethanol solution for 2 min. Thirty buds, on average, were excised from each genotype. After two rinses with sterile distilled water, excision of the scales followed. Then the buds were dipped in freshly prepared 1% aq. sodium hypochlorite solution (20% household bleach) for 10 min. Three rinses with sterile distilled water followed and two to three pairs of primordial leaves were removed. The buds were again dipped in 1% aq. sodium hypochlorite solution (20% household bleach) for 2 min followed by four rinses with sterile distilled water. The rest of the primordial leaves were removed until the meristem was reached. The meristems were then placed onto the MSNB tissue culture medium (Murashige and Skoog basal medium, Sigma Chemical Co. Ltd., 4.4 mg/1; sucrose 20 g/1; NAA 0.1 mg/1; BA 1.0 mg/1; agar 7 g/1; pH 5.7) [16].

All cultures were grown in the growth room (23 ± 2°C, 35–50 μmol/m^2/s, 16-h photoperiod) in 226-ml sterile plastic food containers (Wilsanco Plastic Ltd., N. Ireland) with light provided by white 65/80 W Liteguard fluorescent bulbs (Osram Ltd., UK).

2.3. Bacterial cultures

Pseudomonas aeruginosa was used as the control organism in the sensitivity tests. This, and the newly isolated bacteria, were grown and maintained on TSA (Oxoid Ltd.) and NGA (NA plus 10.0 g/1 glucose) [19] medium. Liquid cultures were grown in TSA (Oxoid Ltd.) and NGA broth (NA plus 10.0 g/1 glucose) [19]) and stored at 4°C and -15°C.

2.4. Identification by fatty acid profile analysis

Fifty-five freshly subcultured isolates were sent to the Central Science Laboratory (Harpenden, Herts., UK) for identification by fatty acid profile analysis (Table 1) [21].

2.5. Sensitivity tests and MBC (minimal bactericidal concentration) determination

The Stokes method was used for the sensitivity tests for 55 isolates [21]. DST medium (Oxoid Ltd.), prepared according to the manufacturer's instructions, and MSNB tissue culture medium, prepared as mentioned above, were the media used in this test. Antibiotic discs of 5 μg/ml and 30 μg/ml of cefotaxime and of 100 μg/ml of carbenicillin (Oxoid Ltd.) were tested.

For the MBC determination, autoclaved TSA broth and liquid MSNB medium were dispensed in sterile plastic universals. Filter-sterilised cefotaxime was added at increasing concentrations to the above media. The tubes were inoculated with 10 µl of bacterial suspension from a 24-h-old bacterial liquid culture and incubated in a shaking water-bath at 28°C for 24 h. Then the minimum inhibitory concentration (MIC) was determined and the tubes that were not cloudy were plated on NGA and TSA medium and incubated at 28°C for the MBC to be determined.

2.6. Meristem culture on cefotaxime-containing medium

Cefotaxime was filter sterilised and added to MSNB medium at 400 µg/ml, which was two to four times the MBC. Contaminated meristem tips were placed and grown on this medium, as mentioned above, for 5 weeks. Then the new growth of the regenerated tissue was transferred to antibiotic-free medium and grown there for another 5-week period. Subculture on antibiotic-free medium followed.

2.7. Bacterial indexing

Plantlets were indexed to confirm the absence of cultivable bacterial contaminants. Pieces of the plantlets or whole plantlets were placed onto five different general bacteriological media, KB, YDC, NGA [19], PYA, TSA (Oxoid Ltd.) at 4-week intervals during subculturing, starting from the first shoots grown on antibiotic-free tissue culture medium.

Plates were incubated at 28°C for 2 weeks and examined for bacterial growth. During subculturing the containers were thoroughly examined for any signs of cloudiness of the medium.

3. RESULTS AND DISCUSSION

3.1. Bacterial contamination of meristem-tip cultures

Sixty-nine percent of the donor plants were contaminated (Table 1). A halo developed in the medium around the infected meristem tips indicating bacterial contamination. Endophytic bacteria, either plant-associated or saprophytes commonly found in soil (e.g. *Enterobacter* spp.), inhibited regeneration from the meristem tip, since the explants died within 5–15 days without showing any growth.

3.2. Sensitivity tests and MBC determination

Most of the bacterial isolates tested were resistant to 100 µg/ml carbenicillin in both DST and MSNB medium. Some of them were either sensitive or moderately sensitive to 5 µg/ml cefotaxime, whereas all the isolates were sensitive to 30 µg/ml cefotaxime. Based on these results, the MBC was determined only for cefotaxime. Most of the isolates showed no growth when cefotaxime, at 200 µg/ml, was added in either TSA or MSNB liquid medium.

Table 1. A list of donor plants, bacterial isolates and bacteria eliminated after antibiotic treatment

Donor plants	Fatty acid profiling	Bacteria eliminated
Species/ssp./cultivar	(Similarity Index)	
H. arborescens	*Enterobacter intermedius* (0.461)[a]	
'Annabelle'	*Enterobacter agglomerans* (0.751)[b]	
'Discolor'	*Enterobacter agglomerans* (0.140)[b]	+
H. argentis		
H. aspera	*Micrococcus kristinae* (0.724)[a]	
aspera		
sargentiana	*Erwinia cypripedii* (0.141)[b]	
strigosa	*Morganella morganii* (0.452)[a]	
	Clavibacter michiganensis (0.520)[b]	
Villosa group	*Escherichia coli* (0.561)[a]	
H. cinerea 'Sterilis'		
H. heteromalla	*Enterobacter intermedius* (0.474)[a]	+
H. involucrata		
'Hortensis'	*Escherichia coli* (0.439)[a]	
H. macrophylla	*Erwinia cypripedii* (0.171)[b]	
macrophylla	*Enterobacter intermedius* (0.557)[a]	+
'Deutschland'	*Enterobacter intermedius* (0.512)[a]	
	Pseudomonas syringae (0.814)[b]	
'Director Vuillermet'	*Enterobacter intermedius* (0.389)[a]	+
'Geoffrey Chadbund'	*Actinobacillus lignieresii* (0.129)[a]	+
	Pseudomonas putida (0.008)[b]	
'Holstein'		
'Kluis Superba'	*Enterobacter intermedius* (0.471)[a]	
'La France'	*Erwinia cypripedii* (0.075)[b]	
'Lilacina'	*Curtobacterium flaccumfaciens* (0.837)[b]	+
	Pseudomonas caricapapayae (0.942)[b]	+
'MDME Emile Moulliéré'	*Erwinia cypripedii* (0.080)[b]	+
'Marechal Foch'	*Clavibacter michiganense* (0.033)[b]	
'Mariesii'	*Enterobacter intermedius* (0.674)[a]	
'Mariesii Perfecta'	*Erwinia cypripedii* (0.103)[b]	
'Nigra'	*Bacillus polymyxa* (0.524)[a]	
	Escherichia coli (0.559)[a]	
	Pseudomonas syringae (0.943)[b]	
	Pseudomonas marginalis (0.855)[b]	
'Sonnengruss'	*Enterobacter nimipresuralis* (0.348)[b]	
'Souvenir de Président Doumer'		
'Superba'		
'Veitchii'	*Erwinia cypropedii* (0.163)[b]	

Table 1 *(continued)*

Donor plants	Fatty acid profiling	Bacteria eliminated
Species/ssp./cultivar	(Similarity Index)	
'Vulkain'	*Enterobacter nimipressuralis* (0.384)[b]	
'Westfalen'	*Morganella morganii* (0.521)[a]	
	Xanthomonas campestris (0.333)[b]	
H. paniculata	*P. tolaasii/P. chlororaphis* (0.748)[b]	
H. petiolaris	*Erwinia cypripedii* (0.118)[b]	
H. prolifera	*Flavobacterium mizutaii* (0.202)[a]	
H. quadricolor		
H. quelpartensis		
H. quercifolia	*Bacillus polymyxa* (0.560)[a]	
	Enterobacter intermedius (0.483)[a]	
H. radiata		
H. sargentiana		
H. scandius chinensis		
H. seemani		
H. serrata 'Acuminata'	*Curtobacterium flaccumfaciens* (0.171)[b]	
'Blue Deckle'	*Pseudomonas tolaasi* (0.729)[b]	
'Diadem'	*Escherichia coli* subgroup B (0.491)[a]	
	P. syringae morspunorum (0.935)[b]	
Preziosa	*Pseudomonas syringae* (0.839)[b]	
	P. syringae morspunorum (0.977)[b]	
H. wilsonii	*X. campestris papervericola* (0.007)[b]	
Hybrid 'Harrows Blue'	*Enterobacter intermedius* (0.465)[a]	+
Hybrid 'La Lorraine'	*Enterobacter intermedius* (0.673)[a]	+
A blue mop (l)	*Escherichia coli* (0.573)[a]	
a late large white mop		
a blue mop (2)	*Escherichia coli* (0.495)[a]	
	Enterobacter agglomerans (0.025)[b]	
pink lacecap	*Escherichia coli* (0.624)[a]	

[a]Non-phytopathogenic bacteria; [b]phytopathogenic bacteria.

4. ELIMINATION OF BACTERIAL CONTAMINANTS

Incorporation of cefotaxime in the tissue culture medium at 400 µg/ml was effective in eliminating several bacteria, without being phytotoxic. The contaminated meristem tips grew very well on the antibiotic-containing tissue culture (MSNB) medium without showing any sign of bacterial contamination and gave rise to uniform plantlets after the incubation period. When the new growth of these shoots was transferred to antibiotic-free medium there was uniform growth of the shoots without any sign of bacterial growth on the tissue culture medium, or on the bacteriological media used for indexing. *Actinobacillus lignieresii, Curtobacterium flaccumfaciens, Enterobacter agglomerans,*

E. intermedius (in five cases), *Pseudomonas caricapapayae, Erwinia cypripedii* (Table 1) were ten randomly selected bacterial contamination cases where elimination of the bacterial isolates from meristem tips occurred. In the case of *H. macrophylla* 'Lilacina' the meristem tips were contaminated with both a Gram-negative and a Gram-positive bacterium (Table 1) and the incorporation of cefotaxime at 400 µg/ml in this case managed to eliminate both.

The diversity of bacterial contaminants isolated in this study [21] suggest that bacterial identification may not be a practical option as the initial step in determination of a choice of antibiotic for *in vitro* use as some advise (e.g. [22,23]; Falkiner, this volume). It may be more practical to go directly to antibiotic sensitivity tests.

That no bacterial contamination emerged in any of the stages of micropropagation after removal from cefotaxime medium indicates the efficacy of this antibiotic in plant tissue culture [7].

ACKNOWLEDGEMENTS

The authors are grateful to Mr. D. Synnott, Director, and his colleagues at the National Botanic Gardens, Dublin; to Mr. B. Cross, 'Lakemount', Cork; to the National Council for the Conservation of Garden Plants and Mr. P. Smith of Derby City Council, UK, for providing plant material. The authors also acknowledge help on the bacterial work from Mr. Don Kelleher, technician in the Plant Science Department of U.C.C.

REFERENCES

1. Trick, I. and Lingens, F. (1985) Appl. Microbiol. Biotechnol. 21, 245–249.

2. Boxus, Ph. and Terzi, J.M. (1987) Acta Hortic. 212, 91–93.

3. Boxus, Ph. and Terzi, J.M. (1988) Acta Hortic. 225, 189–191.

4. Giles, K.L. and Morgan, W. M. (1987) Trends Biotechnol. 5, 35–39.

5. Leifert, C., Waites, W.M. and Nicholas, J.R. (1989) J. Appl. Bacteriol. 67, 353–361.

6. George, E. F. (1993) Plant Propagation by Tissue Culture. Exegetics, Basingstoke, UK.

7. Reuther, G. (1991) in COST 87, Pelargonium Micropropagation And Pathogen Elimination. A Report Of The Pelargonium Group (Applegren M., Hunter C.S., Paludan N., Reuther G. and Theiler-Hedrich R., eds) pp. 17–32.

8. Barrett, C. and Cassells, A. C. (1994) Plant Cell Tissue Org. Cult. 36, 169–175.

9. Bastaiens, L., Maene, L., Jarbaori, Y., Van Sumere, C., Van De Castelle, K. L. and Debergh, P.C. (1983) Med. Fac. Landbouwv. Rijsuniv. Gent. 48, 13–24.

10. Fisse, J., Battle, A. and Pera, J. (1987) Acta Hortic. 212, 87–90.

11. Okkels, F.T. and Pedersen, M.G. (1988) Acta Hortic. 225, 199–204.

12. Eapen, S. and George, L. (1990) Plant Cell Tissue Org. Cult. 22, 87–93.

13. Beauchesne, C. (1974) Landwirtsch. Z. 19, 2881–2882.

14. Preil, W., Kuhne, H. and Hoffman, M. (1978) Nachrichtenbl. Dtsch. Pflanzenschutzdienstes 30, 88–90.

15. Jones, J. B. (1979) in Plant Cell and Tissue Culture (Sharp W.R., Larsen P.O. and Raghavan V., eds) pp. 441–452, University Press, Columbus.

16. Allen, T.C. and Anderson, W.C. (1980) Acta Hortic. 110, 245–251.

17. Stolz, L.P. (1984) HortScience 19, 717–719.

18. Bailey, D.A., Seckinger, G.R. and Hammer, P.A. (1986) HortScience 21, 525–526.

19. Sebastian, T.K. and Heuser, C.W. (1987) Bartr. Sci. Hortic. 31, 303–309.

20. Schaad, N.W. (1988) Laboratory Guide for the Identification of Plant Pathogenic Bacteria. 2nd edition. pp. 1–72, Am. Phytop. Soc., St. Paul, Minnesota.

21. Cassells, A.C. and Tahmatsidou, V. (1996) Plant Cell Tissue Org. Cult. 47, 15–26.

22. Stokes, E.J. and Waterworth, P.M. (1972) ACP Broadsheet No. 55.

23. Falkiner, F.R. (1990) I.A.P.T.C. Newslett. 60, 13–23.

24. Gunson, H.E. and Spenser-Phillips, P.T.N. (1994) in Physiology, Growth and Development of Plants in Culture (Lumsden P.J., Nicholas J.R. and Davies W.J., eds) pp. 379–396, Kluwer Academic Publishers, Dordrecht, The Netherlands.

THE APPLICATION OF ANTIBIOTICS AND SULPHONAMIDE FOR ELIMINATING *BACILLUS CEREUS* DURING THE MICROPROPAGATION OF INFECTED *DIEFFENBACHIA PICTA* SCHOTT.

E. ZENKTELER[1], K. WŁODARCZAK[2] and M. KŁOSOWSKA[1]

[1]*Department of General Botany and* [2]*Department of Microbiology, Institute of Experimental Biology, A. Mickiewicz University, Poznań, Poland*

1. INTRODUCTION

Bacillus cereus are pathogenic bacteria which cause two kinds of food poisoning in humans and other mammalians (diarrhoeal and emetic) [1,2]. No data indicating the infection of plants by this species have been found in the available literature [2,3]. However, in the course of this research, mass isolation of *B. cereus* from a horticultural plantation of *Dieffenbachia picta* was recorded. This indicates that there are pathogenic strains that also infect plants. Therefore, the objective of the present study was to develop an efficient antimicrobial therapy for the infection caused by *B. cereus* in micropropagated *D. picta*.

2. MATERIALS AND METHODS

2.1. Plant material

Three cultivars of *Dieffenbachia picta*: 'Camilla', 'Mars' and 'Perfecta' (VitroPlant Owińska, Poland) were grown and vegetatively multiplied in greenhouses under horticultural conditions. The plants, infected by *Bacillus cereus* (often asymptomatically), were indexed prior to their use in experiments (and tested for the presence of micro-organisms [4,5]). During *in vitro* trials, explants were indexed in every subculture by visual assessment.

2.2. Culture conditions and identification of bacterial strains

Before and after antibiotic treatment, the samples of leaf petioles (whose surfaces were sterilized with 70% ethanol) and water samples used for the watering of the plants were cultured in 100 ml of Brain Hearth Infusion (BHI, Difco USA) at 25°C for 24 h and subsequently subcultured on Nutrient Agar (Difco), plates containing 5% sheep blood, Chapman Agar (Difco) for the isolation of Staphylococci, Levine EMB Agar (Difco) for the isolation of Gram-negative bacteria – Enterobacteriaceae, SS Agar (Difco Lab. USA)

A.C. Cassells (ed.), Pathogen and Microbial Contamination Management in Micropropagation, 183–191.
© *1997 Kluwer Academic Publishers. Printed in the Netherlands.*

for the isolation of Gram-negative bacteria, TCBS Medium (Oxoid England), a selective medium for Vibrios, Aeromonas medium (Difco Lab., USA), a selective agar for the isolation of *Aeromonas* ssp. from a variety of specimens (water), Sabouraud Modified Agar (Difco Lab., USA) for the cultivation yeasts, moulds and aciduric bacteria. The media were incubated at 25°C for 24 h.

Incubation was followed by gross examination of the colonies. Cultures for identification should first be examined microscopically, so films for light microscopy were prepared and stained by Gram's method. The method for staining spores was also used.

The biochemical characters of the isolates were assayed in commercial kits API 20 E and API 20 NE (API-System Bio-Merieux, France) as directed by the manufacturer.

The isolates were identified to the species level by the criteria of Bergey's Manual [3] and according to Collins *et al.* [1]. Antibiotic susceptibility was tested by the Bauer–Kirby Agar disc diffusion method, using discs of 11 antibiotics and Mueller–Hinton Agar II (Becton Dickinson Comp. USA) as directed by the manufacturer.

2.3. Eliminating infection with antibiotics and sulphonamide treatment *in vivo* and *in vitro*

In order to define the treatment, an antibiogram of the *Bacillus cereus* strain isolated from petioles of *D. picta* was made. According to specific tests *B. cereus* revealed susceptibility to the following antibiotics: streptomycin, doxycyclin, tetracycline, amicacin, chloramphenicol, erythromycin, cinoxacin, ophloxacin, gentamycin, tobramycin, and netilmycin. The choice of antibiotics was also made based on good systemic penetration into plant tissues, broad antibacterial spectrum, solubility, stability *in vitro*, lack of side-effects, and their cost [6]. Three bactericidal antibiotics and sulphaguanidine in the form of a drug were used to cure infected plants. Antibiotic dosages generally corresponded to concentrations used by other researchers [4,5]. In order to determine the response of potted plants to watering with antibiotics, ten plants were watered with sulphaguanidine suspension (2 g in 200 ml water for 0.5 l of soil), ten with solutions of gentamycin, ten with streptomycin, and ten with carbenicillin (1 g per 200 ml of water per 0.5 l of soil). All the plants were watered three times with 48-h intervals between treatments. After the excision of leaves, petioles were soaked in the solution of gentamycin or streptomycin (100, 250, 500 and 750 mg/l). The percentage of decontaminated plants was assessed bacteriologically. In order to determine explant multiplication rate in response to antibiotic treatment, 30 leaf petioles were inoculated in culture vessels on MS modified medium containing 100, 250, 500 and 750 mg/l carbenicillin, gentamycin, streptomycin and sulphaguanidine. The multiplication rate was calculated by dividing the number of leaf petioles with buds after the growth period by the number of petioles plated at the beginning of the experiments. The statistical analysis was performed for the subculture data by analysis of variance (F-test) followed by pair-wise comparison (t-test) on the control only.

2.4. *In vitro* culture conditions

After soaking in gentamycin or streptomycin solution for 24 h, excised leaf petioles were surface sterilized with 70% ethanol (1 s), 0.2% $HgCl_2$ (6 min), and washed three times in sterile distilled water. After the dissection of the ends, petioles were cultured on the medium MS [7] supplemented with thiamine 0.4 mg/l, inositol 100 mg/l, 40 g/l sucrose, and 8 g/l agar. Growth regulators such as 2iP and IAA were added at concentrations mentioned in Table 2. The filter-sterilized carbenicillin solution was incorporated to the medium after autoclaving. Explants were grown continuously under 50 μmol/m^2/s of cool white fluorescent light for 24 h daily at 20–22°C. After three subcultures on the medium with carbenicillin, explants free from infection, green and vigorous were selected and transferred to the MS medium without antibiotic and regularly subcultured every 2 weeks on fresh medium with growth regulators. Adventive buds regenerated after 5 weeks of culture were subsequently transferred on a regulator-free medium with a lower sugar content (30 g/l), where small first leaves appeared, followed by the formation of adventitious roots.

2.5. Microscopy

The material was fixed in FAA, dehydrated in graded ethanol–xylene series, and embedded in paraffin. Sections (at 12 μm) were stained with a combination of safranin and fast green.

3. RESULTS AND DISCUSSION

3.1. Identification of the pathogenic bacteria and their source

The strains of *Baciilus cereus* isolated from the petioles of *Dieffenbachia picta* were facultatively anaerobic spore forming bacteria. The spores were oval, central, or subterminal and did not swell the cells. The cells of *B. cereus* were rod-shaped. The colonies had a dull or frosted glass appearance, often with an undulate margin. On nutrient agar plates containing 5% sheep blood, hemolysis was considered present when a zone with hemolysis was detected around the colonies. *B. cereus* produces two hemolysins, one of which, cereolysin, is a potent necrotic and lethal toxin. Phospholipases produced by *B. cereus* may act as exacerbating factors by degrading host cell membranes following exposure of their phospholipid substrates [2]. *Bacillus cereus* causes two types of food poisoning in humans (diarrhoeal and emetic) [1,2]. The virulence factors have been identified for *B. cereus* as a 38–46 kDa protein complex, which in animal models has been shown to cause necrosis of the skin or intestinal mucosa and to be a lethal toxin. This protein complex is responsible for severe *B. cereus* infections [2]. The blanching and necrotic desiccation of leaves observed in the course of the research , most evident in cultivar 'Camilla', is likely to have been caused by these toxins. In cultivars 'Mars' and 'Perfection', whose leaves contained more chloroplasts, symptoms of the infection were weaker. When bacteriosis was fully developed, a strong reduction in the growth rate, yellowish or palish discoloration of leaves, and finally the death of the whole plant were observed. Disease monitoring in greenhouses revealed that all of the three *Dieffenbachia* cultivars were highly susceptible to bacteriosis caused mostly by *Bacillus cereus*, which was the dominant causative agent of infection. Other species found included *Erwinia herbicola, Pseudomonas putida, P. pseudomallei, P. syringae* and *Xanthomonas maltophila*, which were isolated only occasionally (Table 1). Losses due to bacterial infection introduced in the tissue culture with the initial explants were enormous.

Bacteriological examination of the water samples used for watering, taken from the local water supply system and from the River Warta, revealed the presence of *Bacillus cereus* strains. Therefore, the infection of *Dieffenbachia* with *B. cereus* is water-borne. It is well known that bacterial contaminants may originate from unpurified irrigation water [8].

Table 1. Bacterial strains isolated from: (1) leaf petioles of infected stock plants of *Dieffenbachia* , (2) infected explants *in vitro*

(2) March 1994 *Bacillus cereus*[a] 93%	*Xanthomonas maltophila*[a] *Pseudomonas syringae*[a]	4% 5%
(1) March 1995 *Bacillus cereus*[a] 82%	*Erwinia herbicola*[a] *Pseudomonas putida*	9% 9%
(1) April 1996 *Bacillus cereus*[a] 83.4%	*Pseudomonas pseudomallei* 16.6%	

[a]Plant pathogen.

3.2. Antibiotic treatments

The eradication of *B. cereus* in the cultivation of *Dieffenbachia* proved to be extremely difficult due to the presence of mucous substances in plant tissue, which greatly restricts deep penetration of the applied antibiotics. Additionally, the formation by *B. cereus* of endospores resistant to numerous antibiotics, high temperature, and ethanol sterilization [3,6] constitutes another obstacle in effective eradication of the pathogen.

Each antibiotic, which was individually effective against an isolated strain of *Bacillus*, appeared ineffective in treating infected plants (due to poor penetration into plant tissue). Streptomycin, gentamycin, carbenicillin, and a chemotherapeutic (sulphaguanidine) were used for the purpose of watering the potted plants, followed by soaking of the excised petioles, and finally treatment of the explants *in vitro*. It did not appear that *B. cereus* was easily accessible to sulphaguanidine or any other antibiotics used for the watering of the potted plants. After watering with the suspension of sulphaguanidine, the maximum reduction in the presence of *B. cereus* followed exposures of 3 × 48 h and was observed 2 weeks later. The number of *Bacillus cereus*-infected *Dieffenbachia* decreased significantly only after sulphaguanidine treatments. Gentamycin and streptomycin reduced to half the number of infected plants, whose growth was most severely affected after

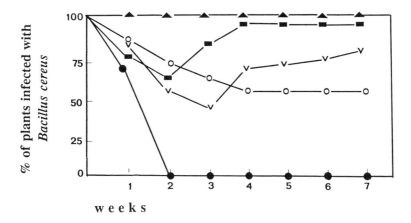

Figure 1. Percentage of *Dieffenbachia* potted plants infected with *Bacillus cereus* after irrigation with three antibiotics and sulphaguanidine for 7 weeks. ▲: untreated *Bacillus cereus*-infected plants; v: irrigated with gentamycin (1 g/l); ○: streptomycin (1 g/l); ■: carbenicillin (1 g/l); ●: sulphaguanidine (2 g/l).

watering. Carbenicillin had a weak bacteriostatic effect on this bacteria. Only sulphaguanidine revealed any inhibitory effect on plant growth, even in high therapeutic doses (Fig. 1).

Generally, 90–100% of the primary explants without antibiotic treatment exhibited contamination within the first week after transfer to the medium. *Bacillus cereus* in tissues of *Dieffenbachia* was not easy to eliminate and required high therapeutic doses during antimicrobial therapy *in vitro*. The multiplication rate of petioles decreased with increasing concentration of the therapeutic (Fig. 2). Treatment with carbenicillin resulted in a reduced multiplication rate only if 500 mg/l was used. The petioles were most severely affected by further regeneration on streptomycin and gentamycin when applied in concentrations of 500–750 mg/l (Fig. 2). Sulphaguanidine was useless due to its incompatibility with the constituents of the medium. After antibiotic treatments, all the explants showed symptoms of phytotoxicity, such as the bleaching of tissues. Part of them turned green during growth on antibiotic-free MS medium.

3.3. Initiation of bud regeneration from leaf petioles

The number of explants capable of bud regeneration depends on 2iP content in the medium and bud formation per explant was directly correlated to 2iP concentration. The presence of IAA together with cytokinin, was found to be indispensable for the induction and formation of buds from petioles of *Dieffenbachia*. The treatment with 30 mg/l 2iP and

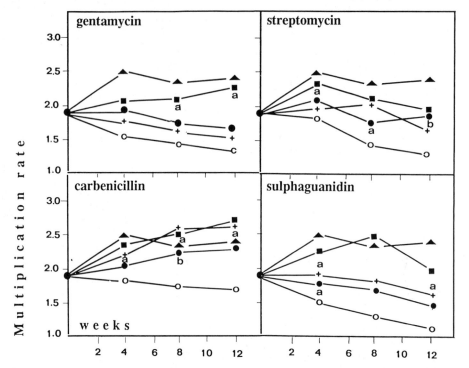

Figure 2. Multiplication rate of *Dieffenbachia picta* petioles individually treated with antibiotics. Concentration of antibiotics (mg/l) ▲: untreated explants, ■: 100; ●: 250; +: 500; ○: 750. **a**: significantly different from control (*P*=0.05); **b**: significantly different from control (*P*=0.001).

6 mg/l IAA gave the most satisfactory result, because of a high percentage of regenerating explants, number of buds formed, and subsequent bud elongation on growth regulator-free medium (Table 2, Fig. 7).

Histological analysis revealed that initiation of meristematic centres in subepidermal cells of the petioles occurred between 2 and 4 weeks on MS medium supplemented with 30 mg/l 2iP and 6 mg/l IAA (Figs. 3 and 4). During four subsequent subcultures, petioles varied in size and had irregular surface often with spherical convexity, which consisted of several clumps. Emergence of the primordium of adventitious buds was visible after 5 weeks (Figs. 5 and 6).

The highest 2iP concentrations (40 mg/l) induced abundant buds and foliar structures in some explants. Further bud development was arrested because most of the explant surface

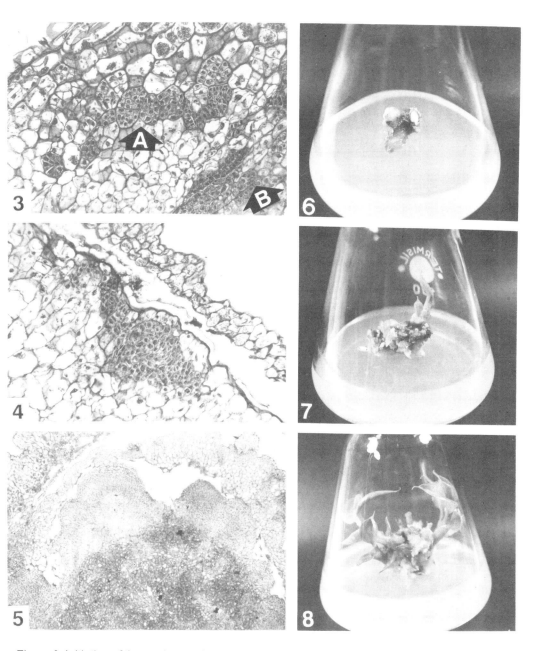

Figure 3. Initiation of the meristematic centres (spherical - A, and linear - B) in subepidermal cells of petiole of *Dieffenbachia picta*. 2 weeks on the modified MS medium with 30 mg/l 2iP and 6 mg/l IAA. ×84. Figure 4. Subsequential stages of development of the meristematic centres 4 weeks on the modified MS medium. ×84. Figure 5. Intensive regeneration of few adventitious buds on the surface of petiole after 5 weeks of culture. ×48. Figure 6. Adventive buds protruded on the surface of leaf petiole after 5 weeks of the culture on the modified MS medium. ×2. Figure 7. First leaves of buds developed on MS medium after 7 weeks of culture. × 1.5. Figure 8. Plantlet formation before excision from primary explant after 9 weeks of culture on MS modified medium. ×1.5.

Table 2. The effect of 2iP and IAA on induction of buds from petioles of *Dieffenbachia picta*

2iP (mg/l)	IAA (mg/l)	No. of explants	Explants with buds (%)	Average no. buds per explant	Explants with plantlets (%)
10	5	30	22.3	0.6	–
10	10	29	35.0	1.3	–
20	10	28	48.4	3.4	1.5
30	6	30	85.2	>7.5	4.4
40	15	27	39.1	1.2	–

> Means significantly different by Duncan's Multiple Range Test $P < 0.05$.

formed callus. A prolonged cultivation of shoots on growth regulator-free medium allowed spontaneous rooting in about 85% of the explants (Fig. 8).

4. CONCLUSIONS

Dieffenbachia picta Schott. is an attractive plant, successfully grown at home. The distribution of plants infected by *B. cereus* may constitute a health risk for their buyers. This is the reason why we developed this antibiotic treatment followed by the micropropagation of healthy plants via adventitious buds regeneration from petiole explants. In the present study, we report the protocol required for a three-stage therapeutic procedure for efficient decontamination of *Dieffenbachia* stock material.

REFERENCES

1. Collins, C.H., Lyne, P.M. and Grange, J.M. (1995) Microbiological Methods. pp. 394-395, Butterworth-Heinemann Ltd., Oxford.

2. Baron, S. (1991) Medical Microbiology. pp. 254, 255, 262, Churchill Livingstone, New York.

3. Claus, D. and Berkeley, R.C.W. (1984) Genus Bacillus in Bergey's Manual of Determinative Bacteriology. Vol. 1 (Krieg N.R., Holt J.G., eds) pp. 1105–1139, Williams and Wilkins Co., Baltimore.

4. Leifert, C., Camotta, H., Wright, S.M., Waites, B., Cheyne, V.A. and Waites, W.M. (1991) J. Appl. Bacteriol. 71, 307–330.

5. Leifert, C., Camotta, H. and Waites, W.M. (1992) Plant Cell Tissue Org. Cult. 29, 153–160.

6. Falkiner F.R. (1991) Newsl. IAPTC 60, 13–23.

7. Murashige, T. and Skoog, F. (1962) Physiol. Plant. 15, 473–497.

8. Seabrook, J.E.A. and Farrell, G. (1993) HortScience 28, 628–629.

9. Cassells, A.C. (1991) in Micropropagation Technology and Application (Debergh P.C. and Zimmerman R.H. eds) pp. 31–44, Kluwer Academic Publishers, Dordrecht, Netherlands.

10. Reed, B.M. and Tanprasert, P. (1995) Plant Tissue Cult. Biotechnol. 1, 137–142.

IDENTIFICATION AND ANTIBIOTIC SENSITIVITY OF BACTERIAL CONTAMINANTS ISOLATED FROM *IN VITRO* CULTURES OF SOME TROPICAL AQUATIC PLANTS

W. KNEIFEL[1], W. LEONHARDT[2] and MARGOT FASSLER[1]

[1]*Department of Dairy Research and Bacteriology, Agricultural University, Gregor Mendel-Str. 33, A-1180 Vienna;* [2]*VitroPlant Ltd., Brunnleiten 17, A-3400 Klosterneuburg, Austria*

1. INTRODUCTION

For a long time, aquarium plants have been imported from tropical regions all around the world. Experience has shown that several micro-organisms and insects are frequently associated with this plant material. For this reason the micropropagation of aquatic plants has become a useful alternative to the conventional procedure. At present, the *in vitro* production of such species has reached more than five million plants per year in Western Europe. It may be expected that *in vitro* produced plantlets should not contain microbial contaminants which are harmful to other plants or animals grown in the aquarium. However, growth of endogenous or hidden micro-organisms can be observed since basic media have been supplemented with coconut water in order to provide special plant nutrients that promote the contaminating microflora.

In this study we describe the isolation and identification of endogenous micro-organisms hosted in six different plant varieties grown in micropropagation on MS standard media for long periods without exhibiting visual signs of contamination.

2. MATERIAL AND METHODS

2.1. Plant material

Four aquatic and two ornamental varieties, *Lobelia* sp. Clone A1 IA, *Lobelia* sp. Clone A1 IIM, *Cryptocoryne wendtii* W8, *Cryptocoryne beckettii* W9, *Echinodorus parviflorus* W10, *Anubia nana* W5/3, were provided by the VitroPlant Ltd. laboratory and taken for microbial examination. The material was grown in plastic containers on solid MS basic medium with 0.2 mg/l IBA.

A.C. Cassells (ed.), Pathogen and Microbial Contamination Management in Micropropagation, 193–199.
© *1997 Kluwer Academic Publishers. Printed in the Netherlands.*

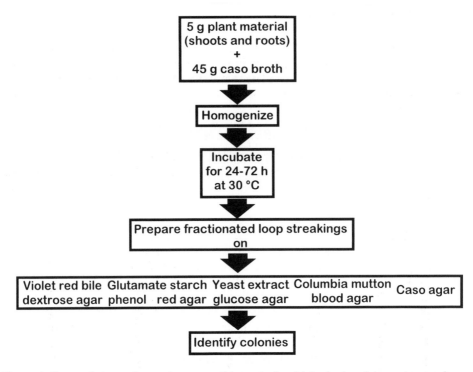

Figure 1. General steps of sample preparation and microbiological enrichment procedures.

Table 1. Additional tests used for the microbiological identification

- Gram staining (Merck staining set)
- Slime formation with 3% KOH (indicator for Gram behaviour)
- Aminopeptidase activity (indicator for Gram behaviour)
- Endospore staining (Malachite Green method; bacilli identification)
- Motility testing (hanging-drop slide)
- Cytochrome c-oxidase activity (indicator for aerobic growth)
- Catalase activity (indicator for aerobic growth)
- Lysostaphin and Nitrofurantoin resistance (differentiation of cocci)
- Slidex Staph clumping test (differentiation of staphylococci)
- Coagulase activity (rabbit plasma test, BioMerieux)
- Growth on starch agar (differentiation of bacilli)
- Polymyxin-egg yolk mannitol thymol blue agar (differentiation of bacilli)
- Calcium caseinate agar (proteolytic activity)
- Caso broth containing different NaCl concentrations (differentiation of bacilli)
- API 50 Chß + API 20E system (identification of bacilli) (BioMerieux)
- API Staph (identification of staphylococci)

2.2. Microbiological examination

Based on microbial enrichment cultures in caso broth (Merck), fractionated repeated loop streaks were performed with the contaminants in order to obtain isolated colonies on different media (Fig. 1). A series of additional tests and kits, as listed in Table 1, was necessary to identify the isolates according to the criteria outlined in the Bergey Manual [1]. All culture conditions were chosen in accordance with the specific properties of the micro-organisms. Antibiotic sensitivity profiles were examined as described earlier [2,3], using commercially available ATP antibiogramme kits (BioMerieux). With two antibiotics, Imipenem (Merck-Sharpe & Dohme, Rahway, USA) and Kathon (Rohm & Haas, Vienna, Austria), the minimum inhibitory concentrations (MIC), indicating the bacteriostatic concentration, and the critical bactericidal concentration (CBC), as exerted against the bacterial isolates, were determined [3,4].

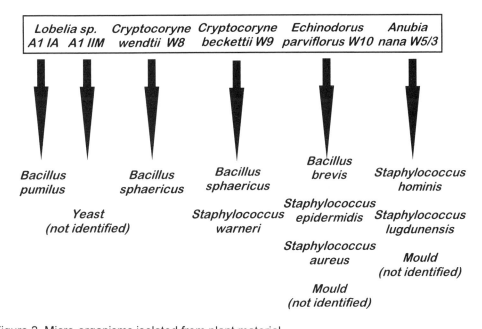

Figure 2. Micro-organisms isolated from plant material.

3. RESULTS AND DISCUSSION

3.1. Bacteriological identification

In Fig. 2 a survey of the microbiological identification results is presented. Three out of six plants contained a mixed contaminating microflora. All bacteria found were Gram-positive. The identification of the fungal contaminants was beyond the scope of this investigation. In Table 2 the results for the most important identification criteria are shown.

Table 2. Identification results of the bacteria isolated from the plant material

Criteria	Isolates								
	A	B	C,D	E	F	G	H	I
Caso broth	+	+	+	+	+	+	+	+
Caso agar	w.c.	y.c.	w.c.	y.c.	w.c.	w.c.	w.c.	w.c.
Columbia blood agar	w.c.	y.c.	w.c.	y.c.	w.c.	w.c.	w.c.	w.c.
VRBD agar	-	-	-	-	-	-	-	-
GSP agar	-	-	-	-	w.c.	w.c.	-	w.c.
YGC agar	-	-	-	-	-	-	-	-
Gram reaction	+	+	+	+	+	+	+	+
Shape	rods	rods	rods	cocci	cocci	cocci	cocci	cocci
Endospores	terminal	central	central	-	-	-	-	-
Motility	+	+	+	-	-	-	-	-
Aminopeptidase	-	-	-	-	-	-	-	-
Oxidase	-	-	-	-	-	-	-	-
Catalase	+	+	+	+	+	+	+	+
Coagulase	n.d.	n.d.	n.d.	n.d.	+	-	-	n.d.
Starch hydrolysis	-	-	-	n.d.	n.d.	n.d.	n.d.	-
ß-Haemolysis	+	-	-	+	+	-	-	-
Mannitol	-	+	+	-	+	-	-	-
Lecithinase	-	-	-	-	-	-	-	-
Proteolysis	+	+	+	-	+	-	-	-
Aerobic growth	+	+	+	+	+	+	+	+
Anaerobic growth	+	+	+	-	-	-	-	-
Growth at	10–37°C	10–37°C	10–37°C	10–37°C	10–37°C	10–37°C	10–37°C	10–37°C
NaCl tolerance	≤ 5%	≤10%	≤5%	n.d.	n.d.	n.d.	n.d.	n.d.
Lysostaphin resistance	n.d.	n.d.	n.d.	n.d.	+	-	-	-
Nitrofurantoin resistance	n.d.	n.d.	n.d.	-	-	-	-	-
Total result including API 50CH + API 20E, API Staph identification	*B a c i l l u s* pumilus	sphaericus	brevis	*S t a p h y l o c o c c u s* warneri	aureus	epidermidis	hominis	lugdunensis

Table 3. Antibiogrammes of the bacterial isolates

Antibiotic	Conc. tested (mg/l)	Bacillus pumilus	Bacillus sphaericus	Bacillus brevis	Staph. aureus	Staph. epidermidis	Staph. hominis	Staph. lugdunensis	Staph. warneri
Penicillin	0.125	–	+	+	–	–	–	+	–
Oxacillin	2	–	+	+	–	–	–	–	–
Kanamycin	8/16	–	+	+/–	–	–	–	–	–
Tobramycin	4/8	–	+/–	+/–	–	–	–	–	–
Gentamycin	4/8	–	–	+/–	–	–	–	–	–
Tetracycline	4	–	–	+	–	–	+	–	–
Minocyclin	4	–	–	–	–	–	–	–	–
Erythromycin	1/4	–	+/–	+/–	–	–	+	+	–
Lincomycin	2/8	–	+/–	+/–	–	–	–	–	–
Pristinamycin	2	–	+	+	–	–	–	–	–
Fosfomycin	32	+	+	+	–	+	–	+	–
Nitrofurantoin	25/100	–	+	+	–	–	–	–	–
Quinolone	1/4	–	+	+	–	–	–	–	–
Rifampicin	0.25/16	+/–	–	+/–	–	–	–	–	–
Fusidinic acid	2/16	–	+	+/–	–	–	–	–	–
Vancomycin	4	–	–	+	–	–	+	–	–
Teicoplanine	4	–	–	+	+	–	–	–	–
Cotrimoxacol	2/8	–	+	+	–	–	+/–	–	–
Imipenem	5	–	+	–	–	–	–	–	–
Kathon	10	+	+	+	–	–	–	–	–

+, bacterial growth in presence of antibiotics at given concentration(s).
–, no bacterial growth in presence of antibiotics at given concentration(s).

3.2. Antibiotic sensitivity of bacterial contaminants

The antibiogramme of the bacterial isolates is shown in Table 3. It can be seen from this list that, in general, the bacilli show a higher resistance to a wide array of antimicrobial compounds. With the exception of Minocyclin, none of the antibiotics was capable of inhibiting the growth of the endospore-formers at given concentrations. Because of experiences described earlier with Imipenem and Kathon [3,4] we chose these substances for the determination of the MIC and CBC levels. Depending on the nature of the bacteria, the MIC of Imipenem ranged from 0.5 to 10 mg/l, that of Kathon from 5 to 20 mg/l; the corresponding CBC ranges were 1–12 mg/l and 5–20 mg/l, respectively (Figs. 3, 4).

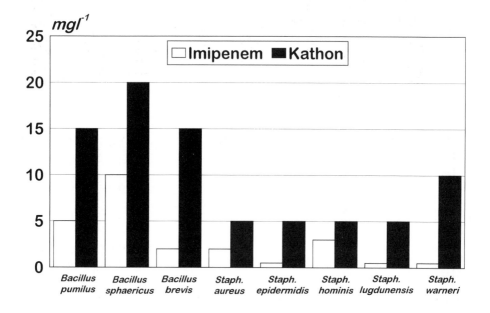

Figure 3. Minimum inhibitory concentrations of Imipenem and Kathon against the bacterial isolates.

Despite their inhibitory effect, only a short-term antibiotic treatment using liquid culture of the plants can be recommended, taking into consideration possible phytotoxicity problems. In general, many staphylococci are well known as being frequently associated with humans. We may, therefore, conclude from the occurrence of staphylococci in the plant material that part of the contamination obviously takes place during the manipulation in the laboratory.

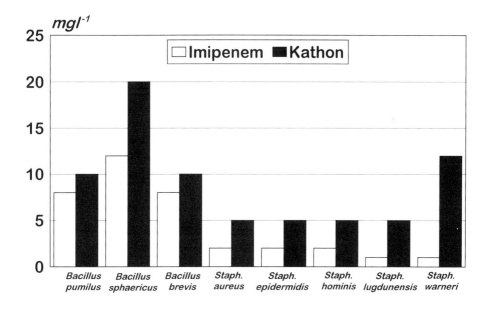

Figure 4. Critical bactericidal concentrations of Imipenem and Kathon against the bacterial isolates.

REFERENCES

1. Sneath, P.H., Mair, N.S., Sharpe, M.E. and Holt, J.G. (eds) (1986) Bergey's Manual of Systematic Bacteriology, Vol. 2, pp. 999–1034 and 1104–1138, Williams & Wilkins, Baltimore, London.

2. Kneifel, W. and Leonhardt, W. (1992) Plant Cell Tissue Org. Cult. 29, 139–144.

3. Fellner, M., Kneifel, W., Gregorits, D. and Leonhardt, W. (1996) Plant Sci. 113, 193–201.

4. Drews, G. (1976) Mikrobiologisches Praktikum, pp. 148–149, Springer Verlag, Berlin.

THE EFFECT OF ACETYLSALICYLIC ACID ON DEVELOPMENT OF BACTERIAL CONTAMINANTS AND GAS EVOLUTION IN APRICOT SHOOT CULTURES

GRAZIA MARINO[1], NANNI ELISABETTA[1], BRUNO BIAVATI[2], MICHELA PESENTI[2] and ANNIE DORO ALTAN[2]

[1]*Dipartimento di Colture Arboree, Via F. Re 6, Bologna, Italy*

[2]*Istituto di Microbiologia Agraria e Tecnica, Via F. Re 6, Bologna, Italy*

1. INTRODUCTION

Bacterial contamination is a serious problem in *in vitro* vegetative propagation of plants. Bacteria can negatively affect shoot growth and, in particular, rooting, sometimes even leading to the death of the young plantlets [1]. While antibiotics have frequently been used to prevent bacterial spread, they can induce phytotoxic effects in plant tissues and lead to the selection of resistant strains [2]. Applied salicylic acid and acetylsalicylic acid (ASA) can influence a number of biological processes in plants, including disease resistance [3]. The present study reports the effect of ASA in culture media on gas evolution, shoot growth and bacterial colony development in contaminated apricot cultures.

2. PROCEDURE

Shoots of cultivar 'San Castrese' contaminated by *Bacillus circulans* and *Sphingomonas paucimobilis*, of cultivar 'Portici' contaminated by *Staphylococcus hominis* and *Micrococcus kristinae* (strains identified by Marino *et al.* [4]), and of both cultivars contaminated by *Sphingomonas paucimobilis* and *Staphylococcus hominis* were grown on MPM8 proliferation medium [5] enriched with 28 and 50 µM ASA, or lacking it (control) and compared for gas exchange, proliferation rate (PR), shoot fresh weight (FW) and bacterial contamination rate. In all treatments, shoots about 15 mm long were grown (8 per jar) in 310-ml glass jars, each containing 50 ml of culture medium and sealed with gas-tight metal screw caps featuring small rubber septa for gas sampling. Culture conditions were $22 \pm 1°C$ and a 16-h photoperiod (60 µmol/m^2/s PAR, OSRAM L 18W/20 Hellweiss cool-white lamps).

2.1. Gas determination

Carbon dioxide, oxygen and ethylene concentrations inside jars were determined in the middle of the photoperiod each day by a Dani DS 86.01 gas chromatograph [6]. Values for gas exchanges are the means of two experiments, except for those presented for both

A.C. Cassells (ed.), Pathogen and Microbial Contamination Management in Micropropagation, 201–206.
© *1997 Kluwer Academic Publishers. Printed in the Netherlands.*

cultivars contaminated by *Sphingomonas paucimobilis* and *Staphylococcus hominis* which refer to a single experiment, each experiment consisting of five jars per treatment and sampling time. Within each time, mean separation was measured by Student's *t*-test and Student–Newman–Keuls (SNK) test (5% level), respectively, for comparisons among two and three ASA treatments.

2.2. Bacterial colony counting

Shoot apexes of proliferating cultures of 'San Castrese' and 'Portici', both contaminated by *Sphingomonas paucimobilis* and *Staphylococcus hominis* but not processed for gas sampling, were collected from each ASA treatment at the end of the subculture, washed with sterile water, and homogenized in an Omni mixer. Living cell count was taken by serial dilutions of homogenates in physiological solutions and plating on Difco Plate Count agar, pH adjusted to 7.0. The number of bacterial colonies was recorded after 48 h at 30°C.

Antimicrobial activity of ASA (concentrations up to 5 mM) was determined for isolated *Sphingomonas paucimobilis* and *Staphylococcus hominis* after Marino *et al.* [4].

3. RESULT AND DISCUSSION

3.1. Gas exchange, shoot proliferation and growth

Data on gaseous exchanges are reported per jar because of the activity of both plants and bacteria, whose development should not be strictly related to shoot growth. Data referring to 'Portici' contaminated by *Staphylococcus hominis* and *M. kristinae* treated with 50 μM ASA are not reported in the figures since the leaf yellowing in many shoots made it impossible to state whether variations in the gas concentrations were due to bacteria or/and to reduced photosynthetic activity likely caused by chlorophyll degradation. According to previous trials [4], contaminated non-ASA-treated cultures showed a gradual CO_2 increase (Figs. 1 and 3). Oxygen correspondingly decreased, at least starting from the second week in culture (Figs. 2 and 4). In ASA-treated cultures CO_2 and O_2 showed variable trends. In particular, in 'San Castrese' shoots contaminated by *B. circulans* and *Sphingomonas paucimobilis* and 'Portici' contaminated by *Staphylococcus hominis* and *M. kristinae* the CO_2 level always remained very low and O_2 tended to increase throughout subculture (up to about 22% and 21.4%, respectively, in 'San Castrese' and 'Portici') (Figs. 1 and 2). In all these shoots bacterial colonies were slightly visible, sometimes only at the end of subculture. In ASA-treated 'San Castrese' and 'Portici', both contaminated by *Sphingomonas paucimobilis* and *Staphylococcus hominis*, CO_2 remained quite low (1–2%) in the former, and increased up to 4–5% (0.4–0.5 mmol/jar produced) in the latter, where contamination was visible a few days after the last transfer. Oxygen thus increased in 'San Castrese' (up to 23–24%) and decreased in 'Portici' (to 16–18%). However, CO_2 and O_2 were always significantly lower and higher, respectively, than in non-treated shoots. Similar trends for gases were found when variations in their concentrations were expressed

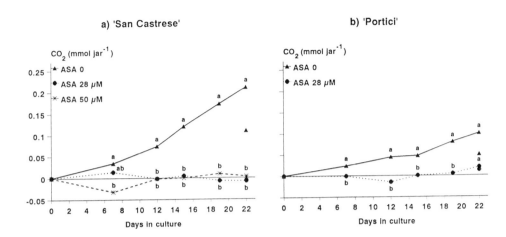

Figure 1. Carbon dioxide variations with respect to concentrations measured at shoot transfer in (a) 'San Castrese' contaminated by *Bacillus circulans* and *Sphingomonas paucimobilis* and (b) 'Portici' shoots contaminated by *Staphylococcus hominis* and *Micrococcus kristinae* and both cultured on proliferation media enriched with 28 and 50 μM ASA or lacking it (control). Separate symbols represent CO_2 variations as per g FW. Mean separation by SNK test (5% level) in 'San Castrese' and Student's *t*-test in 'Portici'.

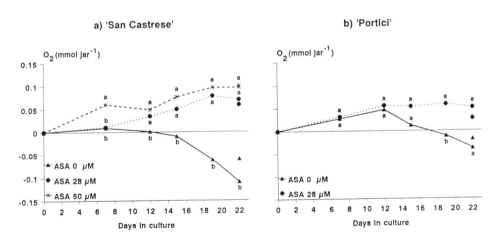

Figure 2. Oxygen variations with respect to concentrations measured at shoot transfer in (a) 'San Castrese' contaminated by *Bacillus circulans* and *Sphingomonas paucimobilis* and (b) 'Portici' shoots contaminated by *Staphylococcus hominis* and *Micrococcus kristinae*, and both cultured on proliferation media enriched with 28 and 50 μM ASA or lacking it (control). Separate symbols represent O_2 variations as per g FW. Mean separation by SNK test (5% level) in 'San Castrese' and Student's *t*-test in 'Portici'.

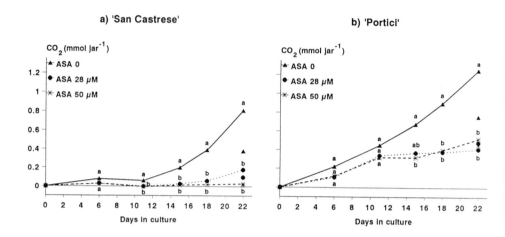

Figure 3. Carbon dioxide variations with respect to concentrations measured at shoot transfer in (a) 'San Castrese' and (b) 'Portici' shoots both contaminated by *Sphingomonas paucimobilis* and *Staphylococcus hominis* and cultured on proliferation media enriched with 28 and 50 μM ASA or lacking it (control). Separate symbols represent CO_2 variations as per g FW. Mean separation by SNK test (5% level).

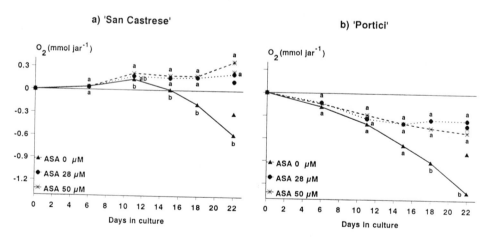

Figure 4. Oxygen variations with respect to concentrations measured at shoot transfer in (a) 'San Castrese' and (b) 'Portici' shoots both contaminated by *Sphingomonas paucimobilis* and *Staphylococcus hominis* and cultured on proliferation media enriched with 28 and 50 μM ASA or lacking it (control). Separate symbols represent O_2 variations as per g FW. Mean separation by SNK test (5% level).

as per g FW, differences among treatments remaining significant for O_2 and CO_2 in all 'San Castrese' cultures, regardless of the identity of contaminating bacteria. No difference was found between the two ASA doses in any of the trials. In agreement with a previous report [7], ethylene accumulation was reduced in ASA-treated cultures, significantly in most cases. Shoot growth and proliferation were also significantly reduced in ASA-treated cultures, except for 'San Castrese' contaminated by *Sphingomonas paucimobilis* and *Staphylococcus hominis* (data on C_2H_4, FW and PR are not reported in the figures).

3.2. Bacterial colony counting

The average number of bacterial colonies was similar in all treatments for 'San Castrese' (approx. 13 and 4×10^6 CFU/g FW in non-treated and ASA-treated cultures, respectively). In contrast, it was 6.8×10^9, 1.3×10^9 and 1.5×10^8 CFU/g FW for 0, 28 and 50 μM ASA, respectively, in the case of 'Portici'. All ASA concentrations tested were ineffective on isolated bacteria.

In previous trials, healthy shoot cultures of 'San Castrese' and 'Portici' clearly showed photosynthetic activity, with oxygen increasing during subculture. In contrast, oxygen quickly decreased and carbon dioxide increased in contaminated shoots [4]. In the present study, CO_2 accumulation, oxygen consumption and ethylene evolution were reduced in ASA-treated cultures of both cultivars, regardless of the identity of the bacteria, suggesting a reduced bacterial proliferation with respect to controls, although shoot growth and proliferation also decreased and some leaf yellowing appeared at the higher ASA concentration. The number of bacterial colonies recorded in the homogenates of ASA-treated cultures was much lower than in controls in the case of 'Portici' and slightly lower in 'San Castrese'. These results support previous findings that ASA can counteract bacterial development, at least in 'Portici' tissues. Further studies should help to determine whether periodic subculturing of shoot apexes of bacterial-contaminated cultures on ASA-enriched media might represent an alternative treatment to antibiotics when these are not satisfactory (i.e. ineffective, toxic to plant tissues, leading to resistant bacterial strains).

ACKNOWLEDGEMENTS

This research was supported by funds from MURST (60%).

REFERENCES

1. Leifert, C., Morris, C.E. and Waites, W.M. (1994) Crit. Rev. Plant Sci. 13, 139–183.
2. Leifert, C., Ritchie, J.Y. and Waites, W.M. (1991) World J. Microbiol. Biotechnol. 7, 452–469.
3. Raskin, I. (1992) Annu. Rev. Plant Physiol. Plant Mol. Biol. 43, 439–463.
4. Marino, G., Doro Altan, A. and Biavati, B. (1996) In Vitro Cell. Dev. Biol. Plant 32, 51–56.

5. Marino, G., Bertazza, G.P., Magnanini, E. and Doro Altan, A. (1993) Plant Cell Tissue Org. Cult. 34, 235–244.

6. Magnanini, E. and Ventura, M. (1994) in 'II Giornate Scientifiche S.O.I. 1994', 22–24 June, San Benedetto del Tronto, Italy, pp. 499–500, S.O.I. Firenze, Italy.

7. Marino, G., Doro Altan, A. and Magnanini, E. (1944) in Abstracts of the VIIIth Int. Congress of Plant Tissue and Cell Culture, 12–17 June, Firenze, Italy, p. 62, IAPTC.

CONTROL OF *BACILLUS* CONTAMINATING DATE PALM TISSUE IN MICROPROPAGATION USING ANTIBIOTICS

Ah. BENJAMA and B. CHARKAOUI

Lab. of Phytobacteriology, INRA-BP, 533 Marrakech, Morocco

1. INTRODUCTION

Microbial contamination is one of the most important causes of losses in *in vitro* culture [1]. Yeast, fungi or bacteria are frequently isolated [2–6]. The most serious contamination is caused by bacteria which reduce the multiplication and rooting speed and can cause the death of tissues [7,8]. To control bacterial contaminants in tissue culture, several trials have been undertaken by different authors using antibiotics [9–13]. Their success, however, depends on the kind of bacteria involved.

In Morocco, it has been shown that date palm tissue culture is contaminated by eight species of endophytic *Bacillus* [3]. Among 40 antibiotics, only Novobiocin and Gentamycin are efficient in pure culture against these bacteria [9]. In this paper, we examine the efficiency of these screened antibiotics on contaminated tissue culture of date palm genotypes.

2. MATERIAL AND METHODS

2.1. Bacterial strains

Eight species of *Bacillus* which are isolated from the tissues of micropropagation date palm were used: *Bacillus sphaericus*, *Bacillus pumilus*, *Bacillus cereus*, *Bacillus brevis*, *Bacillus laterosporus*, *Bacillus subtilis*, *Bacillus circulans* and *Bacillus* sp. [3].

2.2. Determination of minimal inhibitory concentrations

Different concentrations of Novobiocin (Nv) and Gentamycin (Gt) from 0.5 mg/l to 128 mg/l were added to the LPGA medium after autoclaving. Spots of 10^4 bacteria were inoculated on the surface of the medium and incubated at 27°C for 72 h. The lowest concentration that killed all species of *Bacillus* was selected to be tested on contaminated tissue culture in tubes.

A.C. Cassells (ed.), Pathogen and Microbial Contamination Management in Micropropagation, 207–211.
© *1997 Kluwer Academic Publishers. Printed in the Netherlands.*

2.3. Tissues culture of date palm

Two kinds of vegetable material were used. (a) Seedlings (old tissue) of micropropagated genotype Ziz which were very contaminated. The symptoms are characterized by a bacterial layer on the surface of the medium and by a halo around the roots. (b) The second material was composed of the explants produced from cultivars 'Boufegous' (BFG), 'Bouskri' (BSK), 'Tadment' (TDMNT) and from two selected clones 'INRA C10' and 'INRA 3014'. All of these *in vitro* genotypes were young and weakly contaminated by bacteria (halo).

As a first step, the antibiotics were added to tissue culture medium after autoclaving. Then, the contaminated date palm tissues were washed in a solution of the same antibiotic three times, 5 min each time, before their sub-culture in physiological medium with antibiotics. The treated tissues cultures were incubated and controlled every week for 4–5 weeks.

2.4. Index of contamination (ic)

The index of contamination (ic) is established to illustrate the level of contamination in the tubes according to the following scale:

0 = clean *in vitro*-culture
1 = slight contamination around roots of *in vitro* culture
2 = bacterial halo around roots.
3 = bacterial layer on the surface of the tube with a yellow, white or cream colour.

3. RESULTS AND DISCUSSION

3.1. Efficiency of Novobiocin and Gentamycin against strains of *Bacillus* in pure culture

The minimal inhibitory concentration of eight species of *Bacillus* were 1 mg/l for Nv and 3.5 mg/l for Gt.

3.2. Efficiency of the concentrations on *Bacillus* in contaminated *in vitro* culture tissue

The test of minimal inhibitory concentration of Nv and Gt showed that they were not efficient on contaminated tissue. An effect was found when these concentrations were raised 20 times. The progressive increasing of concentration showed that the ic decreases from 3 to 0 (Tables 1, 2) using 20 mg/l for Nv and 30 mg/l for Gt. Fisse *et al.* [12] found that antibiotics are efficient on contaminated tissue when the minimal concentration was raised 1000 times. Cornu and Michel [10] report that for the control of contamination to be successful at least two antibiotics must be used.

Table 1. Effect of increasing concentrations of Novobiocin on contaminated genotype *Ziz*

	Days			
	4	10	21	26
Nv 10 µg/ml	0^1	2.4	2.4	$2.4 (a)^2$
Nv 20 µg/ml	0.4	0	0	0 (b)
Nv 30 µg/ml	0	0	0	0 (b)
Control	0	3	3	3 (c)

[1]Index of contamination (average of ic of 40 tubes).
[2]Numbers followed by the same letter are not significant.

In our study, the results obtained with a concentration of 20 mg/l Nv were highly significant (Table 1). There is no difference of efficiency between the 20 and 30 mg/l concentrations of Novobiocin, but the 30 mg/l was phytotoxic.

The 10, 20 and 30 mg/l concentrations of Gt were significantly efficient (Table 2). The lowest concentration (10 mg/l) was chosen to be combined with 20 mg/l of Nv.

3.3. Effect of Novobiocin and the combination of Novobiocin and Gentamycin using different contaminated genotypes

The incorporation in the medium of the combination of Nv (20 mg/l) and Gt (10 mg/l) showed that, on tissue of genotype Ziz, there was no phytotoxicity and no bacterial growth after 26 days of incubation (Table 3). For other genotypes, the use of this combination allowed 73–100% of clean tissue culture to be obtained, depending on the genotype

Table 2. Effect of increasing concentrations of Gentamycin on contaminated genotype Ziz

	Days			
	4	10	21	26
Gt 10 µg/ml	0^1	1.2	1.2	$1.2 (a)^2$
Gt 20 µg/ml	0.4	1.2	1	1.4 (b)
Gt 30 µg/ml	0.2	1.2	1.2	1.2 (b)
Control	0.4	3	3	3 (c)

[1]Index of contamination (average of ic of 40 tubes).
[2]Numbers followed by the same letter are not significant.

(Table 4). This percentage decreases from 88% to 3% for genotype Ziz (seedling stage) and from 100% to 33% for the other genotypes (explant stage), when the tissue was transferred on medium without antibiotics (Table 4). Other authors observed that contaminations reappeared when the tissues were transferred to media without antibiotics [10], even if these tissues were sub-cultured on medium with a high concentration of antibiotic [12] or maintained for a long time [5].

Table 3. Efficiency of the combination of Novobiocin 20 mg/ml + Gentamycin 10 mg/ml on contaminated genotype Ziz (seedling stage)

	Days			
	4	10	21	26
Tube 1	0	0	0	0
Tube 2	0	0	0	0
Tube 3	0	0	0	0
Tube 4	0	0	0	0
Tube 5	0	0	0	0
Control	0	3	3	3

Table 4. Average of clean *in vitro* culture during (T) and after (R) 40 days of treatment with Nv and Gt on different contaminated genotypes

Genotype	Nv		Nv/Gt	
	T	R	T	R
ZIZ	80.43[a]	3.03	88.37	5.26
BFG	73.33	27.27	100	33.33
BSK	57.89	18.18	81.25	23.07
INRA C10	96.33	30.43	100	35.67
INRA 3014	93.65	27.95	100	39.71
JHL	70.41	19.25	94.62	25.34
TDMT	85.72	20.94	98.85	29.64
Control	0	0	0	0

Nv, Novobiocin; Gt, Gentamycin; T, medium with antibiotic; R, transferred tissue on medium without antibiotic.
[a]Average of safety tissue culture (144 *in vitro* cultures).

The efficiency of antibiotics depends on the physiological stage of the tissue. Young tissues are better controlled than old ones (seedling stage) if the index of contamination is very low. Otherwise cleaning of the tissues in antibiotic solution is necessary before their sub-culturing.

It was noticed that the use of 20 mg/l Nv or the combination 20 mg/l Nv + 10 mg/l Gt gave comparable efficiency. The use of the combination is recommended, however, to prevent bacterial mutation as recommended for other plants [5].

4. CONCLUSION

In order to control *Bacillus* contaminating *in vitro* culture of date palm, we recommend the incorporation of 20 mg/l Novobiocin and 10 mg/l Gentamycin in the tissue culture medium after autoclaving. Moreover, cleaning of the tissue three times in antibiotic solution before sub-culturing is necessary.

Similarly, we observed that the use of antibiotics does not have a negative effect on multiplication rate (Ah. Benjama, unpublished). The results obtained on young tissue (explants) are better than those obtained using physiological old tissue (seedling).

REFERENCES

1. Leifert, C. and Waites, W.M. (1992) J. Appl. Bacteriol. 72, 460–466.

2. Blake, J. (1988) Acta Hortic. 225, 163–166.

3. Benjama, Ah. (1994) Al-Awamia 85, 89–96.

4. Enjalric, F., Carron, M.P. and Lardet, L. (1988) Acta Hortic. 225, 57–65.

5. Leifert, C., Camotta, H., Wright, S.M., Waites, B., Cheyne, V.A. and Waites, W.M. (1991) J. Appl. Bacteriol. 71, 307–330.

6. Leifert, C., Waites, W.M., Camotta, H. and Nicholas, J.R. (1989) J. Appl. Bacteriol. 67, 363–370.

7. Cassells A.C. (1991) in Micropropagation: Technology and Application (Debergh P.C. and Zimmerman R.H., eds), pp. 31–44. Kluwer, Dordrecht.

8. Leifert, C., Waites, W.M. and Nicholas, J.R. (1989) J. Appl. Bacteriol. 67, 353–361.

9. Benjama, Ah., Cherkaoui, B. and Samson, R. (1996) Al-Awamia 92, in press.

10. Cornu, D. and Michel, M.F. (1987) Acta Hortic. 212, 83–86.

11. Falkiner, F.R. (1988) Acta Hortic. 225, 53–56.

12. Fisse, C., Batlle, A. and Pera J. (1987) Acta Hortic. 212, 87–90.

13. Leggat, I.V., Waites, W.M., Leifert, C. and Nicholas, J. (1988) Acta Hortic. 225, 93–102.

MONITORING OF VIRUS DISEASES IN AUSTRIAN GRAPEVINE VARIETIES AND VIRUS ELIMINATION USING *IN VITRO* THERMOTHERAPY

W. LEONHARDT[1], CH. WAWROSCH[2], A. AUER[1] and B. KOPP[2]

[1]*VitroPlant Plant Biotechnology, Brunnleiten 17, A-3400 Klosterneuburg, Austria*
[2]*Institute of Pharmacognosy, Center of Pharmacy, Althanstraße 14, A-1090 Wien, Austria*

1. INTRODUCTION

Vitis vinifera L. (grapevine) is the most important fruit crop grown in Austria. Every year about 250 million litres of wine are produced by Austrian farmers. White wines account for 80% of the total harvest and red wines 20%. Each year some 2 million grapevines are planted in the vineyards of Austria. Until 1992, no work was done in Austria to eliminate virus diseases from the typical Austrian grapevine varieties. As Austria joined the EU in 1995, it is now necessary to produce virus-free plant material for the grape-growing industry according to the EU directions for grapevine trading. Field monitoring had shown that in the region of Niederösterreich more than 40% of the most important variety 'Grüner Veltliner' were infected by the GLRaV1 virus (W. Leonhardt and R. Legin, unpublished). In the grapevine-growing region of Burgenland one-third of the screened samples were virus-positive: 80% of the detected viruses were identified as GLRaV and about 10% each were AMV and GFLV [1].

Besides the possibilities of rapid propagation and long-term storage in gene banks, since the early 1960s tissue culture techniques have focused on the elimination of virus infections [2]. The present study deals with virus elimination in grapevine using *in vitro* thermotherapy. In this, it has to be taken into consideration that the characterization and identification of grape varieties by breeders and nurseries is mainly done by the leaf shape using a series of morphological characteristics which are described in detail elsewhere [3]. Thus, a comparative investigation of conventional vs. *in vitro* propagated grapevine clones using leaf morphology as indicator for somaclonal variation was included in the project. For virus detection inoculation of indicator plants was used to confirm ELISA results as a precaution in case virus titre was reduced in tissue cultures.

2. MATERIAL AND METHODS

2.1. Plant material

Single plants of genotyped vegetatively propagated clones were used. If possible, the use of clones with confirmed positive ELISA results at the location was avoided. From single

A.C. Cassells (ed.), Pathogen and Microbial Contamination Management in Micropropagation, 213–218.
© *1997 Kluwer Academic Publishers. Printed in the Netherlands.*

plants, conventionally propagated stock cultures were established in insect-secure greenhouses. All *in vitro* cultures were started from this material.

2.2. Sterilization method

Pretreatment of shoot tips was done in Chinosol® solution (1 g/l for 3 min) and in 70% EtOH for 3 min. Subsequently the shoot tips (size 5 mm) were sterilized in a solution of Danclor® (4.2% active chlorine) for 10 min.

2.3. Culture media

Establishment of cultures: major and minor salts according to Brendel *et al.* [4], Fe–EDTA, vitamins and sucrose according to Murashige and Skoog [5]. Growth regulators: 0.01 mg/l NAA, 0.03 mg/l BAP; Gelrite 0.25%, pH 5.8.

Propagation, stock cultures and thermotherapy: half-strength MS macro and micro elements, full strength MS $CaCl_2$ and vitamins; 2% sucrose, 0.25% Gelrite, 0.2 mg/l IBA, pH 5.8.

2.4. Thermotherapy

Sterile shoot tips (5 mm) were cultured for 4 weeks at a temperature of 25°C for rooting. Every clone was split up into subclones, each subclone labelled as an individual plant in thermotherapy. Thermotherapy was performed at a temperature of 35°C for a period of 6–8 weeks. This procedure was repeated twice.

2.5. Indexing

Indexing was carried out using the following indicator plants:

- Pinot noir, Cabernet franc — GLRaV
- LN33 — corky bark
- Rupestris St. George — stem pitting and Fleck
- Kober SBB — vein mosaic
- Richter 110 — vein necrosis

The method of indexing is described in detail by Legin [6], Bass [7] and Reustle [8].

2.6. ELISA tests

Fanleaf (GFLV), arabismosaic (AMV) and leafroll associated viruses (GLRaV) were detected according to the protocol of Flack [1].

2.7. Morphogenetic tests

Ten plants of four different varieties originating from conventional cuttings and ten tissue culture derived plants, 1.2–1.5 m in height, were used as source of the leaf samples. Two varieties were GLRaV1-infected whereas the other two were virus-free. According to the guidelines for the data acquisition [3] one fully expanded leaf was collected from each single plant, giving a total of 80 leafs. Of each leaf, 18 morphological characteristics were recorded. Standardization of the data gave values as % of the maximum leaf length.

3. RESULTS AND DISCUSSION

3.1. Infection of field material

Thirty to forty percent of the field material was virus-positive. Only clones with originally negative ELISA results were selected for the program except for a few varieties where only infected clones were available. GLRaV, GFLV and AMV were the most frequent viruses detected. All other important grape viruses seem to be rather rare in Austrian vineyards.

3.2. Sterilization and *in vitro* culture

In the present study 79 white grape varieties, 33 red grape varieties and 15 rootstocks and indicator plants were established in tissue culture (see Table 1). Depending on the variety the sterilization results (10–35% infections) were satisfactory. To avoid possible somaclonal variations shoot tips (5 mm) were used instead of isolated meristems which would require high levels of cytokinins. For the same reason, propagation, stock cultures and heat therapy were done using cytokinin-free media.

3.3. Thermotherapy

All clones including the ELISA-negative ones were subjected to thermotherapy twice. Screening for GLRaV and AMV was done with greenhouse material, after establishing clones in vitro, 8 weeks after thermotherapy in vitro, and 5 months after transfer of plants to the greenhouse. First attempts to cultivate small shoot tips in thermotherapy met with little success. Since nearly all shoot tips showed necrosis, pre-rooted shoots were used for heat therapy with much better results. Two months after thermotherapy nearly all clones were virus-free. ELISA tests were highly sensitive even screening single tissue culture plants. Out of ten originally infected clones with 97 subclones only four subclones (see Table 2) were virus-positive after heat treatments. The fact that infection of four subclones of one clone could be detected by ELISA in all stages seems to indicate that negative ELISA results after thermotherapy were not caused by virus suppression in tissue culture. However, it will be necessary to keep the virus-free clones under further observation in the greenhouse for several years to confirm this result.

Table 1. Grapevine varieties and clones established in tissue culture

Variety	Clones	Variety	Clones	Variety	Clones
White grape		*Red grape*		*Rootstocks*	
Grüner Veltliner	38	Blaufränkisch	13	Aripa	1
Welschriesling	6	Zweigelt	7	Kober 5B B	2
Neuburger	15	Schilcher	4	Riparia	1
Chardonnay	1	Cabernet Sauvignon	2	SO4	1
Frühroter Veltliner	2	Blauer Portugieser	2	*Indicator plants*	
Grüner Sylvaner	1	Pinot noir	3	Cabernet franc	1
Merlot	2	St. Laurent	2	Gamey 4754	1
Müller Thurgau	1			Kober 5 BB	1
Muskat Ottonel	1			LN 33	1
Roter Veltliner	2			Pinot noir	I
Rotgipfler	4			Richter 110	2
Sauvignon blanc	1			Rupestris	1
Weißburgunder	2			St. George 4714	1
Zierfandler	3			Traminer	1

Table 2. ELISA results of two grapevine clones before and after thermotherapy

Before thermotherapy

Clone (no. of subclones)	Greenhouse				*In vitro*		
	GFLV	GLRaV		AMV	GLRaV		AMV
		1	3		1	3	
GV 21/10 (12)	–	12+	–	–	12+	–	–
BA GV 2/1 (10)	–	10+	–	–	10+	–	–

After thermotherapy

Clone (no. of subclones)	*In vitro*					Greenhouse				
	GLRaV				AMV	GLRaV				AMV
	1	2	3	5		1	2	3	5	
GV 21/10 (12)	-	-	-	-	-	-	-	-	-	-
BA GV 2/1 (10)	4+	-	-	-	-	4+	-	-	-	-

3.4. Indexing

To date the indexing of six clones has been completed; none of the clones showed virus symptoms. Due to the extended duration of the process, indexing of the remaining 112 clones is still in progress.

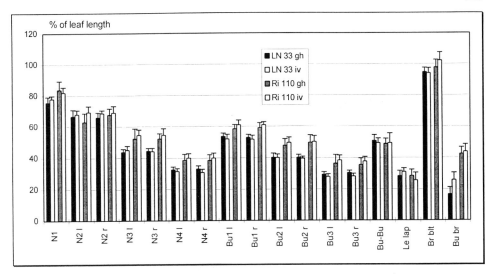

Figure 1. Comparison of leaf morphological characteristics of greenhouse-grown (gh) plants and *in vitro* derived (iv) plants of the virus-free clones LN 33 and Ri I10.

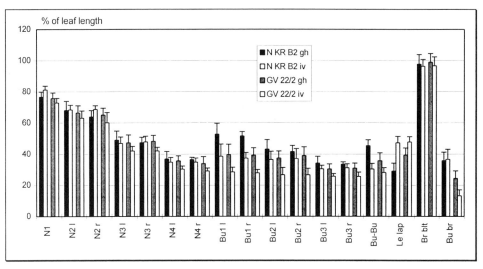

Figure 2. Comparison of leaf morphological characteristics of greenhouse-grown (gh) plants and *in vitro* derived (iv) plants of the leaf roll-infected varieties N KR B2 and GV 22/2.

3.5. Morphogenetic changes

A comparison of greenhouse (cuttings) vs. tissue culture derived plants of both the virus-free indicator clones LN33 and Ri 110 revealed that no significant differences occurred in the morphological characteristics of the leaves (refer to Fig. l): nearly all measured differences lay within the standard deviation. Among the two leaf-roll infected varieties, GV 22/2 and N KR B2, there were differences between the plants propagated from cuttings and those derived from *in vitro* culture, as shown in Fig. 2. The deviations occurred mainly in the characteristics associated with the values of the leaf sinus and were more distinct in GV 22/2 than in N KR B2. Differences in morphology between greenhouse grown and *in vitro* derived-material of the infected varieties may be explained by virus elimination or suppression in the *in vitro* derived plants. This observation has also been made by Legin (personal communication).

ACKNOWLEDGEMENTS

This project is supported by the Austrian Ministry of Agriculture and the Federal Governments of Niederösterreich and Burgenland.

REFERENCES

1. Flak, W. and Gangl, H. (1994) Mitt. Klosterneuburg 44, 163–167.

2. Galzy, R. (1961) C. R. Seances Acad. Sci. (Paris) 253, 706–708.

3. Dettweiler, E. (1991) Preliminary Minimal Descriptor List for Grapevine Varieties, Inst. of Grapevine Breeding Geilweilerhof, D-76833 Siebeldingen, Germany.

4. Brendel, G., Steinmann, P. and Steinmann, K. (1991) Dtsch. Weinbau 2, 58–64.

5. Murashige, T. and Skoog, F. (1962) Physiol. Plant. 15, 473–497.

6. Legin, R., le Gall, O. and Walter, B. (1987) Schweiz. Landwirtsch. Forsch. 26, 313–316.

7. Bass, P. and Legin, R. (1984) Progr. Agric. Vitic. 101, 270–274.

8. Reustle, G., Mann, M. and Heintz, C. (1988) Acta Hortic. 225, 119–127.

HOT-WATER TREATMENT BEFORE TISSUE CULTURE REDUCES INITIAL CONTAMINATION IN *LILIUM* AND *ACER*

MEREL LANGENS-GERRITS[1,2], MARION ALBERS[1,3] and GEERT-JAN De KLERK[1]

[1]Centre For Plant Tissue Culture Research, P.O. Box 85, 2160 AB Lisse, The Netherlands
[2]Bulb Research Centre, P.O. Box 85, 2160 AB Lisse, The Netherlands
[3]Research Station For Nursery Stock, P.O. Box 118, 2770 AC Boskoop, The Netherlands

1. INTRODUCTION

In 1887, the Danish researcher Jensen used a hot-water treatment (HWT) to free plant tissues from pathogens. Ever since, heat treatments have been used to free plants from viruses, bacteria, fungi, nematodes and insects. In agricultural practice, HWTs are used on a large scale with bulbs, tubers and seeds [1]. The advantages of temperature treatments in comparison with chemical treatments are that there are no chemical residues and also that endogenous pathogens are removed. However, a HWT may damage the host plant [2]. The resistance of plant tissues to HWT strongly depends on their physiological condition, for example size, moisture content, vigour, condition of external layers, temperature conditions during growth, dormancy level, age, and genetic constitution [2].

In our lab, HWTs have been examined in tissue culture of *Narcissus* bulbs [3]. The HWT was very efficient and reduced initial contamination from 40–60% to 5%. In this article we report the use of a HWT in tissue culture of another bulbous crop, *Lilium*, and in the woody plant *Acer lobelii*.

2. MATERIALS AND METHODS

2.1. *Lilium*

Field-grown bulbs (circumference 18–20 cm) of *Lilium speciosum* 'Rubrum No. 10' were stored in peatmoss at -1°C. Before use, bulbs were defrosted at 5°C. Bulbs were rinsed with water and roots and brown scales were removed. A HWT of 1 h was given in a waterbath. The bulbs were dried overnight. Scales were surface-sterilized for 30 min in 1% (w/v) NaOCl and rinsed three times for 10 min with sterile water.

A.C. Cassells (ed.), Pathogen and Microbial Contamination Management in Micropropagation, 219–224.
© *1997 Kluwer Academic Publishers. Printed in the Netherlands.*

Tissue culture conditions were as described by Aguettaz *et al.* [4]. Explants of $7 \times 7 \text{ mm}^2$ were placed with the abaxial side on 15 ml medium in culture tubes (ø 2.2; 15 cm high). The tubes were loosely closed with a plastic cap that allowed gas exchange. In one experiment, however, half of the tubes were hermetically closed with parafilm. The medium was composed of MS macro- and microelements [5], 0.4 mg/l thiamine-HCl, 100 mg/l myo-inositol, 30 g/l sucrose, 6 g/l agar (Becton and Dickinson, granulated) and 0.25 µM α-naphthaleneacetic acid. The pH was adjusted to 6.0 before autoclaving for 15 min at 120°C. The explants were cultured at 20°C at a photon fluence rate of 30 µmol/m^2/s (Philips TL 33) for 16 h per day.

Contamination was recorded throughout the culture period. After 14 weeks, the percentage regeneration, the number of plantlets regenerated per non-contaminated explant and fresh and dry weight regenerated per explant were determined. For each treatment, 45 explants were used.

2.2. *Acer*

Branches of *Acer lobelii* were collected in January. Stem segments with two opposing axillary buds were excised. A HWT was given in a waterbath. After that, the segments were rinsed for 1 min in 70% ethanol, surface-sterilized twice in 2% (w/v) NaOCl for 10 min and rinsed three times for 10 min with sterile water. The axillary buds were excised from the segments. With the exception of the two to four inner ones, all scales were removed.

The explants were cultured essentially as described by Marks and Simpson [6] in culture tubes (ø 2.2; 15 cm high) with 10 ml medium. The culture tubes were loosely closed with plastic caps that allowed gas exchange. The medium was composed of MS macro and microelements and vitamins [5], 30 g/l sucrose, 8 g/l agar (Daichin) and 0.01 µM thidiazuron. The pH was adjusted to 5.5 before autoclaving for 15 min at 120°C. The explants were cultured for 4 weeks at 25°C at a photon fluence rate of 30 µmol/m^2/s (Philips TL 33) for 16 h per day. Contamination was scored throughout the culture period. For each treatment, 45 explants were used.

3. RESULTS

3.1. *Lilium*

In the batch of lily bulbs used for the experiment, contamination was high, about 50%. A 1-h HWT resulted in a significant decrease of contamination at all temperatures examined (Fig. 1A; at 40°C: $P<0.01$).

During the first weeks of culture, the explants taken from bulbs treated at 44 or 45°C, became brown and appeared to have deteriorated. From these explants, regeneration started

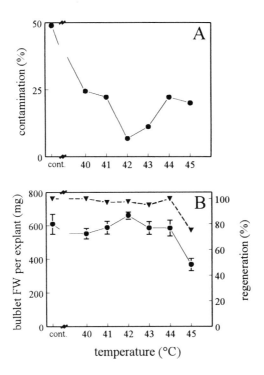

Figure 1. Effect of a 1-h HWT at various temperatures on contamination (A) and regeneration (B) in scale explants of *L. speciosum*. Regeneration was determined as the percentage explants that showed regeneration (▼) and as the total fresh weight (FW ●) of the bulblets regenerated from non-contaminated, regenerating explants.

somewhat later. The number of bulblets per explant and the percentage regeneration were not affected by any of the HWTs with the exception of the treatment at 45°C (Fig. 1B).

In one experiment, the tubes were closed hermetically with parafilm or closed loosely with a plastic cap. The percentage survival is shown in Fig. 2. All *L. speciosum* explants in the hermetically closed tubes died.

3.2. *Acer*

In the batch axillary buds of *Acer lobelii* used in the experiment, contamination reached almost 100%. A HWT at 42.5 or 45°C reduced contamination strongly (Fig. 3). All explants survived the HWT and resumed growth. After the 45°C HWT, some vitrification occurred. Therefore, a temperature of 42.5°C was optimal.

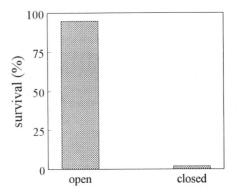

Figure 2. Effect of a 1-h HWT at 43°C on survival of explants of *L. speciosum* cultured in tubes loosely closed with a cap ('open') or hermetically closed ('closed').

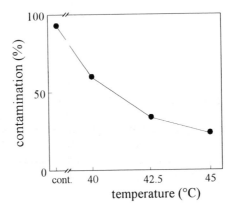

Figure 3. Effect of a 2-h HWT at various temperatures on contamination in *Acer lobelii* axillary buds. All explants survived the HWT.

4. DISCUSSION

The success of a HWT depends on the differential heat sensitivity of pathogen and host. How the HWT affects the pathogens is not known. Various pathogens have different heat sensitivities and the same pathogen may have different sensitivities depending on the host plant [1,2]. This may explain why in *Narcissus* a much higher temperature (54°C) is needed [3] than in *Lilium* and *Acer*.

A HWT may severely damage the plant tissue. In lily, regeneration decreased after a HWT at 45°C. In hermetically closed tubes, all explants of *Lilium speciosum* died after a standard

HWT at 43°C, whereas in tubes that allowed gas-exchange almost all explants survived. An explanation for this is that during and after the HWT metabolism, and therefore need for oxygen, increases (see Ref. 7). During the HWT, the tissue is under partial anaerobiosis because the bulbs are submerged [2]. Oxygen deficiency may lead, for example, to production of acetaldehyde and ethanol [8]. Accumulation of acetaldehyde and ethanol was followed by rapid tissue deterioration in *Prunus* [9]. Ethanol inhibited shoot proliferation and rooting in *Malus* tissue cultures [10]. Thus, death of the explants in hermetically closed tubes may be caused by asphyxiation of the tissue resulting in an accumulation of ethanol and acetaldehyde. An alternative explanation is that in tissues damaged by the HWT a more severe wounding reaction occurred when the explants had been cut. Thus, the production of volatile compounds that occurs after wounding (e.g. ethylene and ethane), might have increased. Ethane production is correlated with the degree of tissue damage [9,11]. High concentrations of ethylene or ethane, accumulating in the tubes, might be toxic.

Occasionally, it has been found that after a HWT contamination rates increase. In the experiment shown in Fig. 1, we also observed higher contamination rates after a 44°C HWT than after a 42°C HWT ($P<0.05$). Due to tissue damage after high temperature treatment, remaining pathogens might be easily released from the tissue. Another possibility is that high temperatures activate spores of pathogens present in the tissue.

As mentioned before, the heat resistance of plant tissue depends, for example, on the dormancy level. Under natural conditions, plants develop dormancy to overcome adverse climatic conditions. Dormant organs (seeds, buds, bulbs, corms and tubers) have an increased resistance to extreme climatic conditions. It was shown, for example, that dormant *Gladiolus* corms are more heat-resistant than non-dormant ones [2]. *Acer* buds taken in January (non-growing) resisted a HWT of 42.5 or 45°C, whereas *Acer* buds that had resumed growth all died after a HWT of only 42.5°C (M. Albers, unpublished).

In conclusion, a HWT was very efficient in reducing endogenous contamination of *Lilium* and *Acer*. Regenerated lily bulblets were not subcultured but, in *Acer*, only occasional contaminations were observed after repeated subcultures.

ACKNOWLEDGEMENTS

One of the experiments was carried out by Dr. Jan Jasik, Comenius University, Bratislava. We thank Dr. P.M. Boonekamp for critical reading of the manuscript.

REFERENCES

1. Grondeau, C. and Samson, R. (1994) Crit. Rev. Plant Sci. 13, 57–75.
2. Baker, K.F. (1962) Phytopathology 52, 1244–1255.

3. Hol, G.M. and Van Der Linde P.C.G. (1992) Plant Cell Tissue Org. Cult. 31, 75–79.

4. Aguettaz, P., Paffen, A.M.G., Delvallée, I., Van Der Linde P.C.G. and De Klerk, G.J. (1990) Plant Cell Tissue Org. Cult. 22, 167–172.

5. Murashige, T. and Skoog, F. (1962) Physiol. Plant. 15, 473–497.

6. Marks, T.R. and Simpson, S.E. (1994) J. Hortic. Sci. 69, 543–551.

7. Van Der Vlugt, C.I.M. (1994) PhD thesis. Agricultural University Wageningen, The Netherlands.

8. George, E.F. (1993) Plant Propagation by Tissue Culture. Part One: The Technology. Exegetics Ltd., Edington, UK.

9. Righetti, B., Magnanini, E., Infante, R. and Predieri, S. (1990) Physiol. Plant. 78, 507–510.

10. De Klerk, G.J., Ter Brugge, J. and Marinova, S. (1997) Biol. Plant. 39, 105–112.

11. Van Aartrijk, J., Blom-Barnhoorn, G.J. and Bruinsma, J. (1985) J. Plant Physiol. 117, 411–422.

ELIMINATION OF BEAN SEED-BORNE BACTERIA BY THERMOTHERAPY AND MERISTEM CULTURE

MATEJA GRUM[1], MARJANA CAMLOH[1], KLAUS RUDOLPH[2] and MAJA RAVNIKAR[1]

[1]National Institute of Biology, Karlovška 19, 1000 Ljubljana, Slovenia

[2]Institut für Pflanzenpathologie und Pflanzenschutz, Georg August Universität, Grisebachstraße 6, D-3400 Göttingen, Germany

1. INTRODUCTION

Common and halo blights caused by *Xanthomonas campestris* pv. *phaseoli*, and *Pseudomonas syringae* pv. *phaseolicola*, respectively, are major world-wide seed-borne diseases of bush bean (*Phaseolus vulgaris* L.) [1]. Seeds contaminated either internally or externally constitute the primary source of the inoculum. In annual plants the main mode of bacterial disease transmission is via the seeds.

The bean cultivar 'Starozagorski' is a commercially interesting low string bush bean in Slovenia, but the yield is very low due to diseases caused by various seed-borne pathogens. During seed health testing we identified *X.c.* pv. *phaseoli* and *P.s.* pv. *syringae*. In addition, a seed-borne potyvirus, the bean common mosaic virus (BCMV), was detected in plants.

To free seeds from bacterial pathogens different heat treatments were used, although a significant decrease in germination especially in the larger seeds of legumes, was observed (for a review, see [2]). To eliminate *P.s.* pv. *phaseolicola* and *X.c.* pv. *phaseoli* from bean seeds dry heat treatment was successfully used [3,4]. Thermotherapy associated with meristem and shoot tip cultures were the main means of producing virus-free plants.

In this study we describe a combination of seed dry heat treatment and thermotherapy of seedlings associated with meristem and shoot tip cultures as a reliable procedure to obtain pathogen-free plants.

2. MATERIALS AND METHODS

2.1. Plant material

Bush bean (*Phaseolus vulgaris* L. cv. 'Starozagorski') seeds were harvested in 1994 and obtained from Semenarna Ljubljana, Slovenia, a commercial producer. Before germination

A.C. Cassells (ed.), Pathogen and Microbial Contamination Management in Micropropagation, 225–231.
© *1997 Kluwer Academic Publishers. Printed in the Netherlands.*

half of them were dry heat-treated at 47–48°C for 3 days. After 5–7 days all seeds were exposed to 25% (v/v) commercial bleach, containing 4.8% NaOCl with a drop of Tween 80, for 1–2 min and then washed several times with water. Each seed was separately soaked in a tube with distilled water for 24 h at room temperature and in daylight. Those seeds which had caused turbidity of the water were eliminated. The other seeds were germinated on a mixture of soil and vermiculite (1:1) for 7 days under the same conditions. Seedlings were then exposed to thermotherapy for 10–14 days at 37°C, 70% relative humidity, and with a photoperiod of 16 h at 65–75 μmol/m^2/s.

To establish tissue cultures, 2-cm seedling segments containing apical and axillary meristems were surface sterilised by soaking for 1 to 2 min in 70% ethanol, then for 15 min in the same solution of commercial bleach as used for the seeds. Explants were rinsed three times with sterile distilled water. As initial explants, meristems 0.6–2 mm in length and in some cases 4- to 10-mm apical shoot tips were used.

2.2. Media and culture conditions

A MS medium [5] with B5 vitamins [6], 1 μM 6-benzylaminopurine (BA), 3% (w/v) sucrose and 0.8% (w/v) Difco–Bacto agar or 0.2% (w/v) gellan gum (Phytagel, Sigma, St. Louis, USA) was used as basal medium (BM). The pH of the medium was adjusted to 5.7–5.8 before autoclaving.

According to publications [7,8], some other bean cultivars regenerated buds from meristems and apical buds on media supplemented with 10–20 μM BA. In our preliminary experiments we tested 1–20 μM BA for culture initiation. Since the best results were obtained with 1 μM BA we used this concentration in further experiments.

Explants were sub-cultured every 2–3 weeks. After the first and all subsequent transfers 5 μM AgNO$_3$ were added to the BM, while BA was reduced first to 0.5 μM, then to 0.2 μM, and after the third subculture it was omitted. The excised buds were elongated on 1/2-strength and rooted on 1/4-strength growth regulator-free BM with 5 μM AgNO$_3$, 0.4% (w/v) agar and 0.15% (w/v) gellan gum. Cultures were kept at 23±2°C with a photoperiod of 16 h at 55–70 μmol/m^2/s (Osram L 65W/20S - cool white lamps).

The regenerated plants were transferred to pots with the same substrate mixture as for seed germination and were maintained under 80% humidity. For the first 3 weeks they were watered with 1/4-strength MS macronutrients.

2.3. Data analysis

After 3 weeks of culture the survival and contamination percentages were determined. The number and the length of buds were also recorded. All experiments were repeated at least twice. The 2×2 chi-squared test (χ^2 test) was used to calculate the levels of statistical significance (P) between the data obtained in different treatments of seeds or explants.

2.4. Isolation and identification of bacteria

2.4.1 Media and culture conditions

One week after initiation of the bean culture, contaminants were transferred with an inoculating loop dipped into the cloudy part of basal media and streaked on Petri plates poured with following media: King's medium B (KB) or yeast dextrose chalk agar (YDC) or nutrient glucose agar (NGA) [9].

Petri plates were incubated at 28°C and colony characteristics were inspected daily. Bacterial isolates were purified by restreaking and subsequently maintained on NGA plates. Single colonies were transferred to YDC slants for short-term storage and to NGA plates from which they were taken as inocula for further characterisation.

2.4.2. Diagnostic tests for bacterial characterisation

Basic diagnostic tests were performed before the hypersensitive reaction test: Gram stain, production of fluorescent pigment on KB, oxidase test, heat test for spore determination [10,11].

2.4.3. Hypersensitive reaction (HR) test

Tomato (*Lycopersicon esculentum* L. cv. 'Lycoprea') and tobacco plants (*Nicotiana tabacum* L. cv. 'Xanthi') were grown under greenhouse conditions, and for the HR test only fully developed leaves were used. Bacteria were prepared by suspending three to five bacterial colonies from 48-h-old cultures in 0.01 M $MgSO_4$. Two millilitres of weakly clouded suspensions containing more than 10^7 cfu/ml were injected into the intracellular spaces of an intact tomato or tobacco leaf with a 27 gauge hypodermic needle [12,13]. A half part of a tomato leaf was used for one bacterial strain. Plants were grown in a greenhouse, and evaluated after 24 and 36 h.

2.4.4. Pathogenicity test

Bean plants cv. 'Red Kidney' were grown under greenhouse conditions until the first trifoliate leaves developed. Bacterial strains inducing a positive hypersensitive reaction were taken from 24-h-old cultures and suspensions in 0.01 M $MgSO_4$ containing 10^7 and 10^5 cfu/ml were prepared. By means of a glass atomizer, the suspensions were sprayed onto the abaxial side of not fully developed trifoliate leaves, so that a faint water-soaking was apparent [13]. The bean plants were grown in a greenhouse and evaluated after 7 and 12 days.

2.4.5. Standard bacterial cultures

Pseudomonas syringae pv. *phaseolicola* NCPPB1321 (Psp), *Xanthomonas campestris* pv. *phaseoli* GSPB1242 (Xcp) and *Xanthomonas campestris* pv. *phaseoli* var. fuscans GSPB271 were used as standards along with the new isolates for comparison.

3. RESULTS AND DISCUSSION

3.1. Germination of seeds and contamination of cultures

After exposure of seeds to dry heat at 47–48°C for 3 days, a significant reduction of seed germination was observed (Table 1). Impaired germination after heat treatment of large-seeded legumes has also been reported by other authors. However, this reduction was more often obtained when heat treatment was associated with water or high humidity (for a review, see [2]). Table 1 also indicates that the contamination rate of explants excised from seedlings of dry heat-treated seeds and cultured for 3 weeks on basal medium was significantly lower when compared to those obtained from seedlings of untreated seeds.

Table 1. The effect of dry heat treatment on germination of bean seeds and contamination of explants excised from bean seedlings originating from treated or untreated seeds after 3 weeks of culture on basal medium. Calculation of the statistical significance between treated and untreated seeds was based on the χ^2 test (n = explants per treatment)

Seed treatment	Germination		Contamination	
	n	%	n	%
47–48°C, 3 days	38	65.8*	120	4.2***
Untreated	43	90.7	119	20.2

*$P<0.05$, **$P<0.01$, ***$P<0.001$.

3.2. Isolation and identification of bacteria

Bacterial growth appeared as a cloudy zone in the basal media around the basal parts of explants excised from seedlings of both untreated and dry heat-treated seeds.

3.2.1. Dry heat-untreated seeds

In the case of untreated seeds, 35 yellow-pigmented bacterial isolates out of 41 isolated in two experiments were suspected to be bean pathogens. Single colonies were transferred to YDC after two-times restreaking on NGA. The colonies were bright yellow, mucoid, convex, and shining. Bacterial isolates causing slight browning of YDC were restreaked on KB, where they produced dark brown, diffusible pigment. The production of brown pigment is characteristic for the fuscous strain of Xcp [14].

Bacterial strains of both types were Gram-negative rods, oxidase negative, and no growth appeared after a heat test for spore determination.

A positive hypersensitive reaction on tomato leaves was induced with both the yellow and brown-pigmented colonies. Tissue collapse within 24 h and formation of desiccated, necrotic areas within an additional 12 h were observed. An injection of bacterial suspensions into tobacco leaves cv. 'Xanthi' resulted only in yellowing of the inoculated areas. Since the hypersensitivity test is predictive of pathogenicity and, in the case of testing xanthomonads, use of certain non-host plants other than the tobacco is recommended [15], tomato plants were used. Induction of hypersensitive symptoms on non-host plants indicated that the bacterial contaminants were plant pathogenic bacteria.

Two fluorescent, Gram-negative, oxidase positive strains were isolated on KB media, but they did not induce HR on tobacco cv. 'Xanthi'.

To confirm pathogenicity of isolated strains on a susceptible host, trifoliate leaves of Red Kidney bean were inoculated with two different concentrations. Inoculations with 10^7 cfu/ml resulted in the development of water-soaked lesions 7 days after spraying. At 10^5 cfu/ml a few inoculated plants showed the same symptoms. Within additional 5 days, larger necroses bordered with yellow haloes were formed at the higher concentration while at the lower concentration water-soaked lesions and small necrotic spots appeared on all inoculated plants. Typical common blight symptoms developed and no differences between yellow and brown-pigmented bacteria were observed. The experiments identified our isolates as *X.c.* pv. *phaseoli* and *X.c.* pv. *phaseoli* var. fuscans.

3.2.2. Dry heat-treated seeds

In the case of treated seeds, bacterial contaminants were white or cream coloured on YDC and NGA. After restreaking on KB, no fluorescence was observed. Since the colony characteristics of these bacteria did not resemble those of bean pathogenic bacteria, we did not do any additional tests for their identification. We concluded that dry heat treatment of bean seeds was sufficient to eradicate pathogenic xanthomonads. Similar dry heat treatment of seeds has also been successfully used for the elimination of the bacteria *Pseudomonas syringae* pv. *phaseolicola* and *X.c.* pv. *phaseoli* from bean seeds [3,4].

3.3. Bean meristem and shoot tip culture

In our preliminary experiments we introduced heat untreated shoot tips into culture but without success. The use of an antibiotic mixture (penicillin–streptomycin–neomycin, Sigma, St. Louis, USA) reduced the bacterial contamination but totally stopped the growth of explants. We then tried to culture meristems and shoot tips of plants which had undergone thermotherapy. This proved to be the best solution for elimination of viruses, but it also partly reduced contamination in cultures (data not shown). Although it is well known that the common bean like other legumes is recalcitrant in culture, we succeeded in regenerating shoots and transferring a few plantlets to soil.

Three weeks after isolation, one to two buds formed on the basal medium. However, further elongation of buds appeared to be the crucial step in plantlets development.

The survival of explants was better on media solidified with Phytagel than with Difco–Bacto agar. Furthermore, on medium solidified with Phytagel a significantly higher number of elongated buds (χ^2 test: $P<0.01$) was obtained (Table 2). Huang and co-workers have shown better regeneration of different species and explants on Gelrite (gellan gum) than agar media. However, lowering the macrosalt content of MS medium significantly lowered gel rigidity of the media solidified with gellan gum [16]. Thus, for elongation and rooting of buds which was performed on 1/2- and 1/4-strength MS medium, a combination of both agar and Phytagel was used. After 3–4 months plantlets with well developed root systems were transferred to soil where they grew vigorously, without any visible disease symptoms. Further evaluation of developed plants is ongoing.

Table 2. The survival and bud development of bean meristems and shoot tips cultured for 3 weeks on basal medium solidified with Difco–Bacto agar or gellan gum (Phytagel)

Gelling agent	Cultured explants (no.)	Surviving explants (no.)	Buds (no.)	Buds longer than 5 mm (no.)
Difco–Bacto Agar	74	40	89	17
Phytagel	72	63	110	53

In conclusion, the dry heat treatment of seeds in combination with thermotherapy of seedlings associated with meristem and shoot tip cultures can be used for successful eradication of different pathogens from the bean cv. 'Starozagorski'. Since substantially lower contamination of tissue cultures and elimination of $X.c.$ pv. $phaseoli$ and its fuscous variant was obtained after exposure of seeds to dry heat treatment, we assume that this step is a critical point of the procedure.

REFERENCES

1. Audy, P., Braat, C.E., Saindon, G., Huang, H.C. and Gilbertson, R.L. (1996) Phytopathology 86, 361–366.

2. Grondeau, C. and Samson, R. (1994) Crit. Rev. Plant Sci. 13, 57–75.

3. Nauman, K. and Karl, H. (1988) Nachrichtenb. Pflanzenschutz DDR 42, 204–208.

4. Koleva, N. (1984) Rast. Zashch. 32, 7–8.

5. Murashige, T. and Skoog, F. (1962) Physiol. Plant. 15, 473–497.

6. Gamborg, O.L., Miller, R.A. and Ojima, K. (1968) Exp. Cell Res. 50, 151–158.

7. Rubluo, A. and Kartha, K.K. (1985) J. Plant Physiol. 119, 425–433.

8. Allavena, A. and Rossetti, L. (1986) Sci. Hortic. 30, 37–46.

9. Rudolph, K. (1990) in Methods in Phytobacteriology (Klement Z., Rudolph K. and Sands D.C., eds) pp. 59–62, Akadémiai Kiadó, Budapest.

10. Schaad, N.W. (1988) in Laboratory Guide for Identification of Plant Pathogenic Bacteria (Schaad N.W., ed.) pp. 1–14 APS, Press, The American Phytopathological Society, St. Paul, MN.

11. Bradbury, J.F. (1970) Rev. Plant Pathol. 49, 213–218.

12. Klement, Z. (1983) Seed Sci. Technol. 11, 589–593.

13. Klement, Z. (1990) in Methods in Phytobacteriology (Klement Z., Rudolph K. and Sands D.C., eds) pp. 96–104, Akadémiai Kiadó, Budapest.

14. Rudolph, K. (1990) in Methods in Phytobacteriology (Klement Z., Rudolph K. and Sands D.C., eds) pp. 65–74, Akadémiai Kiadó, Budapest.

15. Stead, D.E. (1990) in Methods in Phytobacteriology (Klement Z., Rudolph K. and Sands D.C., eds) pp. 252–262, Akadémiai Kiadó, Budapest.

16. Huang, L.C., Kohashi, C., Vangundy, R. and Murashige, T. (1995) In Vitro Cell. Dev. Biol. 31, 84–89.

PREVENTION AND ELIMINATION OF CONTAMINATION FOR *IN VITRO* CULTURE OF SEVERAL WOODY SPECIES

ERIKA SZENDRÁK[1], PAUL E. READ[1] and GUOCHEN YANG[2]
[1]*Dept. of Horticulture, University of Nebraska-Lincoln, NE, USA*
[2]*Dept. of Natural Resources and Env. Design, North Carolina A&T State Univ., Greensboro, NC, USA*

1. INTRODUCTION

Microbial contamination is one of the most critical and frequently occurring problems in plant tissue culture. In previous experiments with several woody genera such as *Castanea*, [1–3] *Ligustrum*, *Spirea*, *Syringa* [4], *Corylus*, *Populus*, *Salix* [5], and others we worked out some successful techniques to establish contamination-free cultures. In most cases the available plant material was in very limited quantity, received from the breeding project of the American Chestnut Foundation [1,2], or as a part of a special plant exchange program with the University of Horticulture and Food, Budapest, Hungary [5]; thus, it was extremely important not to lose any culture because of contamination problems.

2. PROCEDURE AND DISCUSSION

In the most simple case, fresh green growths of greenhouse-grown stock plants are used as an explant source, which provides much cleaner and more vigorous plant material than when explants are taken from field-grown stock plants. In the latter case, culture initiation is generally limited to only a few weeks of the vegetative season. A more advanced method for initiating *in vitro* cultures is to use dormant branches via the "forcing-method". The advantages of this approach are as follows: overcomes bud dormancy; produces fresh, vigorous and clean softwood outgrowth for explants; provides the possibility of delivering growth regulators prior to *in vitro* culture; allows for a very flexible culture initiation time.

The most convenient time for collecting the branches is at the end of winter, when the plants have already overcome the winter bud dormancy. These branches provide a very good initial plant material all year long, if they are stored in a clean cooler at 4–6°C.

Following a short 10 min surface disinfestation in a 10% commercial bleach solution, stems containing 6–10 buds are placed in vessels containing 2% sucrose solution supplemented with 200 mg 8-hydroxyquinoline-citrate(8-HQC)/l at room temperature and under 16 h

A.C. Cassells (ed.), Pathogen and Microbial Contamination Management in Micropropagation, 233–236.
© *1997 Kluwer Academic Publishers. Printed in the Netherlands.*

234

Figure 1. Hybrid chestnut (*Castanea* sp.), left, and corkscrew willow (*Salix matsudana* 'Golden Spiral'), right, twigs forced in solution containing 8-HQC and sucrose ready for the first harvest of explants.

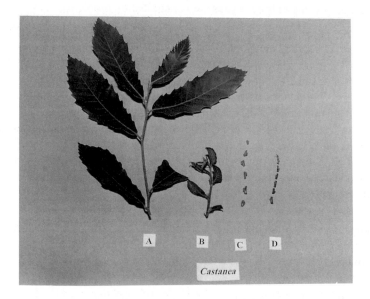

Figure 2. Initial plant material for *in vitro* culture from hybrid chestnut (*Castanea* sp.); A,C: greenhouse-grown branches before and after preparing for disinfestation; B,D: softwood shoots from forcing before and after preparing for disinfestation.

Figure 3. Contamination-free *in vitro* culture of hybrid chestnut (*Castanea* sp.) from apical buds of forced twigs.

illumination per day. Along with the disinfestation, the bleach treatment helps to initiate breaking the dormancy which will be totally overcome using the 8-HQC treatment [1,4]. After 3–4 days the buds start to open and 7–10 days after starting the forcing the first softwood outgrowths appear, which are excellent as an explant source (Fig. 1). Depending on the species, it is also possible to harvest new softwood stems for up to 4–6 weeks after the first harvest.

In both cases — greenhouse or forcing original plant material (Fig. 2) — all the leaves, petioles and stipules are removed from the twigs, and the single-bud final explants are disinfested with a series of bleach and/or ethanol solutions and rinsed in sterile distilled water before placing them in the medium. Usually this method eliminates all the exogenous contamination problems.

When endogenous contamination occurred after the surface disinfestation and subsequent culture, we cultured the excised apical or terminal buds and lateral buds in separate vessels. It is well known that separate *in vitro* culture of apical buds or meristems may help to eliminate contamination caused by endogenous pathogens in many commonly micropropagated plants, but there is less information available regarding woody species [6,7]. From most of the apical or terminal buds (Fig. 3) and sometimes from the first lateral buds we were able to establish clean cultures [1]. In other cases of endogenous contamination, especially when the initial amount of plant material was very limited and it was necessary to use all of the buds, using a traditional microtechnical stain, Malachite Green [8], in the initiation medium at 0.1-1 mg/l concentration as a bacteriostatic agent gave promising results. Using these two different methods, or combinations thereof, we produced cultures free of endogenous contamination without applying any of the more radical antibacterial compounds.

ACKNOWLEDGEMENTS

Special thanks are expressed to the American Chestnut Foundation, the Hungarian State Eotvos Fellowship, the Rotary Club of Omaha, the University of Nebraska-Lincoln and the US–Hungarian Joint Fund for Science and Technology for the support provided for this work.

REFERENCES

1. Szendrak, E., Read, P.E. and Miller, V.I. (1996) Proc. Lippay J. Sci. Symp. in press.

2. Read, P.E. and Szendrak, E. (1995) J. Am. Chestnut Found. IX/1, 50–53.

3. Yang, G., Read, P.E. and Kamp-Glass, M. (1995) HortScience 30(4), 757.

4. Yang, G. and Read, P.E. (1992) J. Environ. Hortic. 10(2), 101–103.

5. Read, P.E., Donald, H., Steinegger, D.H., Szendrak, E., Morrissey, T.M., Schmidt, G. and Hamar, B. (1996) Proc. Lippay J. Sci. Symp. in press.

6. Debergh, P. and Maene, L. (1984) Parasitica 40(2–3), 69–75.

7. Vanderschage, A.M. and Debergh, P.C. (1988) Med. Fac. Landbouww. Rijksuniv. Gent 53(4a), 1763–1768.

8. Zatyko, J. (1992) Proc. Lippay J. Sci. Symp. pp. 377–380, Budapest, Hungary

LABORATORY CONTAMINATION MANAGEMENT; THE REQUIREMENT FOR MICROBIOLOGICAL QUALITY ASSURANCE

CARLO LEIFERT[1] and STEPHEN WOODWARD[2]

[1]*Department of Plant and Soil Science,*[2]*Department of Forestry, University of Aberdeen, Aberdeen, Scotland, UK*

1. INTRODUCTION

Microbial contamination is the single most important cause of losses in commercial and scientific plant tissue culture laboratories [1,2]. However, even in commercial companies the severity and implications of the problem are not recognised or admitted. Many scientific laboratories fail to record contamination losses, and the micropropagation industry often only recognises the sources of contamination after severe losses have occurred. Following a rapid increase in the production of micropropagated plants in the 1980s [3], there has been a steady decline in the number of micropropagation laboratories in the 1990s, which was at least partially caused by the inability of laboratories to reduce contamination losses to a level which allows a predictable production output and quality of micropropagated plants.

A level of contamination losses of not above 2% per subculture is usually seen as the minimum required to guarantee successful production. This level cannot be obtained without the application of regular quality assurance practices and the introduction of a microbiological production control strategy. Such practices are common practice in other industries affected by microbial contamination (e.g. food processing and pharmaceutical companies) [4]. Management/quality assurance strategies such as HACCP (Hazard Analysis Critical Control Points) can easily be applied to tissue culture laboratories. In micropropagation companies which have invested in the establishment of HACCP the additional costs were more than compensated for by the reduction in product losses and quality [1].

The microbial hazards in micropropagation can affect the tissue culture process itself, plant survival during weaning (if plants with fungal contaminants such as *Botrytis cinerea* are transferred to the nursery) and the customer of micropropagation laboratories (if pathogenic bacteria or viruses remain latent until plants have been sold) [5–10]. One important prerequisite for the introduction of HACCP is a production recording system which allows the history of all plants to be traced in detail (including information on the stock plant,

A.C. Cassells (ed.), Pathogen and Microbial Contamination Management in Micropropagation, 237–244.
© *1997 Kluwer Academic Publishers. Printed in the Netherlands.*

media batches, operator(s), flow cabinet(s) and growth room(s) plants have been exposed to). The establishment of HACCP then requires in-depth knowledge of (i) the identity and sources of potential hazards associated with specific stages of the production and distribution process (this will allow routine quality assurance based on the isolation and identification of indicator micro-organisms), (ii) methods for the early detection of hazards and (iii) the development of methods for the treatment of microbiological hazards.

Since the first workshop on bacteria and bacteria-like contaminants in plant tissue cultures was held in Cork 8 years ago [11] considerable advances have been made in both the knowledge and methodology required to establish microbiological quality assurance systems for plant tissue-culture laboratories [1,2,12,13]. This review will critically evaluate recent published literature and personal experiences with the establishment of HACCP procedures in commercial micropropagation.

2. IDENTITY AND SOURCES OF POTENTIAL HAZARDS

Contaminants described to cause severe economic losses in plant tissue culture laboratories include mites [1,14], thrips [1,14], fungi [1,15–18], yeasts [1,19], bacteria [1,11,12,20–22] and viruses [9,10]. Identification of the most common contaminants is relatively simple. The most common fungal and yeast contaminants can easily be identified by macroscopic examination of infected plant tissue cultures [7]. The mycelium/spore characteristics can be used for identification of important fungal/yeast genera associated with specific contamination sources (Table 1). It is also possible to separate the yeasts (yeasts grow vigorously and form white, off-white or red/pink non-translucent growth, and produce a typical yeast odour in the culture container) from bacterial contaminants (bacteria usually produce only very slight growth on tissue culture media or, when more vigorous growth is produced, colonies are slimy and translucent in appearance) based on macroscopic assessment. Very few bacterial species produce characteristic symptoms or growth in plant tissue cultures [20,21] and different bacterial genera, therefore, cannot be separated based on their colony morphology.

However, the main genera/groups of bacterial contaminants can be identified with just a few bacteriological tests and there are several test kit systems available now which can be used to confirm the genus and to identify bacterial contaminants to genus/species level (see Table 2). Genus level identification (which usually does not require the use of relatively expensive test kits) is often sufficient to identify the sources of bacterial contamination. 'Indicator contaminants' which pinpoint specific contamination sources are described in Table 1. The API identification system (bioMérieux sa, 69280 Marcy-l'Etoile, France) requires a range of six standard tests to be performed to select the appropriate test-strip (Table 1).

Table 1. Contaminants indicating specific contamination sources

Contaminant	Likely source for contamination	Comments
Bacteria		
Gram negatives	Inefficient disinfection of explants	often *Pseudomonas*
Gram positives	Inefficient laboratory procedures	and Enterobacteriaceae
Gram-positive cocci (*Staphylococcus epidermidis*)	Poor aseptic techniques in: (i) subculturing plants (ii) pouring of media	*Staphylococcus* spp. are considered obligate inhabitants of animals
Bacillus spp.	Inefficient sterilisation of media	often *Bacillus subtilis* and *Bacillus pumilus*
	Inefficient sterilisation of instruments used for subculture	*Bacillus circulans* can survive in 70% alcohol
Fungi/yeasts		
General increase	Growthroom mite or thrip infestation	Especially when plant-specific growth rooms are affected
Fusarium poe	Growth room infestation with *Sideroptes graminis* mites	*Fusarium poe* forms a white mycelium with a pink base
Grey, black and green moulds (*Botrytis, Aspergillus, Alternaria Penicillium* spp.) *Rhodotorula* spp. (pink yeasts)	High laboratory air contamination Faulty flow cabinets	*Penicillium* and pink yeast are very common air contaminants within buildings
Black moulds	Poor hygiene in growth rooms and plant and media cold stores	Moulds are often found in damp areas of the laboratory walls or grow in areas with frequent condensation. Spores are air transmitted
Cladosporium spp.	Insufficient protection of laboratory against the outside air	*Cladosporium* spores are very common in outside air

The BIOLOG system (Biolog Inc. 3938 Trust Way, Hayward CA 94545, USA), which is based on standard microwell plates, only requires Gram staining of isolates in order to select a Gram-negative, Gram-positive or yeast-specific identification system. Yeast and bacterial contaminants may also be identified by fatty acid profiling [11] and or molecular methods such as plasmid profiling and genetic fingerprinting [23].

3. DETECTION OF ANTAGONISTS

One of the most important sources of contamination is the explant during initiation of plant tissue culture [1,2,24]. The surface disinfection of the explant prior to its placement into a tissue culture test-tube or container is often inefficient. This problem may be due to the disinfectant being inactive or to micro-organisms being protected within the plant tissue used as the explant [11]. Fungal and yeast contaminants are easy to detect when they survive disinfection because they rapidly grow on plant tissue culture media [1]. Bacteria and viral contaminants, on the other hand, may not produce visible growth in the tissue culture medium [9,10,11,25]. Nevertheless, the propagation of such 'latent' infected material can cause severe losses at later stages of tissue culture or after weaning of plants [5,6,9,10,26,27]. Detection of 'latent' bacteria is usually based on 'indexing' of plant tissues [1]. This involves the transfer of pieces of plant material into solid or liquid bacteriological media after disinfection. The media used contain meat, yeast or plant extracts and have been described for sterility testing in other areas, such as the food and water industry and medical microbiology [28–30]. Several 'indexing media' which were developed specifically for the detection of plant tissue-culture contaminants (e.g. medium 523 [32], George and Falkingham's mycobacterium detection medium TT [33] and Leifert and Waites Sterility Test Medium[1]) can now be obtained as commercial products ([31]; Sigma-Aldrich Company Ltd., Fancy Rd., Poole, Dorset BH17 7NH, UK).

The advantage of using indexing media is that they allow growth and detection of a wide range of different bacterial contaminants [1] and many bacteria are detected even when present in very low numbers (101 to 102). However, different bacterial species show different amounts of growth on a particular medium (Table 3) and no bacteriological medium is able to detect all important contaminants [1,13]. Other limitations of 'indexing media' have been described in detail elsewhere [1].

Commercial serological test kits are available for a wide range of plant viruses and specific plant pathogenic bacteria (e.g. *Xanthomonas pelargonii*, *Clavibacter michiganense*, *Pseudomonas solanacearum*, *Erwinia amylovora*, various pathovars of *Erwinia syringae* and *Xanthomonas campestris* and *Erwinia carotovora* pv. *atroseptica*) which may also stay latent *in vitro* ([31]; Adgen Plant Disease Diagnostics, Watson Peat Building, Auchincruive, Ayr KA6 5 HW, Scotland UK.; LOEWE Biochemica GmbH, Nordring 38, Postbox 9, 8156 Otterfing bei Munchen, Germany). Such tests can be very sensitive, but are expensive and only detect one species of bacterium or virus.

Table 2. Identification methods for bacterial contaminants

Classic bacteriological tests used						Suspected genus	API	Biolog
CAT	OF/F	MOB	GRAM	COCC	SPOR			
+	+	−	+	+	−	Staphyloc.	STAPH	GP
+	−	−	+	+	−	Microc.	STAPH	GP
−	+	−	+	+	−	Streptoc.	STREP	GP
+	+	+	+	−	+	Bacillus	50CHB	GP
+	−	+	+	−	+	Bacillus	50CHB	GP
+	+	−	+	−	+	Bacillus	50CHB	GP
+	−	−	+	−	+	Bacillus	50CHB	GP
+	+	+	+	−	−	corynef.	Coryne	GP
+	−	+	+	−	−	corynef.	Coryne	GP
+	+	−	+	−	−	corynef.	Coryne	GP
−	+	−	+	−	−	Lactob.	50CHL	GN
+	−	−	−	+	−	Acinetob.	20NE	GN
+	+	+	−	−	−	Vibrionac.	20NE	GN
+	+	−	−	−	−	Actinob.	20NE	GN
+	−	+	−	−	−	Pseu./Alca	20NE	GN
+	−	−	−	−	−	Flavob.	20NE	GN
+	+	+	−	−	−	Enterob.	20E	GN
+	+	−	−	−	−	Enterob.	20E	GN
	+	+	+	oval	−	Yeasts	ID32C	YT

CAT, catalase; OF/F, anaerobic growth; MOT, motility; GRAM, Gram stain; COCC, coccus shaped as opposed to rod shaped; SPOR, heat-resistant spores formed; Staphyloc., Staphylococcus spp.; Microc., Micrococcus spp.; corynef., coryneform bacteria; Lactob., Lactobacillus spp.; Acinetob., Acinetobacter spp.; Vibrionac., Vibrionaceae; Actinob., Actinobacter spp.; Pseu., Pseudomonas spp.; Flavob., Flavobacterium spp.; Enterob., Enterobacteriaceae.

The suppression of many bacteria and viruses by plant resistance mechanisms and residues of the disinfectants used can result in low inocula being present in the plant pieces tested. Both indexing and serological tests should, therefore, be repeated both *in vitro* and after weaning of plants. Because of the relatively high cost of serological tests, these should only be used to check for organisms which are known to be a problem to particular plant species.

Table 3. Growth (E_{625}) of contaminants on Leifert & Waites medium and tryptone soya broth at different concentrations of nutrients

Test organism	Leifert & Waites medium			Tryptone soya broth		
	FS	HS	1/10S	FS	HS	1/10S
Bacillus subtilis Cot1	0.4	0.4	0.3	0.4	0.4	0.3
Bacillus circulans A11	0.2	0.2	**	0.3	0.3	**
Clavibacter michiganense CM1	**	0.2	0.3	0.2	0.2	**
coryneform H416/3	**		0.2	0.5	0.4	0.3
Lactobacillus plantarum H260/6	0.8	0.8	0.7	0.7	0.6	0.2
Micrococcus kristinae Ho506/4	**	0.2	0.2	0.2	0.3	0.2
Staphylococcus saprophyticus Ch104/5	0.8	0.8	0.3	0.8	0.7	0.2
Acinetobacter calcoaceticus Ho295/35	0.5	0.4	0.3	0.4	0.3	**
Agrobacterium radiobacter Ger840/12	0.9	0.9	0.8	0.4	0.4	0.4
Agrobacterium tumefaciens As001/1	0.8	0.8	0.8	0.4	0.5	0.4
Pseudomonas maltophilia Del603/20	0.7	0.6	0.4	0.8	0.8	0.7
Xanthomonas campestris pv. campestris XCCi	0.5	0.7	0.3	0.5	0.2	0.5
Xanthomonas campestris pv. vesicatoria XCV	0.3	0.4	0.3	0.2	0.2	0.2

FS, recommended concentration; HS, half the recommended concentration; 1/10S, one-tenth of the recommended concentration. **Turbidity not detectable by visual assessment (E_{625}<0.2).

It is important to recognise that latent bacterial contaminants may also be introduced in the laboratory. Indexing plants at the initiation stage may, therefore, not prevent the accumulation of 'latent' contamination when tissue cultures are continuously subcultured. To avoid this problem some laboratories maintain stock cultures of all their production plant lines. These are regularly tested for the presence of latent bacteria (every 2–4 subcultures/months) and 'index-positive' cultures (those showing bacterial growth on bacteriological media) are discarded. Excesses of these stock cultures are regularly used to supplement the production cultures and only remain in production for a specific length of time (up to 2 years). Such production systems have proved to increase reliability, and although they result in additional 'quality assurance' cost, this is usually more than compensated by increased growth rates and lower contamination losses. The factors triggering 'latent' bacteria to become virulent have previously been discussed in detail [1].

4. CONCLUSIONS

Since the last symposium, our knowledge about the sources of contamination has increased considerably and more diagnostic tools are now available. However, we also had to realise that treatment of contamination (for example with antibiotics) is extremely difficult and, with many contaminants, impossible [1,13,34,35]. This means even more emphasis will have to be placed on early detection and prevention of contamination at source. Due to the wide range of sources of contamination in a tissue-culture laboratory it is essential to introduce appropriate quality assurance systems such as HACCP which cover every potential source of contamination. This is not only essential to guarantee reliable production and quality of tissue cultured plants, but also operator safety. Contaminants which can cause disease such as ringworm (*Trichophyton* spp.), oral thrush (*Candida albicans*), gastroenteritis (*Staphylococcus aureus*) can be found in tissue cultures and at least one case where operators became infected from handling *Trichophyton*-infected tissue cultures has been reported [15]. Such systems have now been designed and become common practice in many commercial laboratories. The challenge for the next 8 years should be to perfect such quality assurance systems.

REFERENCES

1. Leifert, C., Morris, C. and Waites W.M.(1994) CRC Crit. Rev. Plant Sci. 13, 139–183.

2. Leifert, C. and Waites, W.M. (1994) in Physiology, Growth and Development of plants in culture (Lumsden P.J., Nicholas J.R. and Davies W.J., eds) pp. 363–378, Kluwer Academic Publishers, Dordrecht, The Netherlands

3. Leifert, C., Clark, E. and Rothery, C.A. (1993) Biol. Sci. Rev. 5, 31–35.

4. Waites, W.M. (1988) Food Sci. Technol. Today 212, 49–51.

5. Cooke, D.L., Waites, W.M. and Leifert, C. (1994) in Plant Pathogenic Bacteria (Lemattre S., Freigoun S., Rudolph and Swings J.G., eds) pp. 183–194, INRA Editions, Versailles Cedex, France.

6. Cooke, D.L., Waites, W.M. and Leifert, C. (1992) J. Plant Dis. Protect. 99, 469–481.

7. Danby, S. and Leifert, C. (1994) Physiology, Growth and Development of plants in culture (Lumsden P.J., Nicholas J.R. and Davies W.J., eds) pp. 379–385, Kluwer Academic Publishers, Dordrecht, The Netherlands.

8. Leifert, C. (1992) Phytomedizin 23, 44–45.

9. Walkey, D.G.A (1985) Applied Plant Virology William Heinemann Ltd., London, UK.

10. Wang, P.J. and Hu, C.Y. (1980) in Advances in Biochemical Engineering 18 (Fiecher A., ed.) pp. 61–99, Springer Verlag, Berlin Germany.

11. Cassells, A.C. (1988) Bacteria and bacteria-like contaminants in plant tissue cultures. Acta Hortic. 225.

12. Cassells, A.C. (1990) in Techniques for Detection and Diagnosis in Plant Pathology (Duncan J.M. and Torrance L., eds) pp. 197–212, Butterworth, London, UK.

13. Leifert, C., Ritchie, J. and Waites, W.M. (1991) World J. Microbiol. Biotechnol. 7, 452–469.

14. Blake, J. (1988) Acta Hortic. 225,163–166.

15. Weller, R. and Leifert, C. (1996) Br. J. Dermatol. in press.

16. Leifert, C., Waites, B., Keetley, J.W., Wright, S., Nicholas, J.R. and Waites, W.M. (1994) Plant Cell Tissue Org. Cult. 36, 149–155.

17. Danby, S., Berger, F., Howitt, D.J., Wilson, A.R., Dawson, S. and Leifert, C. (1994) in Physiology, Growth and Development of Plants in Culture (Lumsden P.J., Nicholas J.R. and Davies W.J., eds) pp. 397–403, Kluwer Academic Publishers, Dordrecht, The Netherlands.

18. Danby, S., Joshi, S., Leifert, C., Epton, H.A.S, Sigee, D.C. and Quernal, L. (1994) Physiology, Growth and Development of Plants in Culture (Lumsden P.J., Nicholas J.R. and Davies W.J., eds) pp. 404–408, Kluwer Academic Publishers, Dordrecht, The Netherlands.

19. Leifert, C., Nicholas, J.R. and Waites, W.M. (1990) J. Appl. Bacteriol. 69, 471–476.

20. Leifert, C., Waites, W.M. and Nicholas, J.R. (1989) J. Appl. Bacteriol. 67, 353–361.

21. Leifert, C., Waites, W.M., Camotta, H. and Nicholas, J.R. (1989) J. Appl. Bacteriol. 67, 363–370.

22. Leggatt, I.V., Waites, W.M., Leifert, C. and Nicholas, J. (1988) Acta Hortic. 225, 93–102.

23. Leifert, C., Berger, F., Steward, G.S.A.B. and Waites, W.M. (1994) Lett. Appl. Microbiol. 19, 377–379.

24. Berger, F., Keetley, J. and Leifert, C. (1994) J. Hortic. Sci. 69, 491–494.

25. Leifert, C. and Waites, W.M. (1992) J. Appl. Bacteriol. 72, 460–466.

26. Long, R.D., Curtin, T.F. and Cassells, A.C. (1988) Acta Hortic. 225, 83–91.

27. Hofferbert, H.-R. (1990) Masters Thesis. University of Gottingen, Germany.

28. Anonymous (1978) in Standard Methods for the Examination of Dairy Products. 15th Edn. American Public Health Association Inc., Washington, DC, USA.

29. Anonymous (1980) in Standard Methods for the Examination of Water and Waste Water. 15th Edn. American Public Health Association Inc., Washington, DC, USA.

30. Anonymous (1982) in The Oxoid Manual. 5th Edn. Tumergraphic Ltd. Basingstoke, UK.

31. Anonymous (1995) in Sigma Plant Culture Catalogue Supplement . Sigma-Aldrich Company Ltd., Dorset, UK.

32. Viss, P.R. (1991) In Vitro Cell Dev. Biol. 27P, 42.

33. George, K.L. and Falkinham, III, J.O. (1986) Can. J. Microbiol. 32, 10–14.

34. Leifert, C., Camotta, H., Wright, S.M., Waites, B., Cheyne, V.A. and Waites, W.M. (1991) J. Appl. Bacteriol. 71, 307–330.

35. Leifert, C., Camotta, H., and Waites, W.M. (1991) Plant Cell Tissue Org. Cult. 29, 153–160.

MICROBIAL COMMUNITIES ON HUMAN TISSUES; AN IMPORTANT SOURCE OF CONTAMINANTS IN PLANT TISSUE CULTURES

RICHARD WELLER

University Department of Dermatology, Edinburgh EH3 6HY, UK

1. INTRODUCTION

Human skin has several defences against microbial colonization. The hair and squamous keratinocytes of the outer skin layers are composed of keratin, a substance unfavourable to microbial growth. Continual shedding of the stratum corneum and surface nitric oxide release limit epidermal invasion. Nonetheless, the lipids, proteins, minerals and hormones on the skin surface support a diverse flora. Some inhabitants of the skin surface are resident and complete their life cycle there. Keds, mites, lice, and *Demodex* spp. live on the hair or in hair follicles. *Staphylococcus* spp., *Micrococcus* spp., *Propionibacterium* spp. and aerobic coryneforms are the major residents in the outer layers of the outer stratum corneum and hair follicles. Transient organisms spend only a limited time on the skin, mainly to feed, and induce an immune response which limits their presence.

Wound infection from skin commensals is a potential complication of surgical procedures. Contamination with human commensals is a major problem in plant tissue culture systems. Techniques developed for surgical disinfection may be appropriate to plant micropropagators. Removal or killing of transient bacteria is relatively easy. Surgical disinfection involves the reduction or elimination of resident bacteria and is more difficult. The disinfection technique used and compliance of subjects significantly affects infection rates.

2. ANATOMY OF THE HUMAN SKIN

The skin is the largest organ in the human body. It is a complex and continually changing structure which serves many purposes. It maintains and regulates body temperature; serves as a sensory organ to external environmental stimuli; processes antigens, and is central to communication and sexual attraction. The skin provides a barrier for the body from harmful stimuli such as ultra-violet radiation, chemicals and dehydration. It provides the first line of defence against infection by bacteria, fungi, viruses and parasites. In addition to being a defence against infection, the skin maintains a stable microbial ecosystem.

A.C. Cassells (ed.), Pathogen and Microbial Contamination Management in Micropropagation, 245–257.
© *1997 Kluwer Academic Publishers. Printed in the Netherlands.*

An understanding of the structure of the skin is important in appreciating the nature of its resident flora. Skin is divided into two layers, the outer epidermis and the inner dermis. The predominant cell type in the epidermis is the keratinocyte. These arise from the basal layer and differentiate as they rise through the epidermis. There is a gradual loss of the nucleus, and thus of cell viability, and a flattening, drying and cornification of the keratinocytes, producing the corneocytes of the stratum corneum in the very outermost layer. Organelles called lamellar bodies in the middle layers (stratum granulosum and spinosum) of the epidermis secrete lipids into the intercellular spaces, completing the barrier that controls water permeability and hydration. The secreted lipid acts partly as a substrate to bacteria on the skin, but also has antimicrobial properties [1].

Corneocytes form skin scales measuring approximately $30 \times 30 \times 3–5$ μM and with the average skin area of an adult being 1.75 m^2, around 10^8 cover the body. A complete layer of scales is lost and replaced every 4 days, resulting in about 2.5×10^7 scales being lost each day [2]. At rest about 10 mg of skin are deposited in the clothes each 2 h. Washing the hands with soap will increase the number of skin squames released by 18-fold or more [3] and walking movements will release about 10,000 scales/min [4]. Scales once shed are carried up in the airstream around the body [5]. 'Flying saucers' of squames, often carrying microbes, make up a large part of airborne debris, having been recorded in 10% of the collected airborne particles in the London Underground [6]. This is the predominant means of dispersal of the skin flora and any organism on the skin can be so disseminated [7].

The dermis lies beneath the stratum corneum and provides support for the epidermis. It is made up of connective tissue, elastic fibres, fat and collagen. Blood vessels, nerves, and specialized structures such as sebaceous, apocrine and eccrine glands are situated here and hair follicles invaginate the dermis from the epidermis (Fig. 1).

Various adnexal structures penetrate the epidermis. Hair follicles are distributed over the entire body surface with the exception of the palms, soles and dorsum of the foot. Anatomically they appear as an invagination of the epidermis into the dermis with inner and outer root sheaths surrounding the developing hair.

Sweat glands are of two types, both with tightly coiled glands lined with secretory cells in the dermis. In the axillae and perineum apocrine sweat glands discharge a viscous sweat which is acted on by aerobic coryneforms and micrococci to create the distinctive musty odour. The glandular section of the apocrine gland is surrounded by smooth muscle-like myoepithelial cells which are richly innervated with sympathetic and parasympathetic nerves. Eccrine glands are smaller, more extensively distributed and have a single set of innervating fibres. They cover most of the body surface and secrete a hypotonic solution to increase cooling during heat stress [8]. The degree of effective cooling produced by sweating has been questioned [9] and it may be that sweat plays other roles than the purely

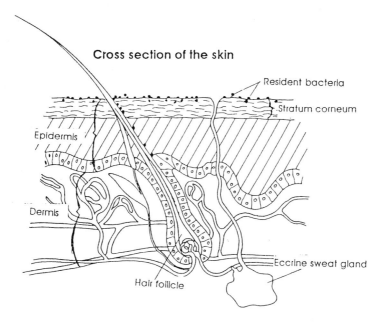

Figure 1. Cross-section of the skin.

thermoregulatory one. Work we have performed suggests that one of these may be as a non-specific surface defence [10].

Sebaceous glands discharge lipid rich sebum to the hair follicles on the face, upper chest and upper back and also directly to the skin surface in hairless areas.

3. THE FLORA OF THE HUMAN SKIN

The skin is sterile at birth, and acquires the normal adult flora at rates varying between bacterial species. Staphylococci are the first to colonize the skin, but it is not until puberty that the full adult flora is gained [11]. The majority of cutaneous bacteria are confined to two sites — the outer layers of the stratum corneum and within the hair follicle. Bacterial densities vary in different parts of the body and different bacteria predominate at different sites. Areas very heavily colonized with bacteria are the axillae, the perineum, the toe web spaces and the nose; heavily colonized, the face and upper chest; moderately colonized, the limbs and torso; lightly colonized, the palms and soles of the hands and feet (Figure 2). In heavily colonized areas bacterial densities may reach $10^7/cm^2$. Bacteria are not spread evenly across the skin surface, but live in discrete microcolonies varying in size between 10^2 and 10^5 cells. The dispersion of microcolonies in the skin results in only 10% of shed squames carrying micro-organisms [12].

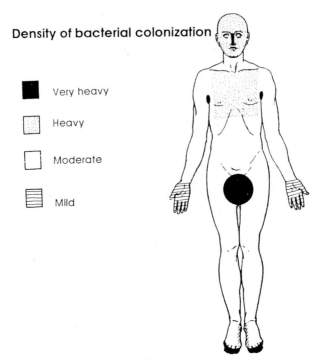

Figure 2. Density of bacterial colonization.

Factors controlling the density of colonization in the stratum corneum and hair follicle are poorly understood. Physico–chemical variables within the hair follicle such as altered oxygen tension and pH contribute to varying microbial proliferation [13]. *Propionibacterium* species are thus favoured in the hair follicle close to the surface of the skin, while *Staphylococcus* spp. colonizes the deeper part of the hair follicle.

Anatomically there are wide variations in types and densities of organisms in different sites, governed by the environmental conditions in these areas. In the pilosebaceous rich skin of the upper torso and face *Propionibacterium* spp. are found; in the occluded, moist skin of the axillae are large populations of staphylococcal and coryneform bacteria; and in the toe web spaces are large numbers of bacteria, in particular Gram-negative organisms, and an increased risk of infection with dermatophyte fungi.

4. PHYSICAL FACTORS AFFECTING SKIN COLONIZATION

The skin is covered with an 'acid mantle' of a pH varying between 4 and 6. Eccrine sweat glands are largely responsible for the acid pH of the normal skin by exchanging H^+ and bicarbonate ions across the membranes of the cells lining the sweat gland; at higher production rates the pH of sweat rises [8]. It has been suggested that the low pH of the skin limits bacterial growth accounting for the apparent reduction in bacterial density in more acidic sites, but alternatively raising the pH of the skin may change the electrostatic charge holding bacteria to the skin surface, thereby producing a higher harvest of the resident flora [14].

Recently, we demonstrated the continual production of nitric oxide (NO) from normal human skin (Weller, R. et al., J. Invest. Dermatol., in press). Nitrate in sweat is reduced by bacterial nitrate reductases to nitrite which, in turn, is reduced to NO in the acidic conditions on the skin surface. Such a system has been shown to be an effective microbicide in the human gut [15] and work we are currently performing is showing a similar sensitivity of skin organisms to NO produced in this way, in particular P. acnes and Candida albicans with Streptococcus spp. being relatively resistant. It would thus seem that a combination of pH and sweat production rather than pH alone is important in limiting bacterial growth.

The stratum corneum at the outer layer of the skin is an impermeable physical barrier created by overlapping cornified keratinocytes connected to one another by junctions called desmosomes. Experimentally it has been demonstrated that a break in the stratum corneum is an absolute requirement for inducing a streptococcal infection [16] . The epithelium is turned over very fast, the transit time from basal layer to shedding at the surface being in the order of 14 days. Organisms on the skin surface are thus sitting on a 'moving staircase' of shedding squames and any pathogen has a limited time in which to invade [17]. Skin surface lipids containing fatty acids are themselves antimicrobial against Streptococcus pyogenes and Staphylococcus aureus, although oleic acid promotes growth of P. acnes and other coryneforms.

Oxygen in the epidermis is at a lower partial pressure than in the atmosphere, while carbon dioxide is present at higher partial pressures [14]. This may be accounted for by the metabolic activities of the keratinocytes. The flora of the skin have varying oxygen requirements with Propionibacterium spp. being anaerobes, Brevibacterium spp. strict aerobes and staphylococci, coryneforms and several Gram-negative bacteria facultative anaerobes. The presence of Propionibacterium acnes so superficially in the stratum corneum is paradoxical for an anaerobe, but a possible explanation is its existence in a biofilm of mixed organisms, producing their own micro-environment by the secretion of metabolites and diffusion of gasses and nutrients [18].

The temperature of the surface of the body varies from the 37°C at the core by up to 20°C with the greatest variation being in the hands and feet. Anatomical site, environmental temperature, clothing and exercise all affect skin surface temperature. Skin commensals such as staphylococci will grow between 20 and 40°C, but some pathogens such as dermatophytes and the fungi causing chromomycosis have temperature optima of 25°C, accounting for the distribution of disease caused by these organisms to the extremities of the body.

The degree of hydration of the skin is an important determinant of microbial density. Occlusion of the skin with a water-impermeable barrier produces an explosive rise in bacterial density [19,20] and a relative increase in Gram-negative rods and coryneform bacteria compared to coccal forms. The water content of the skin derives from the balance of sweat production — predominantly by the eccrine glands — and transepidermal water loss. Transepidermal water loss involves the penetration of the lipid and corneocyte barriers of the skin by water and is used as a measure of the integrity of the barrier function of the skin [21]. It is increased during inflammation of the skin by exogenous irritants [22] or in skin diseases such as psoriasis [23]. Water loss from the skin is predominantly by evaporation and is reduced by high humidity and in occluded sites. Areas of the body such as the axillae, perineum and toe web spaces are thus highly hydrated and these areas have high bacterial colony counts [24].

Ultraviolet light is known to kill micro-organisms *in vitro* and is used for this purpose commercially [25]. *Malassezia furfur* and *Candida albicans* are more sensitive to UV radiation than *Staphylococcus epidermidis* and *S. aureus*, but a small study comparing the flora of sunbathers and non-sunbathers has shown no significant difference in the flora between the two groups [26]. Any use of UV radiation in vivo as a microbicide would be limited as evidence continues to accumulate showing the association between UV exposure and skin cancer [26-29].

Any organism that colonizes the skin must first become attached to the epithelial surface, the ability of an organism to colonize being proportional to its capacity to adhere to the skin [30]. Bacterial cell walls possess surface antigens called adhesins which strongly bind the microbe to the host cell [31]. Separate bacterial species bind with varying avidity to different epithelial surfaces which partly explains the variation in resident microflora on different sites [30]. Skin derived Group A streptococci bind more strongly to the cutaneous epithelium than to the oral mucosa [32]and *Pityrosporum* yeast adheres selectively to stratum corneum cells. *Streptococcus viridans* and *Candida albicans* on the other hand bind poorly to the intact skin [33], paralleling the presence of Group A streptococci and *Pityrosporum* species, but not viridans streptococci or *Candida albicans* as resident organisms of the skin.

Staphylococcal and streptococcal cell walls contain teichoic acids, which are thought to be skin epithelium adhesins. The complementary receptor on the epithelial cell is thought to be fibronectin. An absence of available fibronectin on the undamaged skin surface may account for the poor adherence of *Staphylococcus aureus* and streptococci to bind to undamaged skin rather than eczematous skin, where the surface is already disrupted by the disease process, exposing fibronectins [34]. Pathogenic bacteria must displace the normal flora and this may be achieved by binding more strongly to epithelial cells. A second possible adhesin, protein A, has been identified in *S. aureus* [34], a pathogen frequently involved in exacerbations of eczema, and binding to fibrinogen as well as fibronectin receptors has also been observed [17].

5. THE RESIDENT FLORA OF THE HUMAN SKIN

Only a limited number of organisms are able to tolerate the conditions of the skin environment, and these form the resident microflora. The resident flora is largely made up of coagulase-negative staphylococci, coryneform bacilli, and anaerobic *Propionibacterium* and *Pityrosporum* (*Malassezia*) spp.

Coryneform bacteria are Gram-positive, non-branching, non-spore forming, pleomorphic rods [35]. They may be aerobic or anaerobic and are a large, ill-defined group of bacteria. Historically they were simply divided into lipophilic and non-lipophilic bacteria, but more recent taxonomic classification has been dependent on cell-wall composition. Nonetheless, routine identification of coryneform isolates is not widely carried out clinically. Numerous bacteria have been included with the coryneforms: *Actinomycetes*, *Arachnia*, *Arcanobacterium*, *Arthobacter*, *Bacterionema*, *Bifidobacterium*, *Brevibacterium*, *Cellulomonas*, *Corynebacterium*, *Eyrsipelothrix*, *Eubacterium*, *Kurthia*, *Listeria*, *Mycobacterium*, *Nocardia*, *Oerskovia*, *Propionibacterium*, *Rhodococcus* and *Rothia*. Sixty percent of the aerobic coryneforms on the normal human skin are *Corynebacterium* and 20% *Brevibacterium* [36]. Both lipid-dependent and lipid-independent corynebacteria are found most prevalently in the anterior nares, perineum and toe interspace. They are found less densely over the scalp, forehead and cheeks, areas rich in sebaceous glands. Humid areas of the body thus favour corynebacteria, even, paradoxically, if lipid dependent when a sebum rich area might have been thought to be more favourable.

Anaerobic coryneforms are made up of *Propionibacterium* spp., of which *Propionibacterium acnes* is the most common. It is found in areas rich in sebaceous glands and in the hair follicles and plays an integral part in the development of acne vulgaris [37]. The density of *Propionibacterium* is closely correlated to the amount of lipid secreted by sebaceous glands, and also the amount of free fatty acid on the skin surface, propionibacteria generating this by hydrolysing triglycerides in sebum.

Staphylococci are aerobic, Gram-positive cocci which microscopically are grouped in grape-like clusters . They are classified according to the scheme of Kloos and Schleifer [38]; their principal habitat is the skin. They have developed an increased resistance to penicillin in the last 20 years. *S. aureus* is carried into the anterior nostrils of 30% of the healthy population with some differences in carriage rates between population groups [39]; less frequently it is carried in the axillae, groin and toe web spaces. In addition to racial differences in carriage rates, altered anatomy of the nose increases the carriage of staphylococci [40]. *S. aureus* is uncommon on the skin of healthy people, but is found with increasing frequency on those with an underlying disease such as diabetes or intravenous drug abuse [41] and also on hospital workers [42]. S. aureus is particularly common on patients with eczema; its carriage appears to be proportional to the severity of eczema [43], although whether it contributes to the disease or is a consequence of it remains uncertain. Adherence of *S. aureus* to the skin of subjects with a tendency to eczema appears to be increased. *S. aureus* is a potent pathogen, and exerts its effects by secreting a range of toxins and enzymes [44]; diseases it causes include: boils, furuncles, impetigo, scalded skin syndrome and toxic shock syndrome.

Micrococci are Gram-positive cocci, formerly classified with the staphylococci, but more recently considered closer taxonomically to the coryneforms. They are slower to colonize the skin of children than staphylococci but are part of the resident flora. Micrococci rarely cause infection except in the immunosuppressed, other than pitted keratolysis caused by *Micrococcus sedentarius.*

The coagulase-negative staphylococci *S. epidermidis* and *S. hominis* are the commonest organisms of the normal flora [31]. *S. epidermidis* preferentially colonizes the upper part of the body, particularly the face and trunk, while *S. hominis* predominates over the drier sites such as the arms and legs. At least 18 species of coagulase-negative staphylococci have been identified on the skin, including *S. epidermidis* and *S. hominis*. The coagulase-negative staphylococci appear to be among the first organisms to colonize the skin after birth. They may cause septicaemia in patients with indwelling catheters or cannulas, but otherwise are rarely pathogens.

Streptococci are Gram-positive aerobic and facultatively anaerobic bacteria arranged in chains or in pairs. They are usually transient organisms and die rapidly if placed on normal, unbroken skin [16]. Nasopharyngeal carriage of groups A, C and G haemolytic streptococci occurs in between 6 and 16% of the healthy population, however [45]. During outbreaks of streptococcal infection where the community carriage is high, a large enough innoculum to produce skin infection is more likely to be encountered. Groups B, C and G haemolytic streptococci are found in faeces and may be the source for infection of abdominal and perianal wounds. Non-haemolytic streptococci are of low pathogenicity, but carriage rates are high in the mouth and faeces.

Streptococci may cause local infections such as pyoderma, cellulitis, abscesses, erysipelas and necrotizing fasciitis. The most serious sequelae, however, are the post-streptococcal conditions glomerulonephritis and rheumatic fever. In addition, guttate psoriasis is strongly associated with preceding streptococcal throat infections. Epidemiologically, streptococcal carriage and disease remains widespread. Semmelweiss, over a century ago found that rates of puerperal fever could be dramatically reduced by the simple expedient of washing the hands between deliveries. Nonetheless, outbreaks of infection continue to occur in hospitals worldwide where insufficient attention is given to hygiene. Outbreaks of streptococcal skin infection are also recorded in rugby football teams, police training centres, residential schools and prisons and in meat handlers [45]. In addition to direct physical contact and droplet spread from nasopharyngeal carriage, shared equipment such as coconut matting and climbing ropes used in gymnasia has been found to be abrasive to the skin and contaminated with *S. pyogenes* and a possible instrument of transmission [45].

Gram-negative bacteria, other than *Acinetobacter* spp., are infrequently found on the skin surface, probably because of its desiccation. *Acinetobacter* is found in 25% of people, particularly in the toe web spaces, axillae and other moist areas. *Acinetobacter* is found more frequently in the summer months when sweatier bodies make the cutaneous environment less hostile [46]. *Klebsiella*, *Proteus*, *Enterobacter* and *Escherichia* have been cultured from the toe web spaces of healthy individuals.

The only resident fungi on the skin are the *Pityrosporum* yeasts which are thick-walled fungi inhabiting the superficial stratum corneum. There is some debate over whether there is one species of *Pityrosporum* with varying morphology, or two species, *P. ovale* and *P. orbiculare*. *Pityrosporum* yeasts are most prevalent in young adults and in sebaceous-rich areas such as the scalp, chest and upper back. They produce a lipase which splits fatty acids in sebum, and on invasion of the stratum corneum induce the production of azelaic acid that alters the ability of melanocytes to form pigment. *Pityrosporum* yeasts are responsible for pityriasis versicolor, where pigmentary abnormalities are part of the disease process, as well as seborrhoeic dermatitis and pityrosporum folliculitis [47]. The hair follicle mite *Demodex* is extremely widespread and harmless in the adult population.

6. PATHOGENIC FLORA OF THE SKIN

Many of the bacteria discussed under the heading of commensals of the skin may be pathogenic in certain circumstances. Other organisms are always pathological. Gram-negative organisms may superinfect eczematous skin, or skin damaged by fungi or extreme wetness. 'Immersion foot' is most commonly due to *Pseudomonas aeruginosa* or *Pseudomonas mirabilis* and occurs in conditions of high humidity. Candidal infections occur where there is reduced bacterial competition due to concurrent antibiotic administration, or reduced resistance as a result of diabetes or immunosuppression. Once again, wet conditions favours the growth of candida — chronic paronychia is most commonly caused by *Candida albicans* and most frequently occurs in those whose hands are continually wet for occupational or domestic reasons. Various dermatophyte fungi infect human skin, hair and nails. The commonest fungi in the UK are *Trichophyton rubrum* followed by *T. mentagrophytes* var *interdigitale*. Other dermatophytes include *Epidermophyton floccosum*, *T. tonsurans*, *Microsporum canis* and *Maudouinii*. Typically dermatophytes cause the erythema and scaling of ringworm or athletes foot (tinea corporis and tinea pedis). We have reported a case of a *T. interdigitale* infection transmitted from one horticulturist to another via an intermediate plant host (Weller R., Leifert C. Transmission of *Tricophyton interdigitale* via an intermediate plant host. Br. J. Dermatol., in press). Zoophilic fungi such as *T. verrucosum* and *T. mentagrophytes* var *erinacei* may cause florid reactions such as kerions.

Several arthropods are capable of infesting man. Some, such as ticks, are not harmful in themselves, but may act as vectors for a number of rickettsial illnesses. Mites such as Sarcoptes scabei cause scabies with zoophilic *Cheyletiella* species also capable of causing disease in man. Two species of lice parasitize man, *Pediculus humanus* and *Pthirus pubis*, the head and pubic louse. Infection with the human flea *Pulex irritans* is now rare, but cat and dog fleas *Ctenocephalides felis* and *C. canis* will readily bite man, the bites producing intensely itchy papules, usually situated below the knee. Mosquitoes and flies feed on humans, and some members of the *Diptera* family — mostly situated in the Americas — spend their larval stage in mammalian tissue, causing miasis in man. Not all insects are able to penetrate the human epidermis; thrips, although an agricultural pest, have no effect on man.

7. TRANSIENT FLORA OF THE SKIN

The transient flora are defined as those organisms which are deposited on the skin, but do not colonize or multiply there [48]. Transient microflora can be removed relatively easily from the skin, often with soap and water. Any organism found in nature may be transient on the skin: bed bugs, stable flies, group A streptococci, *Bacillus* spp. and viruses are some of the commoner examples.

8. CONTAMINANTS OF PLANT TISSUE CULTURES AND MEANS OF DISINFECTION

Contamination with micro-organisms is considered to be the single most important cause of loss of plants during *in vitro* culture; and, of contaminants, bacteria have been most frequently described [49]. The frequency of infection with common skin organisms of *Staphylococcus* spp. and *Micrococcus* spp. and the increased percentage of infection with serial subcultures implies contamination from the human skin [50].

Sterilization of the human skin is not possible. Hygienic hand disinfection with soap and water is usually effective at removing transient organisms if performed diligently. [51]. Washing with bar soap, however, increased the shedding of skin squames and viable bacteria 18-fold, while the use of a surgical solution had no effect on shedding of bacteria [3]. There are thus theoretical benefits in the routine use of antibacterial solutions such as chlorhexidine or povidone-iodine. Such solutions will reduce the resident bacterial population, and a 2 min scrub is as efficient as a 10 min scrub [52]. The benefit of any disinfectant process is lost if taps or dirty towels have to be touched. Sterile towels and elbow taps are used in hospitals. Over-frequent washing can itself cause problems. Nurses have been observed to wash their hands up to 100 times in a single shift, with resultant irritation and cracking of the hands. An increase in the bacterial count before and after chlorhexidine use was then observed in this group [53]. Wearing of sterile surgical gloves will further reduce the exposure of any material handled to skin organisms, but at the cost of increasing bacterial colony formation on the underlying hands.

Skin squames are the major route of dispersal of cutaneous organisms to the environment. The wearing of tightly woven ventile operating suits has been shown to reduce this [54], but they are too hot to be comfortable for extended use. Laminar air flow systems are used in orthopaedic operating theatres, particularly for surgery involving the insertion of artificial prostheses, where absolute asepsis is required. Airborne organisms are thus reduced to a minimum.

REFERENCES

1. Miller, S.J., Aly, R., Shinefield, H.R. and Elias, P. (1988) Arch. Dermatol. 124, 209–215.

2. Noble, W.C. (1975) Br. J. Dermatol. 93, 477–485.

3. Meers, P.D. and Yeo, G.A. (1978) J. Hyg. 81, 99–105.

4. Sciple, G.W., Riemensnider, D.K. and Schleyer, C.A.J. (1967) Appl. Microbiol. 15, 1388.

5. Lewis, H.E., Foster, A.R., Mullan, B.J., Cox, R.N. and Clark, R.P. (1969) Lancet 1, 1273–1277.

6. Clark, R.P. and Shirley, S.G. (1973) Nature 246, 39–40.

7. Benediktsdottir, E. and Hambraeus, A. (1982) J. Hyg. (Camb.) 88, 487–500.

8. Boysen, T.C., Yanagawa, S., Sato, F. and Sato, K. (1984) J. Appl. Physiol. 56, 1302–1307.

9. Jenkinson, D.M. (1973) Br. J. Dermatol. 88, 397–406.

10. Weller, R., Pattullo, S., Smith, L., Golden, M., Ormerod, A.D. and Benjamin, N. (1996) J. Invest. Dermatol. 107.

11. Leyden, J.J., McGinley, K.J., Mills, O.H. and Kligman, A.M. (1975) J. Invest. Dermatol. 65, 379–381.

12. Noble, W.C. and Davies, R.R. (1965) J. Clin. Pathol. 18, 16.

13. Cunliffe, W.J. (1989) Acne. Martin Dunitz, London.

14. McBride, M.E. (1992) in The Skin Microflora and Microbial Skin Disease (Noble W.C., ed.) pp. 73–101, Cambridge University Press, Cambridge, UK.

15. Dykhuizen, R.S., Frazer, R., Duncan, C., Smith, C.C., Golden, M., Benjamin, N. and Leifert, C. (1996) Antimicrob. Agents Chemother. 40, 1422–1425.

16. Leyden, J.J., Stewart, R. and Kligman, A.M. (1980) J. Invest. Dermatol. 75, 196–201.

17. Kinsman, O.S. (1982) Semin. Dermatol. 1, 127–136.

18. Hamilton, W.A. (1987) in Ecology of Microbial Communities (Fletcher M., Gray T. R.G. and Jares J.G., eds) pp. 361–385, Cambridge University Press, Cambridge, UK.

19. Bibel, D.J. and LeBrun, J.R. (1975) Can. J. Microbiol. 21, 496–500.

20. Aly, R. (1982) Semin. Dermatol. 1.

21. Pinnagoda, J., Tupker, R.A., Agner, T. and Serup, J. (1990) Contact Dermatitis 22, 164–178.

22. Tupker, R.A., Pinnagoda, J., Coenraads, P.J. and Nater, J.P. (1989) Contact Dermatitis 20, 108–114.

23. Rogers, S. (1993) Clin. Exp. Dermatol. 18, 21–24.

24. Leyden, J.J., McGinley, K.J., Nordstrom, K.M. and Webster, G.F. (1987) J. Invest. Dermatol. 88, 65s–72s.

25. Zelle, M.R. and Hollaender, A. (1955) in Radiation Biology (Hollaender A., ed.) pp. 365–400, McGraw-Hill, New York.

26. Holman, C.D.J., Armstrong, B.K. and Heenan, P.J. (1986) J. Natl. Cancer Inst. 76, 403–414.

27. Elwood, J.M., Whitehead, S.M., Davison, J., Stewart, M. and Galt, M. (1990) Int. J. Epidemiol. 19, 801–810.

28. Weinstock, M.A., Colditz, G.A., Willett, W.C., Stampfer, M.J., Bronstein, B.R., Mihm, M.C.J. and Speizer, F.E. (1989) Pediatrics 84, 199–204.

29. Kricker, A., Armstrong, B.K., English, D.R. and Heenan, P.J. (1995) Int. J. Cancer 60, 489–494.

30. Gibbons, R.J. and van Houte, J. (1975) Annu. Rev. Microbiol. 29, 19–44.

31. Roth, R.R. and James, W.D. (1988) Annu. Rev. Microbiol. 42, 441–464.

32. Alkan, M., Ofek, I. and Beachey, E. (1977) Infect. Immun. 18, 555–557.

33. Aly, R., Shinefield, H.R. and Maibach, H.I. (1981) in Skin Microbiology: Relevance to Clinical Infection (Maibach H., ed.) pp. 171–179, Springer-Verlag, New York.

34. Cole, G.W. and Silverberg, N.L. (1986) Arch. Dermatol. 122, 166–169.

35. Leyden, J.J. and McGinley, K.J. (1992) in The Skin Microflora and Microbial Skin Disease (Noble W.C., ed.) pp. 102–117, Cambridge University Press, Cambridge.

36. Pitcher, D.G. (1977) J. Med. Microbiol. 10, 439–445.

37. Holland, K.T. (1989) in Acne (Cunliffe W.J., ed.) pp. 178–210, Vol. 1, Dunitz, London.

38. Kloos, W.E. and Schleifer, K.H. (1975) J. Clin. Microbiol. 1, 82–88.

39. Noble, W.C. (1974) Acta Dermatol. Venerol. (Stockholm) 54, 403–405.

40. Noble, W.C. (1992) in The Skin Microflora and Microbial Skin Disease (Noble W.C., ed.) pp. 135–152, Cambridge University Press, Cambridge.

41. Berman, D.S., Schaefler, S., Simberkoff, M.S. and Rahal, J.J. (1987) J. Infect. Dis.

42. Tuazon, C.U. (1984) Am. J. Med. 76, 166–171.

43. Goodyear, H.M., Watson, P.J., Egan, S.A., Price, E.H., Kenny, P.A. and Harper, J.I. (1993) Clin. Exp. Dermatol. 18, 300–304.

44. Noble, W.C. (1992) in The Skin Microflora and Microbial Skin Disease (Noble W.C., ed.) pp. 153–172, Cambridge University Press, Cambridge.

45. Barnham, M. (1992) in The Skin Microflora and Microbial Skin Disease (Noble W.C., ed.) pp. 173–209, Cambridge University Press, Cambridge.

46. Kloos, W.E. and Musselwhite, M.S. (1975) Appl. Microbiol. 30, 381–395.

47. Hay, R.J. (1992) in The Skin Microflora and Microbial Skin Disease (Noble W.C., ed.) pp. 232–263, Cambridge University Press, Cambridge.

48. Price, P.B. (1938) J. Infect. Dis. 63, 301–318.

49. Leifert, C., Ritchie, J.Y. and Waites, W.M. (1991) World J. Microbiol. Biotechnnol. 7, 452–469.

50. Leifert, C., Morris, C.E. and Waites, W.M. (1994) Crit. Rev. Plant Sci. 13, 139–183.

51. Lowbury, E.J.L., Lilly, H.A. and Bull, J.P. (1964) Br. Med. J. ii.

52. Aycliffe, G.A.J., Coates, D. and Hoffman, P.N. (1984) Chemical Disinfection in Hospital. PHLS, London.

53. Ojajarvi, J., Makela, P. and Rantasalo, I. (1977) J. Hyg. (Cambridge) 79.

54. Mitchell, N.J. and Gamble, D.R. (1974) Lancet, 1133–1136.

CONTAMINATION BY MICRO-ARTHROPODS IN PLANT TISSUE CULTURES

JOHAN PYPE, KOEN EVERAERT and PIERRE DEBERGH
Laboratory for Horticulture, University of Gent, Coupure Links 653, 9000 Gent, Belgium

1. INTRODUCTION

Micro-arthropods are, although not always acknowledged, a serious source of contamination in tissue cultures. The damage that is caused by mites has been severely underestimated in past decades and only recently laboratory managers have admitted that mites can be responsible for severe losses. A review in the UK [1] revealed that one-third of the laboratories had major, and another fifth minor, problems with micro-arthropods. Personal communications with different laboratory managers indicate that these figures are certainly not overestimated.

Though the direct harm caused by *Acari* in tissue cultures is limited to suction, their most important damage is caused as vectors for bacteria and fungi. However, as contrasted with other pathogens in tissue cultures such as viruses, bacteria and fungi that are spreading passively by carry-over during manipulations, turbulence or by specific vectors, micro-arthropods are actively spreading in culture rooms by crawling from one container to another.

Because most of the life-cycle of the mites takes place in the protected environment of a tissue culture container, it is very unlikely that a population can be eradicated by a simple pest control in the culture room.

Different approaches and methods for the control and eradication of mites in tissue cultures are discussed.

2. LIFE CYCLE AND SPREADING OF *ACARI* IN PLANT TISSUE CULTURES

Micro-arthropods, more commonly called mites, can only occasionally be seen by the naked eye. Fungal spores from the sporotheca, cavities in the abdomen where spores are collected, of *Acari* commonly found in tissue cultures, contribute to the transfer and dispersion of

A.C. Cassells (ed.), Pathogen and Microbial Contamination Management in Micropropagation, 259–266.
© *1997 Kluwer Academic Publishers. Printed in the Netherlands.*

fungi. Hence, rapid spreading of fungal contamination is a first indication of contamination by mites.

Micro-arthropod contamination is often accompanied by bacterial contamination. Populations of *Erwinia* sp. were found in media and on plant remains of infected cultures. The determination of the micro-arthropods found in tissue cultures is confusing because no adequate references are available. Much of the determination is actually based on the work by Krczal [2]. Several genera have been observed in tissue cultures. These include *Tyrophagus*, *Siteroptes* and *Pyemotes*. All species observed are plant-specific *Acari* and cannot be confused with house- or dust mites.

Since contaminating micro-arthropods require a high relative humidity, an optimal environment is provided in tissue culture containers. A normal life cycle under these protected conditions takes 10–20 days. Each female will generate an offspring of 100 to 200 mites of which 98% are female. The first mite born is of the male sex. It will attach itself to the swollen abdomen of the female parent and copulate with the successive female descendants during their birth. However, if the swollen abdomen, or hysterosoma, is crushed a progeny of 50% males and 50% females will mature from the eggs.

Once fertilised, the young mites will spread in the cultures. They can leave and penetrate containers through the smallest cracks, actively looking for plant material. Even the use of parafilm will not prevent them from spreading. Young mites will attach to plants by means of modified chelicera (styleta) and will feed on cell sap. Gradually the abdomen or hysterosoma will swell due to the developing eggs. Nine days after their birth, the first mites of a next generation will brood. Hatching will continue for about 7 days.

A tissue culture container infected with a single mite can theoretically bear 25,000 descendants after 30–40 days and more than 5 million after 60 days. Young mites will spread in the culture rooms and infect nearby containers. Gradually the complete environment will be infected. Hence, infected containers — providing shelter and an ideal environment for mite propagation — remaining in the culture rooms for longer periods can cause an invasion of micro-arthropods resulting in a complete loss of the cultures.

3. PREVENTION AND CONTROL OF MICRO-ARTHROPODS

3.1. Pesticide use in tissue culture laboratories

In order to prevent epidemic spreading of micro-arthropods in tissue cultures, the application of pesticides can be considered. However, the selection of appropriate products can be erratic.

Frequently observed micro-arthropods attacking horticultural crops *in vivo* are *Panonychus*, *Brevipalpus* and *Tetranychus*, and selection of commercial acaricides is limited to products efficient against these harmful species. Therefore, they are not necessarily efficient against the species occurring *in vitro*. Small spectrum acaricides should therefore be avoided and preferably broad spectrum ones, or even broad spectrum insecticides should be applied.

Since the preferably selected pesticides are also toxic to humans, special attention should be paid to the safety of the personnel working in the locations where the products have been applied or to those manipulating the cultures with residues. Priority should be given to protect the personnel from these hazards. If complete safety cannot be guaranteed, application should be prohibited.

3.2. Application of pesticides in tissue culture media

Pesticides can be incorporated in the media during preparation. If this is done prior to autoclaving, the temperature stability of the active ingredients should be evaluated. In our experiments the temperature stability as well as the photodegradation of dicofol and broompropylate were analysed. A concentration of 10 ppm broompropylate was reduced by 15–40% during autoclaving while 10 ppm of dicofol was decreased by 15–55%. Although a reduction in active ingredients occurred, the concentration was still sufficient to control micro-arthropod infection. After 1 month the concentration was measured a second time. No further breakdown was observed. In some objects the concentration increased even slightly due to water loss by evaporation.

Since technical data on fenbutatin oxide indicate that it is extremely stable to heat, light and atmospheric oxygen, breakdown was not measured.

Phytotoxicity of dicofol (5, 50, 500 and 5000 ppm), broompropylate (5, 50, 500 and 5000 ppm) and fenbutatin oxide (0.275, 2.75, 27.5 and 275 ppm) was gauged on *Rosa* spp., *Ficus cyathistipula* and *Calathea roseopicta*. For rose, no growth reduction or toxicity was observed but instead some growth promotion took place. Compared to the control, fresh weight increased and leaves were greener. Only the application of 5000 ppm dicofol was moderately toxic for *Calathea*. Concentrations of more than 50 ppm dicofol and of more than 500 ppm broompropylate were toxic for *Ficus*.

In vitro, acaricides can also be applied in a double layer system [3], a thin film of a few millimetres thick on top of the agarified medium. This method allows application independently of the medium preparation or in populations that were subcultured earlier.

The results indicated that the phytotoxic effects of the latter method were more pronounced compared to the incorporation of acaricides in the medium. For *Calathea* strong toxicity with yellowing leaves and pronounced growth reduction already took place at concentrations of 500 ppm dicofol.

To avoid phytotoxic reactions, acaricides can also be applied as saturated cotton plugs. In this procedure spreading of micro-arthropods is controlled by the introduction in the tissue culture container of a sterile cotton plug that is soaked in a sterile solution of the active ingredient. Acaricides that work by direct contact with the micro-arthropods will be ineffective in this system. Hence, products should be selected that are active by their partial pressure.

Application of pesticides in culture containers does not stop contamination immediately, but prevents epidemic propagation of mites which penetrate new containers. Even if only a single mite penetrates a culture container, the culture will be contaminated by fungi.

3.3. Physical control of spreading

When ineffective products are applied, room treatments with these acaricides can have a dispersing rather than a curtailing effect. Indeed, if wire netting is used for the shelves micro-arthropods will only suffer from a temporarily knock down and tumble massively down to reach the cultures on lower shelves.

From the above it can be concluded that the shelf structure is an important factor in the dispersion of micro-arthropods in a laboratory. If contamination is observed it is indicated to provide physical barriers, even if this would imply reduced aeration. Moreover, to reduce horizontal dispersion, culture containers should be placed distant from each other to prevent direct contact. To slow down vertical dissemination of micro-arthropods a paper layer can be placed on the shelf surfaces.

To be more effective the paper can also be drenched in a pesticide solution or an acaricide can be sprayed on the paper before cultures are placed on it. The use of a paper layer allows easy removal and renewal afterwards and prevents residue accumulation.

An efficient method to prevent their spreading in tissue culture rooms is the utilisation of adhesive surfaces. For this purpose, a paper sheet coated with a dry type of glue is put on the shelves. In our experiments an acrylic glue was applied (Casco Nobel 7268, Denmark), but other dry type of glues or hot melts can also be considered. Micro-arthropods scattering from a contaminated container are instantly immobilised when they reach the glued paper. If no direct contact between different containers occurs, dispersion is effectively prevented.

3.4. Application of attractants or repellents

In the course of our experiments it was observed that certain crops were more infected than others, even if placed in a contaminated environment. This lead to the conclusion that some plants release gaseous substances that attract or repel micro-arthropods.

The repellent activity of menthol is well known in apiculture to prevent *Varroa jacobsoni* infection in honeybee hives. However, menthol cannot be applied in tissue cultures due to prevailing phytotoxic effects. Already 1 h after subculture of *Spathiphyllum* and *Ficus* on a medium with 0.02% menthol, wilting was observed and 48 h later the plants were completely yellow and had died off.

3.5. Development of a bioassay

The development of a bioassay can be of utmost importance to assess the efficiency of pesticides or other actions undertaken. The aim of the assay was to culture mite populations free of fungal contamination, which could then be used for further experiments.

According to Suski [4] sporotheca of *Siteroptes* spp. are not always filled with fungal spores. To obtain a sterile population, blown up physogastric female micro-arthropods were transferred three times. Before transfer they were submerged in alcohol for a few seconds. A gradual decrease of spore concentration in the sporotheca of a sterile population was expected. However, no sterile populations could be obtained. Possibly more subcultures are required. Furthermore, it is indicated to use micro-arthropods with a larger hysterosoma and more developed eggs. Indeed, when transferred, the mites are not able to move and cannot attach to the new substrate for feeding, although this is required for the subsequent development of immature eggs.

To control the development of moulds the incorporation of a fungicide in the feeding substrate was considered. Since previous experiences with imazalil in tissue cultures were successful, it was selected as an additive to a modified medium of Murashige and Skoog [5].

The addition of imazalil to a feeding substrate was adequate to reduce the development of fungi. Micro-arthropods were not immediately killed, but their activity was drastically reduced. Since the application of imazalil has positive effects on the growth and propagation of various plants in tissue cultures [6], its application can be beneficial when contamination with micro-arthropods is suspected.

4. DISCUSSION

In vitro cultures are extremely sensitive for ravaging micro-arthropod populations. This is especially true for large mono-cultures in commercial laboratories that are often left unattended for some weeks. Consequently, the control of micro-arthropods is required.

Although the infection of cultures is probably accidental, a rapid increase of the population is stimulated in the protected environment with high relative humidity and abundant food. Spreading occurs by active crawling from container to container. If inappropriate extermination methods are applied this will lead to an expansion rather than a decrease of the pest.

When pesticides are applied for the control of micro-arthropods attention should be paid to the selection of effective products. Since the species involved in contamination of tissue cultures are not causing any important economic damage to other field or greenhouse crops, the effects of new active ingredients on these micro-arthropods are not evaluated in the screenings of phyto-pharmaceutical companies.

If the use of acaricides or insecticides is considered, temperature stability, light degradation and efficiency should be evaluated. When applied, extreme care should be taken for the personnel handling the cultures.

The incorporation of an acaricide in the medium is the most cost-efficient and technically feasible method. If chemicals can be selected that are effective against the species frequently occurring in tissue cultures and that are neither toxic to humans nor to plants, systematic use can be considered for a limited period. Also the application of the fungicide imazalil, can slow down the rapid increase of micro-arthropod populations.

A good alternative is the installation of physical barriers. If a micro-arthropod infestation is perceived, direct contact of culture containers should be avoided by placing them distant from each other and perforated shelves should be covered with a non-perforated sheet. The application of a paper layer with a coating of acrylic glue will stop natural spreading instantly. However, this method will not eliminate the origin of the problem.

Although there is probably an affinity for certain environmental conditions, e.g. appropriate relative humidity or the plant considered, not much is known about these preferences. To evaluate the preferences qualitative and quantitative tests are required. These experiments can only be accomplished if clean cultures of micro-arthropods can be obtained. Factors to be evaluated are the effects of repellents or attractants, the efficiency of pesticides or the effects of culture conditions.

One of the most important factors to control micro-arthropod infestations is the general tidiness and hygienic standards of the laboratory. Probably all laboratories occasionally have cultures infected with mites. Well-managed labs are able to control these outbreaks by their standard procedures. If contaminated containers are regularly removed, a dramatic increase in the population can be avoided. In commercial laboratories the follow-up of cultures is sometimes insufficient. At least twice a week all cultures should be controlled for contamination. Especially cultures that require long subculture periods should be followed up since they allow an abundant growth of the population. Contaminated cultures and plant remains from initiations or subculture should be removed from the laboratory on a daily basis. Although these measures should be standard procedures in a tissue culture laboratory, they are often neglected.

Micro-arthropods species that are harmful for *in vitro* cultures are sensitive for environmental conditions in the culture room and often seasonal fluctuations are observed. Generally they prefer warmer conditions and a high relative humidity. If the relative humidity is lowered, they are less active and spreading and propagation is reduced. However, changing the environmental conditions has an important effect on the cultures.

5. CONCLUSION

The review of Blake [1] and personal information from several researchers and managers clearly indicate that an important number of laboratories suffer from micro-arthropod contamination. Hence, management should always be alert for this problem and take appropriate intervention when it is observed. Neglecting the problem will lead to dramatic losses.

General cleanness and tidiness as well as the application of standard procedures will effectively reduce the risk of an epidemic spread.

If, however, a problem arises, several options for the control of the pest are possible. In the infected area physical barriers can be created to avoid scattering of the micro-arthropods. Combined with the application of an acrylic glue, spreading is effectively controlled.

To prevent multiplication of micro-arthropods, an acaricide or insecticide can be incorporated in the medium or a cotton plug drenched in a volatile pesticide can be added in the culture container. Attention should be paid to poisonousness, stability, phytotoxic effects and efficiency of the product.

Although at present no useful repellents or attractants have been identified, they could offer an opportunity for the control of micro-arthropods.

ACKNOWLEDGEMENT

This research was supported by the I.W.O.N.L., the Institute for Scientific Research in Industry and Agriculture, and by private tissue culture laboratories.

REFERENCES

1. Blake, J. (1988) Acta Hortic. 225, 163–166.

2. Krczal, H. (1959) in Beitrage zur Systematik und Okologie Mitteleuropaischer Acarina (Stammer H.J., ed.) pp. 385–625, Academische Verlagsgesellschaft, Geest & Portig K.-G. Leipzig.

3. Maene and Debergh (1985) Plant Cell Tissue Org. Cult. 5, 23–33.

4. Suski, Z. W. (1973) Annales zoologici 17, Polska Akademia Nauk, Instytut Zoologiczny, Warszawa, 27 p.

5. Everaert, K. (1994) Bestrijding van micro-arthropoden in in vitro culturen, Thesis, Universiteit Gent.

6. Werbrouck, S.P.O. and Debergh, P.C. (1995) J. Plant Growth Regul. 14, 105–107.

PHOTOAUTOTROPHIC MICROPROPAGATION — A STRATEGY FOR CONTAMINATION CONTROL?

ROGER D. LONG

Green Crop Ltd, Littlewood, Coolkenno, Tullow, Co. Carlow, Ireland

1. INTRODUCTION

Photoautotrophic micropropagation is a tissue culture technique whereby a chlorophyllous explant is placed in environmental conditions that induce it to photosynthesise, grow and multiply. This requires cultivation of the micropropagated material under conditions of carbon dioxide enrichment and high photosynthetic photon flux densities. This reduces, or removes, the requirement for the addition of sucrose, or any other carbohydrate source to the medium. Photoautotrophic micropropagation has been quite widely discussed in the literature for the past decade There is little evidence as to its use as a commercial micropropagation system throughout the world. It has been developed and applied in Ireland since 1990 and is now routinely used for the production of microplants to produce pre-basic seed potatoes. Approximately 80% of pre-basic seed potatoes in Ireland are produced via photoautotrophic micropropagation [1]. Commercial laboratories utilising photoautotrophic micropropagation are known to exist in Ireland, Africa and the Middle East. There are reports of its use on an experimental basis from many laboratories and for many species (see reviews by Kozai, etc.).

Microbial contamination is often cited as one of the major reasons for losses and high costs in micropropagation [2–4]. There are three generally recognised sources of contamination that occur: (1) epiphytic micro-organisms on the explant, (2) endophytic micro-organisms in the explant and (3) micro-organisms that can be either pathogenic or saprophytic that are introduced into the culture by poor operator technique or invasion of the culture vessel. For the micro-organism to be able to survive and grow in culture, it requires a source of nutrients and carbohydrates. Both epiphytes and endophytes derive these from the plant in nature but, under the nutrient-rich conditions of *in vitro* culture, some species can grow out from the plant tissue and swamp or spoil the culture. Similarly, organisms unwittingly introduced into the culture vessel often find the carbohydrate-rich medium ideal for growth, with concomitant rapid spoilage of the culture. It is also possible that some epiphytic organisms conditioned to surviving in a low nutrient status and harsh phylloplane environment, do in fact fail to grow under nutrient-rich *in vitro* culture conditions. Given the survival mechanisms often present in such organisms, it is probable that these can at least remain viable on the plant material in culture.

A.C. Cassells (ed.), Pathogen and Microbial Contamination Management in Micropropagation, 267–278.

Photoautotrophic micropropagation takes place in the absence of sucrose in the medium, with the consequence that the medium does not support the growth of many of the potentially contaminating organisms concerned. Due to this fact, it is often postulated that photoautotrophic micropropagation is a possible means of, at best, contamination control or, at least, damage limitation.

2. PHOTOAUTOTROPHIC MICROPROPAGATION

Photoautotrophic micropropagation requires the growth of chlorophyllous explants under the appropriate environmental conditions that include carbon dioxide enrichment (CDE) and a higher level of photosynthetic photon flux density (PPFD) than is common in many growth rooms.

There are numerous systems reported in the literature which can be broadly categorised as shown in Table 1.

Table 1. Types of photoautotrophic micropropagation system

	Exogenous carbohydrate	Medium/substrate	Container	Ventilation
(i)	None	Agar/Gelrite	Conventional	Passive
(ii)	None	Liquid with various matrices	Conventional	Passive
(iii)	None	Agar/Gelrite	Custom-made	Forced
(iv)	None	Liquid with various matrices	Custom-made	Forced

The first type is perhaps closest to conventional micropropagation in that the level of sucrose in the medium is reduced or completely eliminated to encourage photosynthetic activity. This is the system that has been reported on most widely and is the one most likely to be used in that it does not require very much additional investment. Indeed, it can successfully be achieved by the use of conventional growth rooms with the addition of CDE, which is relatively inexpensive.

The second type of system is very similar but uses an alternative substrate to agar. Here the rationale is that in order to get the cultures/plantlets to act in as 'normal' a manner as possible you require not only a normal gaseous environment but also a normal root environment. Attempting to grow material in Murashige and Skoog medium with agar is about as abnormal as you can get! The ideal matrix for root growth and nutrient uptake has to have an adequate air/water volume percentage ratio and the nutrient solution used should be balanced and with a reasonable conductivity. A variety of substrates have been used experimentally from peat moss, rockwool, cellulose to polyurethane foam [5].

The third system uses large custom-made containers and an umbilical cord system to maintain the gaseous environment [4], and the final system is very similar except that, since the medium is liquid and a porous substrate is used, there is the possibility of pumping fresh medium into the container as well as maintaining the gaseous environment (microponics).

3. THE ENVIRONMENTAL CONDITIONS REQUIRED

Of necessity the conditions required can only be briefly mentioned here. For a full review see Kozai [4].

Generally, photoautotrophic micropropagation requires conditions of high photosynthetic photon flux density (PPFD) and carbon dioxide (CO_2) enrichment. Under the growth room conditions obtaining in most commercial laboratories the photosynthetic rate of explants/plantlets is restricted by the low CO_2 concentration in the culture vessel during the photoperiod; consequently, the cultures are either heterotrophic or mixotrophic. The CO_2 concentration inside the culture vessel often reduces to the compensation point during the photoperiod and only rises due to dark respiration [6]. By increasing the CO_2 concentration inside the vessel during the photoperiod the net photosynthetic rate increases. However, the presence of sugar in the medium has been shown to inhibit the photosynthetic activity of plantlets in vitro. With single node cuttings of potato the net photosynthetic rate has been shown to be eight to ten times greater in the absence of sugar in the medium at CO_2 concentrations in the range 500 to 1000 µmol/mol [7]. With CO_2 enrichment the main issue is whether to enrich the room or the individual vessel. To enrich the room is easier but the concentration inside the vessel is then a function of the gas exchange characteristic of the vessel. To enrich the individual vessel implies a forced ventilation system, which may confer some benefit with respect to growth rates [4] but is expensive to set up and carries contamination risks.

Photosynthetic photon flux density is a measure of the amount of photosynthetically active radiation being emitted from the light source being used, i.e. between 400 and 700 nm. Growth rooms in commercial laboratories, in my experience, have low levels of PPFD, in the region of 20–60 µmol/m^2/s at the shelf surface. The type of container and closure used obviously reduces the level at the leaf surface even further [8]. The light compensation point for chlorophyllous tissue in vitro is thought to be about 5 µmol/m^2/s. The relationship between PPFD and CO_2 concentration in terms of maximising the net photosynthetic rate in vitro has still to be studied on a wider range of species. In practical terms most photoautotrophic growth rooms operate with a PPFD in the region of 150 µmol/m^2/s and a CO_2 concentration of 1000 µmol/mol. Attempts at growing cultures at higher PPFD values lead to problems with dissipating the heat generated by the lights unless novel and expensive lighting systems are used [4].

4. TYPES OF CULTURE AND THE RANGE OF SPECIES

The type of explant to be used in photoautotrophic micropropagation has, by definition, to contain chlorophyll. The most successful type of explant is a leafy single node or multinode cutting. Since nodal sections and axillary bud culture is the most widely used in commercial micropropagation at the multiplication phase, this is not a problem. Regeneration from leafy explants via adventitious buds has been achieved with relatively easy to propagate species, e.g. *Begonia* (see Sayegh and Long, this volume). Regeneration from callus, cells and protoplasts is not possible but germination and growth of somatic embryos is a possibility.

Even with nodal cultures the hormonal requirements are often quite different to those required under hetero- or mixotrophic conditions. Plants such as potato that are normally grown with no hormones and whose multiplication is dependent on elongation of the stem axis and nodal increase, grow much faster under photoautotrophic conditions and produce plants that are stronger, establish faster and mature earlier (R.D. Long, unpublished results). Plants that rely on hormones to stimulate multiple axillary bud development often require different hormone combinations to achieve comparable or better multiplication rates. This is particularly so in systems that do not have agar as the support substrate.

Photoautotrophic micropropagation has been used on an experimental basis at least with *Cymbidium* [9,10], *Pelargonium* [11], potato [12,13], strawberry [14] carnation [15], *Eucalyptus* [16,17] and *Theobroma cocoa* [18].

5. ADVANTAGES OF PHOTOAUTOTROPHIC MICROPROPAGATION

Photoautotrophic micropropagation has many advantages over conventional methods. Kozai [4] cites the advantages listed in Table 2. The quality of plantlets produced, the ease with which they can be weaned and the lack of physiological problems with plants derived from photoautotrophic cultures are attributes with which anyone who has experience of photoautotrophic micropropagation can easily concur. From the point of view of this paper, however, we need to consider claims 7, 8 and 11 in a little more detail.

6. OCCURRENCE OF CONTAMINATING MICRO-ORGANISMS IN PHOTOAUTOTROPHIC CULTURES

Contamination in micropropagation arises from one of two sources, either carried over in or on the plant tissues or introduced into the culture by faulty laboratory procedures or some other factor such as faulty seals on the tissue culture vessel.

Table 2. Advantages of photoautotrophic micropropagation over heterotrophic micropropagation (after Kozai [4])

---

1. Growth and development of shoots/plantlets *in vitro* are promoted.

2. Physiological/morphological disorders and mutation are reduced and plantlet quality is improved.

3. Relatively uniform growth and development is expected.

4. Leafy single or multinode cuttings can be used as explants.

5. Procedures for rooting and acclimatisation are simplified.

6. Application of growth regulators and other organic matter can be minimised.

7. Loss of shoots/plantlets due to contamination is reduced.

8. A larger vessel can be used with minimum contamination.

9. Then, environmental control of the vessel becomes easier.

10. The control of growth and development by means of environmental control becomes easier.

11. Asepsis in the vessel is not required if no pathogen is guaranteed.

12. Automation, robotisation and computerisation become easier.

The literature, as far as I am aware, contains no references to contamination occurring in photoautotrophic cultures. Theoretically and practically, however, contamination by micro-organisms can and does occur in the same manner as in heterotrophic cultures but the consequences are less obvious and usually ignored. With no readily utilisable carbohydrate source in the tissue culture medium, growth of these micro-organisms is at least significantly reduced or at best stopped.

7. EPIPHYTIC AND ENDOPHYTIC SOURCES

In the absence of any literature base on contaminating micro-organisms occurring in photoautotrophic cultures, I am forced back on anecdotal evidence based on 8 years experience of utilising photoautotrophic methods. To date it has been the practice, in many of the reports in the literature and in commercial reality, for photoautotrophic cultures to be set up from material derived from heterotrophic cultures. This is either purely as a matter of convenience or because the material has been derived from a meristem culture programme to obtain elite material. When Stage I is carried out heterotrophically, many of the epiphytic micro-organisms present and that are capable of growth on the high sucrose and high salt concentrations usually present in heterotrophic medium, have already been eliminated purely through such cultures being discarded. Any micro-organisms that do not have the ability to express themselves and grow under such conditions are probably not going to express themselves under photoautotrophic conditions. In situations where plant material is collected from mother plants, disinfected and placed into culture, exactly the same rules apply as for heterotrophic cultures. Sterilisation procedures are the same; however, the main organisms of concern are those which are potentially pathogenic rather than the saprophytic organisms that in heterotrophic culture will over-run the medium. In this case, the support substrate can play a role. I have seen a case where material contained an unidentified fungal contaminant that quickly over-ran an agar-based medium but was not observed to grow when the same material was placed on a liquid medium with a polyurethane foam support. It was postulated that the agar source contained contaminating carbohydrates and perhaps amino acids that allowed the growth of the contaminating organism.

Similarly with endophytic micro-organisms, if they are not expressed under heterotrophic conditions, they are probably even less likely to express themselves under autotrophic conditions. Not all contaminants express themselves by over-running the culture and either killing the explant or rendering it unfit for further sub-culture. It is possible for them to survive as a non-observed, latent contaminant. These have been shown in some cases, in heterotrophic cultures, to affect the productivity of the culture or the subsequent performance of the progeny plants [19,20]. There is no reason why they could not have a similar affect on photoautotrophic cultures. Depending on the source of the starting material for Stage II culture, as is found in heterotrophic culture, the plant material can be found to be 'axenic' or to carry a large payload of micro-organisms.

8. INTRODUCED CONTAMINANTS FROM FAULTY LABORATORY PROCEDURE

The most frequently observed contamination in photoautotrophic cultures arises from faulty laboratory procedure. One feature of the transfer step from heterotrophic to photoautotrophic conditions is that the explant often carries over with it carbohydrates that can act as a substrate for the growth of micro-organisms. Consequently, contaminants introduced due to faulty laboratory procedure can colonise in a ring around the explant, either due to small amounts of medium transferred with the explant or possibly leakage of carbohydrate-rich solutes from the cut end of the explant inserted into the medium. These are rarely, if ever, observed when support matrices other than agar or Gelrite are used, presumably because the carbohydrates diffuse rapidly away and in the totality of the volume of medium are diluted to an extent which renders them of no consequence as a substrate for the growth of micro-organisms. In a gelled substrate, however, diffusion is slow and allows a high localised concentration sufficient to support the growth of micro-organisms.

A distinction should be drawn between the types of support matrix used and the level of cleanliness used in setting up the cultures. Generally, agar-based medium used in photoautotrophic culture is far less susceptible to exhibiting symptoms of contamination caused by faulty laboratory procedure than a sucrose-containing agar-solidified medium used for heterotrophic culture; but it is not totally exempt from showing symptoms. When support matrices other than agar are used the probability of observing contaminants on the surface of the support is virtually nil. This is not to say that the micro-organism is not present. If the mineral salt mix that makes up the nutrient solution is capable of supporting growth of the organism, such growth might be detected by turbidity in the liquid medium. In practice this has never occurred in my experience, presumably because such growth is either extremely slow or is just not supported by a medium without the presence of sucrose and various other organic ingredients.

The level of cleanliness used in initiating and handling Stage II cultures is obviously critical. On agar-based photoautotrophic medium, the handling procedures and containerisation are often the same as for heterotrophic cultures (except for the better quality and more leafy explants). Hence, the mind set of the laminar flow operator is exactly the same and they follow exactly the same (hopefully good) procedures. When operators do not see their normal contamination rates appear, it is easy to become blasé and for standards to slip. It is important that this does not happen because, whilst there may be no obvious contamination occurring, it is possible that latent contaminants, which may be a problem at some stage, will be building up in the plant material. When the medium is not agar based but some other support matrix is used, this slippage can happen even more quickly because the operator is very aware that this is a different system. This is even more the case when large, non-standard, containerisation is used.

With non-agar-based medium in large, non-standard, containers the largest problem that I have experienced has been with pathogenic fungi. The diseases experienced have been damping off (*Pythium*) and a root rot (*Phytophthora*); both appeared well into Stage II multiplication and were assumed to be introduced by faulty sterilisation of the container or by the laboratory operative. Fortunately, both were identified and dealt with early. One laboratory I know routinely includes a fungicide in its nutrient solution in an attempt to prevent this type of problem occurring; this is not to be recommended.

One feature of photoautotrophic micropropagation is the requirement for carbon dioxide enrichment. In conventional culture vessels this requires caps to be kept very loose, a membrane filter system, e.g. Suncaps, or the use of gas permeable films [21]. Obviously, any method that breaches the integrity of a sealed system carries risk although, depending on the source of the planting material, the substrate and the micro-organism(s) concerned, any potential contaminant may or may not be detectable and may or may not be deemed a risk.

When the ventilation system is forced, the potential for introducing contaminating micro-organisms obviously increases. This is even more the case if containers are linked in series and particularly so where there is the ability to circulate the nutrient solution as well. I know of no commercial system utilising such methodology but it has been suggested in the literature.

9. LESSONS FROM HYDROPONICS

Methods such as those suggested immediately above are reminiscent of hydroponic systems and would appear to carry the same type of risk. In the early days of development of NFT, the perceived disease risk was one of the largest impediments to its wider use. The literature on plant disease in hydroponics is limited but it is true to say that the incidence is generally less than was originally anticipated. It appears that host stress, pathogen ecology and the presence of non-pathogenic competing organisms around the roots all have a role to play in the susceptibility or non-susceptibility of the plant to pathogenic micro-organisms in hydroponically grown plants [22,23]. The same general principles apply in microponics where there is forced circulation of the gaseous environment and the nutrient solution. Using technology derived from bioreactor cultures, it should be possible to achieve the pumping and handling of solutions and gases in a sterile manner. Undetected contaminants in one container or on one plant could spread and cause large losses in such systems. This is one of the reasons why such systems are not in commercial use.

10. WHITHER AXENIC CULTURES?

This brings us to the point where we can ask the questions: do we really require axenic cultures or is it even possible to achieve axenic cultures? The answer to both questions is, almost definitely, no. Many researchers have questioned the ability of disinfestation procedures to remove contaminating micro-organisms from the explants used in micropropagation. Such organisms have been shown to exist intercellularly in virtually every plant tissue [24], particularly so in plants that have been vegetatively propagated by conventional means for decades. In such plants, is there any evidence that such an endophytic microflora is harmful to the plant? The evidence is very often to the contrary; many plants have beneficial associations with their microflora, particularly in their rhizosphere but undoubtedly in their phyllosphere too. As pointed out by Herman [25], there is increased interest in engineering specific micro-organisms associated with plants to modify and improve aspects of plant growth and crop production.

In plant tissue culture the effect of the presence of cultivable organisms has not been widely examined. Long *et al.* [19] illustrated that visually undetected bacterial contaminants can indeed be harmful to growth rates and subsequent performance *ex vitro*. Similarly, there is plenty of anecdotal evidence of 'contaminated' cultures that wean and grow perfectly acceptably [25], or in fact have enhanced growth, (see Nowak, this volume). It is now quite widely accepted that a truly axenic plant tissue culture is a quite rare or unique occurrence [25,26].

In the production of rooted cuttings for pot plants and hardy nursery stock using conventional glasshouse-based technology, whilst pathogen testing is, or at least should be, rigorous no one would be foolish enough to suggest that the cuttings should contain no cultivable organisms. The major reason that these organisms have become such an issue for micropropagation technologists is the fact that in heterotrophic micropropagation they can over-run the cultures and if this happens late in the multiplication cycle the losses can be considerable. Of course there are other reasons as well, such as the inadvertent dissemination of potential pathogens in supposedly 'axenic' germplasm that is distributed between laboratories.

Now that we have developed technology where the presence of an endophytic microflora is not an issue as far as productivity and quality of microplants is concerned it should become of less concern to the industry. However, we come back to the question raised by de Fossard [27] as to the whether or not all micro-organisms are potentially harmful to *in vitro* systems. We need to be assured that in photoautotrophic systems the naturally occurring microflora does not contain pathogens or vitropaths, organisms capable of damaging or killing the cultures under the *in vitro* conditions [28].

11. SCHEMATICS FOR QUALITY ASSURANCE

Any micropropagation programme utilising a photoautotrophic stage or stages should, as far as is possible, screen for potential pathogens and vitropaths. Unfortunately, information on vitropaths will only be built up in the light of experience. My experience has been that the only organisms of concern are those that are known pathogens of the plant being cultured. Even in situations where photoautotrophic methods have been used to rescue contaminated and even 'over-run' heterotrophic cultures, those micro-organisms have been of no concern provided the explant is relatively uncontaminated by debris from the heterotrophic medium and the material is placed in a support matrix other than agar with a 'simple' nutrient solution (see Sayegh and Long, this volume).

For most photoautotrophic micropropagation programmes a simple screening schedule is proposed (Fig. 1).

Irrespective of the route to be followed, screening of the mother plants for both pathogenic organisms and promiscuous endophytes should be employed. In the presence of pathogens, steps must be taken to eliminate them, which in practice means searching for 'clean' mother plants or, in the event that they are not available, meristem culture to eliminate pathogens. In the event that promiscuous endophytes are present it then becomes a judgement as to whether these will be detrimental to the cultures. In the event of meristem culture and heterotrophic multiplications, they may become a complicating factor by over-running the cultures. In the case of photoautotrophic culture, if the organism is very aggressive or if the plant tissue gives out large amounts of exudate then the decision as to whether to use agar or some other support substrate, such as Hortifoam, will have to be taken.

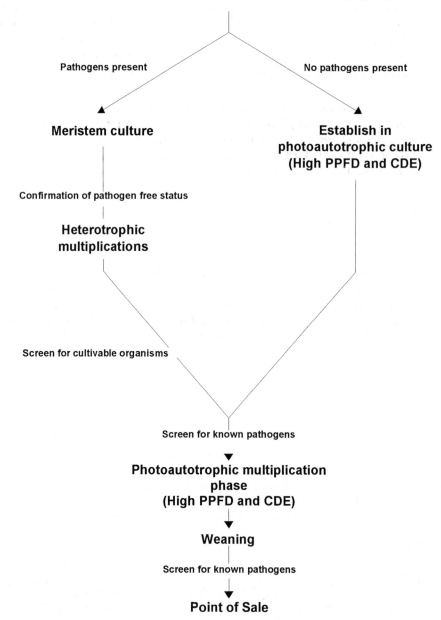

Figure 1. Quality control schematic for photoautotrophic micropropagation.

After meristem culture the normal confirmation of pathogen-free status will have to be carried out. Often there are a number of cycles of heterotrophic culture following meristem culture, to build up the plant numbers sufficiently for testing and passing on to the multiplication stage. If this is the case, it is probably wise to screen for cultivable organisms since they can pick up contaminants easily, even in the best managed laboratory. Whilst their presence may not be deleterious in photoautotrophic culture, it is well to know the microflora load being carried, if only to emphasise the need for keeping heterotrophic and photoautotrophic material well separated whilst being handled by the laboratory operatives.

Once material is transferred to photoautotrophic culture, the only required screening is for known pathogens. Since under photoautotrophic conditions, and particularly on substrates other than agar, plant growth more closely approximates to that seen under natural conditions, the probability of detecting pathogens by their normal symptom expression is high. Disease indexing using an appropriate technique and sampling rate can also be used.

12. CONCLUSIONS

Photoautotrophic micropropagation offers a number of possibilities including non-axenic culture of plants, reducing the problems experienced in contamination of cultures through faulty laboratory procedures and the rescue of valuable contaminated heterotrophic cultures. Although many micro-organisms that are vitropathic under heterotrophic conditions are no longer a problem in autotrophic conditions, it does not necessarily mean that they are not present. Transferring material back from autotrophic to heterotrophic conditions is notoriously difficult. Accordingly, photoautotrophic cultures should be handled separately to heterotrophic cultures in the laboratory to reduce the risk of cross-contamination.

Photoautotrophic micropropagation is a major step forward in reducing the risk of loss of cultures to both plant-borne and laboratory introduced micro-organisms. It has the other major advantage of the production of superior quality plants that are easier to wean, grow faster and mature earlier. The costs of converting to photoautotrophic conditions are not high, particularly when balanced against the losses caused by contamination in most laboratories. Its one major disadvantage remains the amount of work still to be done to achieve higher multiplication rates with many of the commercially desirable species.

REFERENCES

1. Roche, T.D. *et al.* (1996) Acta Hortic. 440, 515–520.
2. Constantine, D.R. (1986) in Plant Tissue Culture and its Agricultural Applications (Withersand L.A. and Alderson P.G., eds.) pp. 175–186, Butterworths, Guildford, UK.
3. Leifert, C. *et al.* (1991) World J. Microbiol. Biotechnol. 7, 452–469.

4. Kozai, T. *et al.* (1992) in Transplant Production Systems (Kurata K. and Kozai T., eds.) pp. 247–282, Kluwer Academic Publishers, Dordrecht, The Netherlands.

5. Brouchard, P. (1991) in Biotechnology in Agriculture and Forestry. Vol. 17. High Technology and Micropropagation (Bajaj Y.P.S., ed.) pp. 270–284.

6. Pospisilova, C. *et al.* (1988) Photosynthetica 22, 205–213.

7. Nakayama, M. *et al.* (1991) Proceedings of 11th Meeting of Plant Tissue Culture Society, Japan, pp. 186–187.

8. Fujiwara, K. *et al.* (1987) J. Agric. Meteorol. 43(1), 21–30.

9. Honjo, T. *et al.* (1988) J. Agric. Meteorol. 43(3), 223–227.

10. Kozai, T. *et al.* (1990) Plant Cell Tissue Org. Cult. 22, 205–211.

11. Reuther, G. (1991) Acta Hortic. 300, 59–75.

12. Kozai, T. *et al.* (1988) Acta Hortic. 230, 121–127

13. Cournac *et al.* (1991) Plant Physiol. 97, 112–117.

14. Desjardins, Y. *et al.* (1988) Acta Hortic. 230, 45–53.

15. Kozai, T. *et al* (1988) Acta Hortic. 230, 159–166.

16. Kirdmanee, C. *et al.* (1995a) In Vitro Cell Dev. Biol. Plant 31, 144–149.

17. Kirdmanee, C. *et al.* (1995b) Environ. Control Biol. 33(2), 123–132.

18. Figuera, A. *et al.* (1991) J. Am. Soc. Hortic. Sci. 116(3), 585–589.

19. Long, R.D. *et al.* (1988) Acta Hortic. 225, 83–92.

20. Leifert, C. *et al.* (1989) J. Appl. Bacteriol. 67, 353–370.

21. Cassells, A.C. and Walsh, C. (1994) Plant Cell Tissue Org. Culture 37, 171–178.

22. Holderness, M. and Pegg, G.F. (1986) in Water Fungi and Plants (Ayers P.G. and Boddy L., eds.) pp. 189–205, Cambridge University Press, Cambridge.

23. Evans, S.G. (1979) Plant Pathol. 28, 45–48.

24. Bastiens, L. (1983) Meded. Fac. Landbouwwet. Rijksuniv. Gent 48, 1–11.

25. Herman, E.B. (1990) Acta Hortic. 280, 233–238.

26. Cassells, A.C. (1991) in Micropropagation (Debergh P.C. and Zimmerman R.H., eds) pp. 31–44. Kluwer Academic Publishers, Dordrecht, The Netherlands.

27. de Fossard, R.A. and de Fossard, H. (1988) Acta Hortic. 225, 167–176.

28. Herman, E. (1987) Agricell Rep. 9, 33–35.

THE USE OF HORTIFOAM SUBSTRATE IN A MICROPROPAGATION SYSTEM. COMPARATIVE STUDY OF SUGAR-BASED AND SUGAR-FREE MEDIUM

A.J. SAYEGH[1] and R.D. LONG[2]

[1]*Plant Technology Ltd, IDA Industrial Estate, Dublin Road, Enniscorthy, Ireland*
[2]*Green Crop Ltd, Littlewood, Coolkenno, Tullow, Co. Carlow, Ireland*

1. INTRODUCTION

Micropropagation has established itself as an integral part of the horticultural industry. It responded to the needs of the nurseryman in rapid release of new products, cleaning of the mother stocks and as a breeding tool. As with all production processes, plant micropropagation needs to look carefully at the cost elements in order to be competitive against the conventional processes. One of the problems which is increasing the costs of microplants is that of culture contamination at different stages of the production process [1].

Kozai [2] put forward the proposition that a fundamental reason for high production costs in micropropagation is the presence of sugar in the propagation medium. The presence of sugar, in addition to being a cause of certain physiological disorders, makes the medium suitable for the growth of bacteria and fungi. This leads to the use of smaller vessels, and increased cost, to avoid larger losses in cases where contamination occurs.

Sources of contamination have been suggested as either carried with the original explant or introduced to the culture through faulty procedures [3]. The contaminating organism(s) may be either endogenous or exogenous on the explant [4]. Endogenously occurring bacteria are not necessarily detectable on the ordinary tissue culture medium, but they can show signs of their presence as in the case of black brackets in marantha [5]. The source of introduced contamination could be faulty operational procedures [3,6], movement of Arthropods such as thrips [7] and mites [8]. The other major source of introduced contamination is associated with the air temperature differential between day and night and the greenhouse effect within the culture jar. Culture jars are not sealed completely, which allows for the ingress of contaminating micro-organisms, the degree of which is dependent on the type of vessel (degree of sealing), the air temperature differential and the cleanliness of the air.

A.C. Cassells (ed.), Pathogen and Microbial Contamination Management in Micropropagation, 279–285.
© *1997 Kluwer Academic Publishers. Printed in the Netherlands.*

In our experience with a wide range of plant species, the two most commonly occurring problems are so-called latent endogenous contaminants that express themselves at some stage of the micropropagation process or introduced contamination. The conventional approach to eliminate contamination by discarding contaminated cultures [9], isolation of the contaminant and test for antibiotic sensitivity [10–12] either due to limited efficiency or higher costs, is impractical under most commercial conditions.

Kozai [2] suggested photoautotrophic micropropagation as one of the concepts upon which a future micropropagation system should be based, with the emphasis that fundamental techniques for producing micropropagated plantlets on a commercial scale will differ from those developed as research tools. In the photoautotrophic system the micropropagated plant is dependant on its own supply of photosynthates rather than using an exogenous carbon source in the form of sugar.

In our efforts to minimise contamination losses in our production we studied a number of different systems to integrate with what, until this, had been our conventional micropropagation system based on mixotrophic cultures on agar solidified medium. The work included searching for a suitable substrate that is adaptable to larger scale production units than the conventional jars and is also cost effective.

2. MATERIALS AND METHODS

2.1. Experiment 1

This was a comparison of Hortifoam as a support system with agar and liquid stationary culture, using conventional culture medium (containing sugar), for both material showing visual evidence of bacterial contamination and material visually free from bacterial contaminants. The medium used was based on Murashige and Skoog (1962) medium [13] except that the macroelements were used at half-strength, calcium nitrate at 1000 mg/l replaced ammonium nitrate and the medium was supplemented with 20 mg/l ascorbic acid and 80 mg/l citric acid.

Shoots derived from tissue cultures, that were either visibly free from contaminants or visibly contaminated, of *Calathea "Freddie"* and *Alpinia leuticarpa* were trimmed to 5–7 cm in length. These were placed in either stationary liquid culture, in Hortifoam substrate or in agar solidified medium, three explants per jar for the stationary liquid and agar solidified medium and six explants per jar with the Hortifoam substrate. Ten replicate jars were used in each treatment.

The containers used were the Magenta GA7 vessel (Sigma V8505) for Hortifoam and baby food jars (Sigma V9004) for liquid and agar solidified medium. The Hortifoam was supplied by Plant Biotechnology (UCC) Ltd. The containers and medium/substrate were sterilised by autoclaving at 121°C for 15 min. The incubation environment was a 14 h day length at a light intensity of 90 μmol/m^2/s, 24 ± 3°C and a growth room gaseous environment enriched with 1000 ppm CO_2. To enhance CO_2 gas exchange, Suncap membrane (Sigma C692) was used.

The medium used was the laboratory specific multiplication medium developed for that species or cultivar. The medium volume used was 50 ml in magenta vessels and 22 ml in the baby food jars.

2.2. Experiment 2

In this experiment, Hortifoam was used as the support matrix with a sugar-free medium. The medium used was the one developed in the laboratory for photoautotrophic growth of potato. The medium was supplemented with 2 mg/l IBA, 10 mg/l IBA or 10 mg/l BAP and 2 mg/l IBA. The medium was autoclaved after pouring at 105°C for 15 min. The plant material used was a double flowered *Primula* hybrid and *Maranta kerchoveana varigata* from either contaminant free, bacterially contaminated and fungally contaminated cultures. The cuttings of *Primula* were of single shoots with developed leaves. *Maranta* were of clumps of different sizes with at least one partially opened leaf.

2.3. Experiment 3

This consisted of the testing of Hortifoam as a "rescue" medium with a wide range of contaminated plants; these included *Calathea* (*C. ornate*, *C. royale*, *C. rosa picta*, *C. pictorata angenea*, *C. "Freddie"*, *C. sanguina* and *C. openhaninca*), *Lilium*, *Alpina varigata*, *Alpina leuticarpa*, *Opholoides*, *Begonia*, *Hosta*, *Streptocarpus*, *Penstimon*, *Yucca*, *Rhododendron* and *Cordyline terminulis*.

3. RESULTS

3.1. Experiment 1

In conventional agar solidified multiplication medium both the *Calathea* and *Alpina* normally develop a number of basal shoots. The clump developed at the base normally contains three shoots which require a further elongation/rooting cycle to produce a marketable product. In the stationary liquid culture the average multiplication rate increased slightly to four and the shoots developed faster than on agar solidified medium. In addition the *Alpina* developed a thick root that was absent in both Hortifoam and on agar solidified medium. The shoots placed on Hortifoam did not multiply at all on this medium but the shoot developed into an excellent saleable plant within the duration of the experiment (6 weeks).

With the visibly contaminated plants the bacterial growth was very slow on the standard agar solidified medium. This is the normal experience in our laboratory when it is usually possible to go through a number of sub-cultures before the material loses its ability to multiply. When contaminated material was placed in Hortifoam with a sugar based medium, there was rapid growth of the bacterial population which resulted in a decline of vigour and yellowing of the shoots. In stationary liquid culture the bacteria formed a layer at the bottom of the medium in the culture vessel but plant growth continued as in the clean cultures, performing better than the contaminated material on agar solidified medium. When the shoots were transferred to the elongation/rooting medium, those derived from the liquid multiplication medium performed in a superior manner to those from the agar solidified multiplication medium.

3.2. Experiment 2

In this experiment, bacterially contaminated *Primula* and *Maranta* showed continuous growth of the main shoot when placed on Hortifoam and medium containing 2 mg/l IBA, provided the stem length was long enough to have good contact with the medium and they had at least one leaf with a reasonable degree of lamina development. No bacterial growth was observed in the medium, i.e. there was no turbidity, and the plantlets all appeared healthy. In spite of continuous shoot and leaf development, visible root development was slow and did not appear before week eight and sometimes took as long as 12 weeks. (Contaminated plants normally take 6 weeks on agar based medium, non-contaminated plants root in approximately 3 weeks.) At 12 weeks the plant quality was excellent and the number of leaves and leaf area is far larger than the standard plants grown on agar based sugar containing medium.

Plants from cultures contaminated with fungi, normally *Penicillin* or *Trichoderma*, also showed a very favourable response on the Hortifoam. On normal sugar-containing medium both of these non-pathogenic fungi normally swamp the culture, strangling the plants in culture. In this experiment there was no further fungal growth except in cases where there was callus attached to the plant base. As the callus became necrotic, the fungus survived and grew on the exudates, in some cases affecting the developing plant. Generally the plants survived and grew, developing into excellent quality saleable material, as in the case of the bacterially affected cultures.

Growing the material in medium supplemented with 10 mg/l IBA or dipping the cuttings, prior to placing in Hortifoam, into 1000 mg/ml IBA did not have any effect on accelerating root development. Inclusion of BAP in the medium did not induce any multiplication and neither did a 1000 mg/l spray of BAP onto the leaves.

3.3. Experiment 3

The results from Experiment 2 suggested a possible role for Hortifoam cubes in rescuing contaminated material, rather than discarding it. Accordingly we tested many lines that are either in production or at the protocol development stage. The results showed that with one standard medium we can produce well-rooted, well-developed plants from cultures that would otherwise be lost from production in the laboratory. Some plant species showed that Hortifoam could be part of their production protocol as it produced better cuttings or better rooted plants than when grown on a comparable agar based medium. *Onaphaloides*, with which we had a problem initiating roots and elongating plants is one example. Another is *Begonia* which develops large leaves allowing more cuttings to be taken and the system used as a multiplication tool. In *Streptocarpus* the clumps usually only produce a small number of saleable cuttings but using this system larger numbers develop to a saleable size. *Hosta* is another to benefit, with the quality of the cutting being significantly improved.

Plants rescued and rooted on Hortifoam also weaned more easily than the conventional material. The larger leaf are was found not to be disadvantageous, the plants are photosynthetically active and have functional roots that continue to grow. Our experience is that this can significantly reduce or eliminate the need for a weaning stage, depending on the plant species.

4. DISCUSSION

It is not possible or viable to keep all cultures axenic when carrying out large-scale commercial micropropagation. Contaminants are not necessarily visible or easily detectable in routine tissue culture work [5]. These contaminants are not necessarily harmful to the plant culture and they will not necessarily render the culture unworkable in the short-term [14]. In many cases they require a change in the culture environment, such as a change from multiplication to rooting medium, in order to visibly express themselves [15].

The main objective of the production laboratory is to supply cuttings to the nurseryman free of specific pathogens at a competitive cost. Discarding all contaminated cultures not only increases production costs but, in some instances, it may make micropropagation totally non-viable for a particular species. The prospect of "rescuing" contaminated cultures or developing micropropagation systems that eliminate the problems of contamination is extremely attractive and should act to reduce production costs.

From the results presented here it is obvious that contaminated cultures can be "rescued". Removing contaminated material from the production cycle and rooting it in the absence of sugar in the medium worked well for all the species tested. Using Hortifoam substrate was generally slower than rooting in agar but the quality of the plant was much improved and the presence of functional roots in the Hortifoam cubes generally means faster weaning and establishment. In the case of potatoes, not only is establishment faster, maturity occurs earlier and yields are higher (R.D. Long, unpublished data). Whether such a system should become routine in a production laboratory is open to debate. It centres, as do all issues in contamination, on knowledge of the contaminating organism(s). Herman [16] used the term vitropath to describe organisms pathogenic to the tissue culture (*in vitro*) as opposed to organisms that are pathogenic to the plant *in vivo*. Our work demonstrates that the effect of a vitropath can be reduced or eliminated by removing the material to autotrophic conditions in the absence of a sugar in the medium.

As to whether photoautotrophic micropropagation is genuinely an alternative micropropagation system capable of eliminating contamination risks is difficult to say. Certainly by removing sucrose from the medium both endophytic and epiphytic contaminants do not survive and multiply to the same degree. Airborne contaminants entering the vessel either due to poor operator technique or air movement between the internal and external environment, are less likely to establish on the medium and overrun the cultures. However, the organism(s) are still present, and in the case of agar solidified medium can establish and survive, presumably due to the availability of organics and carbohydrate sources in the agar. Generally though, photoautotrophic cultures do not become contaminated to anything like the same degree. In our experiments, penicillin was only a problem when callus clumps became necrotic and the leakage of solutes provided a substrate that could be utilised for growth. Non-axenic systems on sugar free agar based medium for multiplication of *Begonia* have been reported by Bowes [17] where there were no problems with growth of micro-organisms on the medium. The major problem we experience in the utilisation of photoautotrophic methods is in achieving similar multiplication rates as we do in mixotrophic cultures. Species that propagate by nodal cuttings generally perform well but species dependant on other pathways of regeneration tend to be difficult and usually require a totally different medium formulation. However, we have had success with *Begonia* and *Streptocarpus* and many successes have been reported in the literature [2].

Until further work is carried out to improve multiplication rates in photoautotrophic cultures our results suggest that a combination of mixotrophic and photoautotrophic methods can be used, depending on the plant type. With many of the species we produce this means the use of sugar containing medium for the multiplication stage, possibly stationary liquid culture if the explant is large enough, followed by the use of sugar free medium to elongate the shoot and for functional root development of the cutting. In this way the quality of the plants produced is enhanced, weaning is much easier and the contamination risk is reduced in the latter part of the production cycle. If photoautotrophic conditions can be extended for use in the multiplication phase of a wider range of species this will also help in reducing the contamination risk in the laboratory and growth rooms.

REFERENCES

1. Cassells, A.C. (1991) in Micropropagation, Technology and Application (DeBergh P.C. and Zimmerman R.H., eds.) pp. 31–44, Kluwer Academic Publishers, Dordrecht, The Netherlands.

2. Kozai, T. (1991) in Micropropagation, Technology and Application. (DeBergh P.C. and Zimmerman R.H., eds.) pp. 447–469, Kluwer Academic Publishers, Dordrecht, The Netherlands.

3. Leifert, C., Waites, W.M. and Nicholas, J.R.(1989) J. Appl. Bacteriol. 67, 363–270.

4. Hennerty, M.J., Upton, M.E., Harris, D.P. and Eaton, R.A. (1988) Acta Hortic. 225, 129–137.

5. DeBergh, P.C. and Vanerschaeghe, A.M. (1988) Acta Hortic. 225, 77–81.

6. de Fossard, R.A. and de Fossard, H. (1988) Acta Hortic. 225, 167–176.

7. Klocke, J.A. and Myers, P (1984) HortScience 19, 480–481.

8. Blake, J. (1988) Acta Hortic. 225, 163–166.

9. Boxus, P. and Terzi, J.N. (1987) Acta Hortic. 212, 91–93.

10. Fisse, J., Battle, A. and Pera, J. (1987) Acta Hortic. 212, 91–93.

11. Bradbury, J.F. (1988) Acta Hortic. 225, 27–38.

12. Bastiens, L. (1983) Meded. Fac. Landbouwwet. Rijksuniv. Gent 48, 1–11.

13. Murashige, T. and Skoog, F. (1962) Physiol. Plant. 15, 473–497.

14. Long, R.D., Curtin, T.F. and Cassells, A.C. (1988) Acta Hortic. 225, 83–92.

15. Cassells, A.C., Harmey, M.A., Carney, B.F., McCarthy, E. and McHugh, A. (1988) Acta Hortic. 225, 153–162.

16. Herman, E.B. (1990) Acta Hortic. 280, 233–238.

17. Bowes, B.G. (1990) Prof. Hortic. 4,113–120.

PLANT TISSUE CULTURE LABORATORY ESTABLISHMENT UNDER CHALLENGING CONDITIONS

PAUL E. READ[1] and W.K. CHISHIMBA[2]

[1]*Dept. of Horticulture, University of Nebraska-Lincoln, NE, USA*
[2]*National Council for Scientific Research, Kitwe, Zambia*

1. INTRODUCTION

Micropropagation and related plant tissue culture technologies can provide significant scientific and economic benefits to developing and developed countries at a relatively low cost when compared to many other modern technologies. However, there exist many challenges to successful implementation of micropropagation and other tissue culture technologies, especially in rural areas of both developing and developed countries. Inadequate infrastructure may cause many commonly available tools, equipment and supplies to be unavailable, or available only after significant and highly detrimental delays. In some cases, the distance from the supplier combined with inadequate transportation facilities causes severe limitations on establishment and successful operation of the laboratory. Limited or lack of availability of specialized equipment and/or supplies unique to tissue culture can also be a problem.

Climate and weather characteristics may also pose serious challenges to the researcher or entrepreneur wishing to establish a tissue culture laboratory. Rapid and sudden changes in environmental factors, especially temperature, are frequently damaging to the tissues being cultured. In addition, extremely windy conditions, high humidity and high temperatures may exacerbate the problems of contaminants, both exogenous and endogenous.

Many of the aforementioned problems and challenges can be overcome by use of air conditioners, filters, laminar flow hoods, autoclaves and modern construction methods that incorporate sophisticated environmental controls. However, although such equipment and approaches may be readily available in many areas of the world, they may be unavailable or too expensive for other locations. Methods for creating homemade or alternative equipment and materials together with alternative approaches to achieving asepsis and establishment of clean cultures will be presented in the following examples.

A.C. Cassells (ed.), Pathogen and Microbial Contamination Management in Micropropagation, 287–291.
© *1997 Kluwer Academic Publishers. Printed in the Netherlands.*

Figure 1. The necessity of working in buildings in which it is difficult to eliminate contaminants (left) is a common obstacle to successful micropropagation in developing countries. Creation of dust-free clean laboratories such as this one (right) at the Tree Improvement Research Centre in Kitwe, Zambia, facilitates micropropagation and biotechnological research.

2. EXAMPLES

2.1. Tree Improvement Research Centre, Zambia

The physical structure or building in which the tissue culture laboratory is to be established often presents the initial challenge to the success of the venture. Figure 1 illustrates some aspects of this challenge and how it has been overcome. Ideally, one would construct a tissue culture facility from the ground up [1] but this was not possible at the Tree Improvement Research Centre (TIRC). Therefore, it was necessary to clean and retrofit rooms in existing building. Cracks in the walls and between ceiling and walls needed to be sealed — in some rooms, entire floors required replacement or to be newly fabricated — and cleaning and disinfesting followed by painting with a non-phytotoxic paint was required. Entrance of dust was thus prevented. Special care was also required to eliminate mites, insects and larger animals existing in the buildings and to prevent the future entrance of such fauna. Installation of simple room air conditioners where electricity was available was a major step in dealing with environmental variables, especially those associated with seasonal changes. The rainy season brings high humidity and rapid proliferation of soil-,

Figure 2. Establishment of stock blocks in the field can present many difficulties such as the termite mounds in citrus (top) and difficult to propagate native species such as *Ricinodendron* sp. (bottom).

air- and plant-borne organisms that increase potential contamination problems. The dry season leads to high temperatures and serious dust problems.

Contamination of cultures is generally attributed to one of two sources, the presence of organisms on or in the plant part to be cultured (the explant) or as a result of faulty procedures [2–4]. Establishment of stock plants of appropriate genotypes (Fig. 2) and successfully disinfesting the explants generally provides the opportunity to control exogenous contaminants. Culture of apical meristems [5,6] was initiated to aid in resolving problems of endogenous contamination.

2.2 Deejay Enterprises, Bangalore, India

As was noted in the TIRC example, appropriate modifications to an existing building were necessary. Achieving asepsis was again a significant challenge when consulting with Deejay, a major poultry enterprise wishing to diversify by using micropropagation to enter horticultural markets. Several laminar flow hoods had already been purchased, but obtaining pathogen-free explants was a serious problem. Fungal contamination was drastically reduced by introduction of improved hygiene measures and by stock plant selection. Some endogenous problems, essentially all bacterial or bacteria-like, were reduced by use of meristem or shoot apex culture combined with use of bacteriostatic compounds. However, use of bactericides or bacteriostatic compounds alone was ineffective and only useful as a supplement to other approaches, which agrees with other reports [2,3,6]. For several species, e.g. potato, it was necessary to import certified specific pathogen-free *in vitro* cultured nuclear stock plant materials. As was the case for laboratories established in other developing areas, use of locally available materials aided greatly in successful culturing of a number of species (Fig. 3).

2.3 Arlin Acres, Hemingford, Nebraska

Arlin Acres tissue culture laboratory was begun in a farm kitchen with limited facilities, but ultimately evolved into an enterprise capable of producing over a million mini-tubers of potato per year. Central to the success of this business was the owner's ability to use existing, low cost equipment combined with construction of homemade equipment. Home pressure canners were used on a stove top, in lieu of an expensive autoclave, for preparing medium and sterilizing tools. Small alcohol lamps used in conjunction with dipping tools in alcohol and flaming off the alcohol proved to be an adequate means of sterilizing forceps, scalpels and other small tools.

Home-constructed individual transfer hoods eliminated the need to purchase expensive laminar flow cabinets. Made of plexiglass pieces cut to form and glued together and depending only on a simple low cost fan and a small piece of HEPA filter, these hoods cost less than US $50 to construct. Numerous other simple and inexpensive devices have enabled this small enterprise to flourish in a highly competitive market, that of producing nuclear stock mini-tubers for farmers producing certified seed potatoes.

3. CONCLUSION

Although many difficulties confront the researcher or entrepreneur, it is possible to overcome most obstacles and to develop a successful tissue culture laboratory under challenging conditions. Prospects for tissue culture and biotechnology to make significant

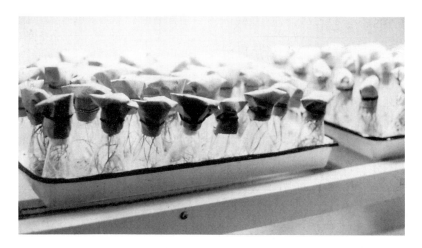

Figure 3. Use of locally available materials such as seen in this rural laboratory can help overcome problems posed by inaccessibility of materials commonly found in more developed locations.

contributions to the economics of rural, developing and developed areas of the world are promising [7,8] and consideration of some of the approaches discussed will help bring such contributions to fruition.

REFERENCES

1. Mageau, O.C. (1991) in Micropropagation: Technology and Application (Debergh P.C. and Zimmerman R.H., eds) pp. 15–30, Kluwer Academic Publishers, Dordrecht, The Netherlands.

2. Cassells, A.C. (1991) in Micropropagation: Technology and Application (Deberg P.C. and Zimmerman R.H., eds) pp. 31–44, Kluwer Academic Publishers, Dordrecht, The Netherlands.

3. Debergh, P. and Maene, L. (1984) Parasitica 40, 69–75.

4. Kyte, L. (1987) Plants From Test Tubes: An Introduction to Micropropagation. Timber Press, Portland, OR, USA.

5. Szendrak, E., Read, P.E. and Yang, G. (1996) this volume.

6. Bastiaens, L., Mains, L., Harbaoui, Y., Van Sumers, C., Vandecasteele, K.L. and Debergh, P.V. (1983) Med. Fac. Landbouww, Rijksuniv. Gent 48/1, 1–11.

7. Murashige, T. (1974) Annu. Rev. Plant Physiol. 25, 135–166.

8. Prakash, J. (1990) in Progress in Plant Cellular and Molecular Biology (Nijkamp H.J.J., Van Dier Plas L.H.W. and Van Aartrijk J., eds) pp. 789–794, Kluwer Academic Publishers, Dordrecht, The Netherlands.

IN VITRO CALLUS FORMATION ON WATERMELON COTYLEDON EXPLANTS IS INFLUENCED BY THE PRESENCE OF ENDOPHYTIC ORGANISMS

S. KINTZIOS, K. TSERPISTALI, A. KALOGIROS, J. DROSSOPOULOS and C. HOLEVAS

Department of Plant Physiology, Faculty of Agricultural Biology and Biotechnology, Agricultural University of Athens, Iera Odos 75, 11855 Athens, Greece

1. INTRODUCTION

Endophytic micro-organisms, apart from being a common source of plant tissue culture contamination, frequently interfere with the process of tissue morphogenesis *in vitro*. A negative effect is usually demonstrated indirectly, i.e. by observing the enhancement of callus induction and organ/plant regeneration after the addition of antibiotic agents to the culture medium. For example, application of cefotaxime (cephalosporin) at 500 mg/l could effectively induce somatic embryogenesis on carnation leaf explants [1], while penicillin G and carbenicillin also induced somatic embryogenesis but with lower frequencies. Callus growth and shoot regeneration in barley and apple tissue culture were stimulated by cefotaxime (at 60–100 and 250 mg/l, respectively) and the number of calluses with shoot primordia was increased up to 75% over the control [2,3]. Incorporation of different antibiotics such as carbenicillin, cefotaxime and streptomycin sulphate enhanced plant regeneration from somatic embryos of finger millet [4]. On the other hand, a direct effect of plant micro-organisms on tissue morphogenesis *in vitro* has also been reported in some cases, such as the growth suppression of *Acer macrophyllum* callus by co-cultured *Cryptodiaporthe hystrix* isolates [5] and the stimulation of maturation and plant regeneration from carrot somatic embryos by *Synechococcus* sp. and *Rhizobium* sp. exudates [6,7].

In the present study we investigated the possible effect of endophytic micro-organisms on explant necrosis, callus induction and somatic embryogenesis from watermelon cotyledon and stem explants cultured *in vitro*. Several genotypes were tested in respect to this explant–microorganism interaction under the application of various culture treatments.

A.C. Cassells (ed.), Pathogen and Microbial Contamination Management in Micropropagation, 293–298.
© *1997 Kluwer Academic Publishers. Printed in the Netherlands.*

2. MATERIALS AND METHODS

2.1. Preliminary experiments – microscopical observations

In a preliminary attempt to establish embryogenic callus cultures, cotyledon and stem explants received from aseptically grown seedlings of the watermelon *(Citrulus lanatus* L.) cultivar 'Tresor' were cultured on a Murashige and Skoog (MS) [8] basal medium supplemented with 3% sucrose, glucose or maltose, 0.8% agar and different plant growth regulators (PGRs) (2,4-dichlorophenoxyacetic acid (2,4-D), *p*-chlorophenoxyacetic acid (pCPA), α-napthylacetic acid (NAA) and/or kinetin (Kin)) at various concentrations (0.5–2.0 mg/l). A11 cultured explants developed extensive necrotic areas and turned brown 2–6 days after culture initiation. Since browning of cultured explants is frequently indicative of the presence of oxidized polyphenols, we made longitudinal and transverse sections of cotyledon explants cultured for 1–4 days in darkness on a solid medium supplemented with 3% glucose and 10 mmol NAA ('standard culture medium'), which were subsequently stained either in acidic phloroglucinol or in safranin. Abundant populations of endophytic bacteria were microscopically observed in sections stained in safranin, although sections stained in acidic phloroglucinol also showed a positive reaction. Therefore, we conducted a number of experiments in order to determine whether the presence of either oxidized polyphenols, endophytic micro-organisms, or both, was affecting normal explant dedifferentiation. Unless otherwise stated, a 'standard culture medium' (MS basal medium supplemented with 3% glucose and 10 μmol NAA) was used in all experiments.

2.2. Prevention/reduction of microbial contamination

- Explants were surface sterilized in 10% or 20% sodium hypochloride for 3, 5, or 7 min.
- Explants were cultured on 'standard culture medium' supplemented with various antibiotics (either penicillin, streptomycin or miconazole) at 3–20 mg/l.
- Explants were cultured in a liquid 'standard culture medium' with a lower pH value (pH 4.8 or 4.0, instead of 5.8).

2.3. Prevention/reduction of oxidized phenol accumulation

- Explants were cultured on a medium supplemented with 1% (w/v) activated charcoal (absorption of toxic phenolic derivatives).
- Explants were cultured on a medium supplemented with 10 % (w/v) ascorbic acid (reduction of the oxidization of phenolic compounds).
- Explants were cultured in a liquid 'standard culture medium' (increased diffusion of oxidized phenols).
- Explants were pre-soaked in 0.015 % (w/v) ascorbic acid for 3 h before inoculation.
- Explants were regularly transferred to fresh medium.
- Explants were initially cultured in a liquid medium for 1–3 days, then transferred to a solid medium with activated charcoal for another 1–3 days and finally cultured on solid 'standard culture medium'.

2.4. Other culture treatments

- Explants were received from seedlings of different ages (5, 15 and 20 days old).
- Explants were cultured on an MS basal medium supplemented with either 2,4-D, NAA, 4-CPA, BA and Kin, alone or in all possible auxin/cytokinin combinations, at various concentrations (0.5–2.0 mg/l).
- Different sugars (sucrose, glucose, maltose) were supplemented as the carbon source to the medium, at 1–3% concentration.
- Explants were cultured either under a high photosynthetic proton flux density (PPFD) or in the dark.
- Explants were cultured on a medium supplemented with 25 µM silver nitrate (an ethylene inhibitor).
- Apical, middle and basal hypocotyl sections were used as explants instead of cotyledons.

2.5. Genotypic effects

In order to investigate the effect of the genotype on the explant response to culture, cotyledon explants of eight watermelon cultivars ('Bruno Silva', 'Crimson Sweet', 'Royal Sweet', 'Royal Charleston', 'Super Galaxy', 'Super Galaxy II', 'Imperial' and 'Rodeo') were inoculated on 'standard culture medium' supplemented or not with 20 mg/l miconazole.

3. RESULTS

3.1. Prevention/reduction of microbial contamination

Surface-sterilization of the explants in 10 or 20% sodium hypochloride for 3–7 min had no effect on explant browning, and 90–100% of the explants declined within 3 days in culture. However, normal development and callus induction were observed on co-cultured squash explants, so that the observed results could not have been attributed to inadequate surface sterilization of the watermelon explants. Explant decline could not be prevented by culturing them in a liquid medium with a low pH value; on the contrary, when the initial pH value of the liquid culture medium was 5.8, this was gradually reduced to 3.6 during the first 4 days after culture initiation and parallel to explant decline. Explant necrosis was not prevented either by the addition of 3–20 mg/l streptomycin or penicillin to the culture medium: over 90% of the cultured explants developed necrotic symptoms after 4 days in culture. The addition, however, of miconazole to the culture medium significantly delayed explant browning. This effect was more apparent at higher antibiotic concentrations, and callus induction was observed on 85% of the explants cultured in 20 mg/l miconazole after 7 days in culture.

3.2. Prevention/reduction of oxidized phenol accumulation

None of the treatments employed to reduce the accumulation of oxidized phenols in explants could effectively prevent explant necrosis, although explant decline was delayed. Initial signs of tissue dedifferentiation were observed (both macroscopically and microscopically) on most explants during the first 4–6 days after culture initiation. After this period, however, an extensive explant chlorotic discoloration was observed. Since the necrotic symptoms were not associated with explant browning, we assume that explant decline was partially due to the accumulation of oxidized phenols, even though this did not seem to be the main cause of explant death.

3.3. Other culture treatments

Modification of the culture conditions did not remarkably improve explant survival and cultured tissues turned brown and died after 7 days in culture. However, very significant differences ($P < 0.0001$) were observed among different treatments within the first 3 days after culture initiation. The decline of the explants was delayed when cotyledon explants were received from older seedlings (at least 18–20 days old) and cultured in darkness on a solid medium supplemented with 3% glucose and 10 µmol α-napthylacetic acid ('standard culture medium'). Under these conditions, 80% of the cultured explants demonstrated initial stages of dedifferentiation during the first 3 days in culture.

3.4. Genotypic effects

Each watermelon cultivar responded differently to tissue culture, indicating a direct genotype × treatment interaction. None of the tested cultivars could effectively produce a callus without the addition of miconazole to the culture medium. Even when the medium was supplemented with this antibiotic, cotyledon explants of the cultivars 'Royal Sweet' and 'Bruno Silva' declined and died within 3 weeks from culture initiation. Only explants from 'Super Galaxy II', 'Imperial' and 'Rodeo' did not decline after 12 weeks in culture. Several immature somatic embryos (at the globular stage) were formed on calli of the cultivars 'Rodeo' and 'Imperial' (at a frequency of 100% and 75% of the explants, respectively). An extended rhizogenesis was observed on every callus received from explants of 'Super Galaxy II'.

4. DISCUSSION

The results of the present study indicate that callus induction from watermelon cotyledon explants might have been negatively affected by endophytic micro-organisms (probably bacteria), which have been observed microscopically in sections of declined explants. This hypothesis is further supported by the fact that only addition of miconazole (a broad-spectrum antibiotic) to the culture medium, among many different treatments, could effectively promote callus induction and morphogenesis. To our knowledge, this is the first time that somatic embryos were induced from mature cotyledons derived from watermelon seedlings. So far, somatic embryogenesis in watermelon has been achieved only from cotyledon explants excised from immature zygotic embryos [9]. On the other hand, the possibility cannot be excluded that explant decline is also caused, at least partially, by the accumulation of oxidized phenols, since the various treatments applied in order to prevent phenol oxidization also caused a remarkable reduction of explant browning (although explant death was not prevented).

Considering the observed results, we propose the following model in order to describe the *in vitro* interaction between the endophytic micro-organisms and the cultured explants.

Upon explant removal from the donor plant and culture initiation, the bacteria (possibly stimulated by an ingredient of the culture medium, e.g. a growth regulator) begin to grow at the expense of the host tissue and the nutrient medium. Plant cells respond to the infection by overproducing polyphenolic compounds (some of which may act as phytoalexins, but can be toxic to the host cells at excessive concentrations). Accumulation of these compounds may lead to a drastic reduction of the pH value of the culture medium, which in turn can accelerate cell damage. The final explant decline may result both from the direct parasitic action of the bacteria on the host cells and from the damaging effect of the oxidized phenols. The observed explant–microorganism interaction could be genotype dependent, as demonstrated by the results of experiments with several watermelon cultivars. Previous experiments with cefotaxime on barley callus cultures [2] also indicated an interaction between the antibiotic and both the genotype and the growth regulator concentration. It is also possible that explants from older seedlings are less affected by phenol accumulation because of their larger size, which allows for a delayed diffusion of the toxic compounds produced as a result of the tissue lysis by the bacteria. This model is schematically presented below:

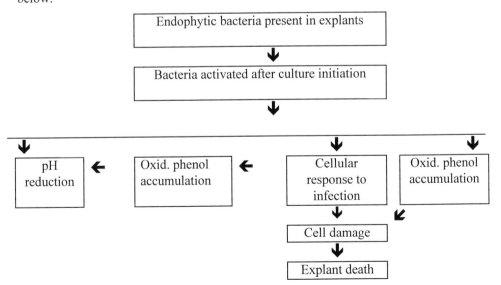

REFERENCES

1. Nakano, M. and Mii, M. (1993) J. Plant Physiol. 141, 721–725.

2. Mathias, R.J. and Mukasa, C. (1987) Plant Cell Rep. 6/6, 454–457.

3. Yepes, L.M. and Aldwinckle, H.S. (1994) Plant Cell Tissue Org. Cult. 37, 257–269.

4. Eapen, S. and George, I. (1990) Plant Cell Tissue Org. Cult. 22, 87–93.

5. Sieber, T.N., Sieber-Canavesi, F. and Dorworth, C.E. (1990) Mycologia 82, 569–575.

6. Wake, H., Akasaka, A., Umetsu, H., Ozeki, Y., Shimomura, K. and Matsunaga, T. (1992) Plant Cell Rep. 11, 62–65.

7. De Jong, A.J., Heidstra, R., Spaink, H.P., Hartog, M.V., Meijer, E.A., Hendriks, T., Lo Schiavo, F., Terzi, M., Bisseling, T., Van Kammen, A. and De Vries, S.C. (1993) Plant Cell 5, 615–620.

8. Murashige, T. and Skoog, F. (1962) Physiol. Plant. 15, 472–497.

9. Compton, M.E. and Gray, D.J. (1993) Plant Cell Rep. 12, 61–65.

EPIPHYTIC BACTERIA: ACTIVITIES, RISKS AND BENEFITS

H.A.S. EPTON

*School of Biological Sciences, University of Manchester, Manchester,
M13 9PL, UK*

1. INTRODUCTION

Epiphytic bacteria have many features which are of special relevance to plant tissue culture. This review will focus on (i) the assessment and location of epiphytic bacterial populations, (ii) aspects of adhesion and features of the plant's environment that influence bacterial adhesion, (iii) features that influence population levels and fluctuations in populations, and (iv) the development and distribution of antibiotic-resistant strains of epiphytic bacteria, as their occurrence creates problems with their elimination.

Certain features of epiphytic bacteria may be of use, or may represent a hazard, during the culture of plants. For example, in order to minimise the risk of the establishment of plant pathogenic epiphytic populations, particular consideration should be given to the establishment of beneficial epiphytes during weaning of plants into soil. The frost tolerance of plants may also be greatly influenced by the composition of the established epiphytic population, quite apart from variations between cultivars determined by physiological parameters. Similar considerations apply to the tolerance by some plants of low environmental humidity.

Concerning the host plant, variations in the response of microbial resistance mechanisms of specific cultivars to the establishment of epiphytic bacterial populations may also influence the survival of plants under natural conditions, due to interactions between the resistance mechanisms and pathogenic or ice-nucleating-active bacteria.

2. ESTIMATION OF POPULATIONS OF EPIPHYTIC BACTERIA

2.1. Population studies

Many population studies of bacterial epiphytes have been approached with an objective of observing changes in a species or a group of isolates which have a particular physiological,

A.C. Cassells (ed.), Pathogen and Microbial Contamination Management in Micropropagation, 299–308.
© *1997 Kluwer Academic Publishers. Printed in the Netherlands.*

biochemical or pathological effect. Fewer studies are aimed towards giving an overall picture of bacterial population composition and fluctuations, on a particular plant species. The complexity of such studies, if undertaken over a long time-scale, and the difficulties in the identification of bacteria from 'environmental' rather than 'medical' sources probably explain their rarity. One such study, by Ercolani [1] of mesophilic bacteria of olive leaves, covered a 6-year period. Being an evergreen plant, on which new leaves unfold throughout the year, there was the additional factor of leaf age to be considered. From leaf washings at 3-month intervals, a collection of 1701 representative strains was selected on the basis of colony characteristics. Using cluster analysis of the results from 210 phenotypic tests, about 75% of the strains were identified to generic or lower levels (Table 1).

Table 1. Estimation of occurrence of phylloplane bacteria on olive leaves over a 6-year period

Genus/species	%
Pseudomonas syringae	51
Xanthomonas campestris	6.7
Erwinia herbicola	6
Acetobacter aceti	4.7
Gluconobacter oxydans	4.3
Pseudomonas fluorescens	3.9
Bacillus megaterium	3.8
Leuconostoc mesenteroides subsp. *dextranicum*	3.1
Lactobacillus plantarum	2.8
Curtobacterium plantarum	2.2
Micrococcus luteus	2.2
Arthrobacter globiformis	1.4
Klebsiella planticola	1.2
Streptococcus faecium	1.2
Clavibacter sp.	0.98
Micrococcus sp.	0.82
Serratia marcescens	0.81
Bacillus subtilis	0.57
Cellulomonas flavigena	0.4
Erwinia sp.	0.37
Zymomonas mobilis	0.3
Bacillus sp.	0.29
Alcaligenes faecalis	0.27
Erwinia carotovora	0.08
Pseudomonas aeruginosa	0.04

Adapted from Ercolani, G.L. (1991) Microb. Ecol. 21, 35–48.

This extensive study illustrates that decisions necessitated by practical considerations of the scale of the experimental effort must also restrict the data obtainable. For example, the use of a single, high-nutrient medium, in this instance yeast extract/tryptone/glucose agar, will have resulted in a selection of those organisms able to grow on such a high nutrient source. Some organisms well adapted to the low nutrient levels existing in the epiphytic environment may not have been able to survive on such a medium. Isolation of organisms at a single temperature (Ercolani used 26°C) selects against bacteria favouring a psychrophilic range, while a short incubation period selects against slow-growing bacteria.

The limitations of taxonomic methods also create difficulties, perhaps leaving a large residue of unidentifiable strains, or grouping together in a single species, such as *Pseudomonas syringae*, a vast number of pathovars or non-pathogenic strains, which have the capacity for a multiplicity of interactions with their hosts.

2.2. Non-culturable epiphytes

A further dimension to the difficulties of studying epiphytic populations has been clarified by Wilson and Lindow [2] who have demonstrated that viable population assessments of epiphytes by conventional viable colony counting methods are unreliable when measuring populations of older bacterial cells, resulting in 2- to 4-fold under-estimation of populations. Viable, but non-culturable bacteria were assessed by a direct viable count method, based on a technique used to study bacteria in aquatic systems, involving the measurement of elongation of bacterial cells during incubation in a nutrient solution. These non-culturable bacteria, usually in populations that were older than 80 h, appeared to have lost culturability due to low nutrient availability on the plant surface, similar to a starvation-survival state that has been observed with *Pseudomonas* spp. in aquatic systems [3,4]. Such conditions are also likely to be encountered on plant surfaces.

2.3. Strategies for sampling and counting

Most population studies of bacterial epiphytes have been undertaken because of the particular effects of an organism, either as a plant pathogen, as a potential biocontrol agent or due to some other interaction with the environment of the host plant such as ice-nucleation activity. Recent work by Kinkel *et al.* [5] has focused on sampling scales for assessing viable populations, comparing the objectives of sampling of leaf discs, entire leaflets or whole plants, depending on the desire to observe localised sites on leaflets, variability between leaflet environments, or whole-plant differences in bacterial habitats. The study provides a very useful comparison of the different sampling strategies and of the considerations that need to be applied to the statistical treatment of such data.

The variations that can be observed in epiphytic populations depend on the frequency of sampling the population and the statistical treatment of the data obtained. Thus, at one extreme Hirano *et al.* found that epiphytic population size of *P. syringae* on bean leaves varied with the time of day [6], while Ercolani's studies [1] required that samples were taken only every 3 months. Awareness of short-term variations is important if statistically dependable data are to be obtained in long-term studies.

When assessing total bacterial cell populations, a recent study in Manchester [7], observing leaves and wax surfaces, showed the difficulties of direct electron microscopic observations of surfaces. Because of physical difficulties, relating to magnification scales, sample size or surface features, total counts below approximately 4×10^4/cm^2 are very difficult to measure.

3. LOCATION OF EPIPHYTES

Observations of the location of epiphytic bacteria clearly show considerable variation in distribution on different species of plant. Work by Brown [7] on the location of bacteria (*Listeria monocytogenes*) which had been applied to leaf surfaces showed substantial differences between the niches occupied by bacterial cells on the surface of cabbage and lettuce leaves, with bacteria being particularly associated with wax particles on cabbage and with depressions at the margins of epidermal cells on lettuce. Even on a single leaf, bacteria may occupy a variety of niches, as demonstrated by Mariano and McCarter [8] when examining the survival of *Pseudomonas viridiflava* on tomato plants. Bacteria were found to survive as microcolonies in depressions between epidermal cells, around trichomes, along veins, and around stomata. Knowledge of such features, or species variations, may provide guidance towards the most appropriate removal/sterilisation techniques to employ.

4. ADHESION OF EPIPHYTIC BACTERIA

While many studies have been made of the adhesion of bacteria to surfaces, few of these relate to epiphytic species or to plant surfaces. Such studies take on particular relevance when the removal of bacteria from plants or sterilisation of plant surfaces needs to be achieved. Awareness of these mechanisms may indicate how detachment techniques can be improved, or contact between the bacteria and sterilising agents be enhanced.

4.1. Physico–chemical mechanisms

Work in Manchester by Brown [7] has shown that with the, luckily occasional, epiphyte *Listeria monocytogenes*, adhesion of bacterial cells to hydrophobic surfaces is greatly influenced by the mineral salt content of the suspending fluid. Similar relationships were found to apply to the very waxy surfaces of white cabbage leaves. It has also been found [9] that heavy metals (Cd, Cu, Hg) increased adhesion of both Gram-negative and Gram-positive bacteria, to solid surfaces when cells were suspended in aqueous media. It is possible that the adhesion of rain-splashed soil-borne bacteria to epiphytic surfaces may be enhanced by the mineral salt content of the suspending soil water, resulting in differences in epiphytic populations on plants obtained from locations with different soil conditions. Adhesion may also be affected by leaching onto the leaf surface of minerals from within plants grown in soils with differing mineral salt content.

4.2. Bacterial surface structures

In an investigation of adhesion of the common epiphyte *Pseudomonas fluorescens* [10], adhesion-deficient mutants were obtained which were found to lack flagella and flagellum-related proteins. Although in this instance studies involved a soil isolate and adhesion to sand grains, it is possible that the same mechanism may relate to the epiphytic attachment of this species.

A further form of adhesive mechanism, more usually seen in biofilms on fluid-bathed surfaces such as the teeth, involves the production by bacteria of gum-like extracellular materials. Among epiphytic organisms, *Xanthomonas campestris* is known to produce gums which provide both adhesion to the plant surface and protection from desiccation under the low humidity conditions which frequently occur in this environment [11].

5. FEATURES THAT INFLUENCE EPIPHYTIC BACTERIAL POPULATIONS

The composition of epiphytic bacterial populations is influenced by a variety of features related to the environment of the host plant, the microenvironment of the location on the plant, interactions between component micro-organisms of the population or due to characteristics of the host plant.

5.1. Effect of plant nutrients

In a study of the epiphytic population of *Xanthomonas campestris* pv. *vesicatoria* (causal organism of Bacterial Spot of Tomato) as influenced by the application of nitrogen and potassium fertilisers [12], it was found that, early in the season, increasing the rate of potassium application reduced the epiphytic population of *X c. vesicatoria*, with sharper reductions when nitrogen levels were also high. At fixed levels of potassium application, increasing the level of nitrogen fertilisation also reduced epiphytic populations of the pathogen. Regarding defoliation of plants resulting from the disease, this was reduced by increases in nitrogen, but was unaffected by the potassium levels applied. This evidence of the influence of plant nutrition on epiphytic populations of a bacterium, and on disease levels in tomato, indicates that plant nutrition can interact with the level of harmful, or of beneficial bacterial epiphytes, which may be of particular importance when weaning plants into non-sterile environments.

5.2. Antimicrobial compounds

The significance of siderophores in competition between epiphytic micro-organisms was demonstrated by McCracken and Swinburne [13], and subsequent studies have reinforced their role in the inhibition of micro-organisms on and around plants [14,15]. There are reports of the effect of bacterial siderophores in the control of soil-borne bacterial pathogens, and of deleterious soil bacteria. Epiphytic bacteria have been shown to control fungal pathogens by siderophore production, but evidence is still sought for the inhibition of bacteria on plant surfaces.

A recent study of *Pseudomonas fluorescens* and *P. chlororaphis* [16] suggested that certain siderophores appear to act as stimulants of the growth of some fungal plant pathogens, indicating that an assumption of inhibition is unsafe if based solely on chemical evidence of siderophore production by a bacterium, additionally requiring that biological evidence of inhibition is presented.

Blakeman [17] maintains that while it is easy to demonstrate antibiotic production by epiphytic bacteria in agar plate cultures, because of the low nutrient levels on leaves, it is unlikely that there would be the resources available for sufficient production to influence interactions between micro-organisms on the phylloplane. However, Leifert *et al.* [18] working with epiphytic bacteria on low nutrient media showed that a strain of *Serratia plymuthica* increased production of antimicrobial compounds in response to reduction in the nutrient levels of media. Additionally, epiphytic sites such as floral parts or fruit surfaces may have higher nutrient levels than the phylloplane, and recent work in Manchester provides substantial evidence for the involvement of antibiotic production in the antagonism of *Erwinia herbicola* towards *E. amylovora* in pear fruit and hawthorn flowers (M.H. El Masry, T.A. Brown, H.A.S. Epton and D.C. Sigee, unpublished). It is also of significance that where antibiotic production by micro-organisms is possible, even under restricted conditions, scope exists for the development of antibiotic resistance in associated microbial populations, which may have practical considerations for the culture of plant tissues.

5.3. Micro-environmental factors in population fluctuation

In an evaluation of the biological and physical variables which interact with epiphytic bacterial populations, O'Brien and Lindow [19] examined population changes of 19 bacterial strains on seven plant species under various light intensities and humidities. Comparing *Pseudomonas syringae* with non-epiphytes, strains of *P. syringae* were found to be superior in establishing and maintaining epiphytic populations, the difference being most marked under dry, high light intensity conditions, when the *P. syringae* strains achieved populations at least 25-fold higher than the most prolific non-pseudomonad, although no such distinction was evident under low light wet conditions. Thus, survival under low humidity and high light intensity, rather than ability to grow on leaves, were concluded to be major determinants of the epiphytic habit among bacteria. Tolerance of humidity fluctuation will be of particular importance where wetting and drying of plants occurs on a daily basis, as a result of dew, mist or rain.

5.4. Host preferences of epiphytic bacteria

The influence of plant species on epiphytic populations of *P. syringae* has been found to be substantial, with as much as 17-fold differences from one plant species to another [19]. It was found that generally plants with waxy cuticles, such as corn, oats and pea tended to have lower populations than those with rough, trichome-bearing leaves such as bean, tomato and cucumber. In this study it was found that pathogenic and non-pathogenic strains of *P. syringae* on a particular plant species did not differ in their ability to establish populations, or to persist. However, previous reports of host preference among strains of pathogenic epiphytic bacteria are at variance with this finding [20–22] and O'Brien and Lindow [19] concluded that the conditions during their study may not have been equivalent to those of previous studies.

Georgakopoulos and Sands [23] found significant differences between the *P. syringae* epiphytic populations of barley lines and cultivars observed over entire growing seasons. Their observations over 2 consecutive years indicated that plant genotype had a significant role in the determination of epiphytic population size. However, statistically significant differences between the rates of population increase on various lines were not observed consistently over the 2 years of the experiment, although the ranking of barley lines according to their average epiphytic populations were very similar in both years.

Interaction between the host plant, strain pathogenicity and environmental conditions in the establishment of epiphytic populations has been investigated by O'Brien *et al.* [24] by comparing the fate of the root pathogen, *Rhizomonas suberifaciens*, the successful epiphyte *Pseudomonas syringae*, and a bacterium not usually associated with plants *Escherichia coli*, after inoculation onto leaves of lettuce (host of *R. suberifaciens*) and barley (non-host). Plants were kept at high humidity, close to 100%, and at low light intensity, for 72 h, followed by 72 h at low humidity (50%) and high light intensity. Survival of the epiphyte *P. syringae* was only slightly reduced under the low humidity/high light regime, but the more notable finding was that the root pathogen, while surviving under high humidity, rapidly declined under low humidity, high lighting conditions, but declined more markedly on the non-host barley, than on the leaves of the host plant, lettuce. Other workers have also shown that some bacteria survive better on some plants than others, for example, pathovars of *P. syringae* attacking cherry and pear were only epiphytic on their respective hosts [20].

Some of the above features may be relevant to the bacterisation of cultured plants. In particular, expectations of the formation of an epiphytic population of a particular bacterial species or strain must take into account cultivar-specific interactions with the epiphyte.

6. INTERACTIONS OF EPIPHYTIC BACTERIA RELATED TO DAMAGE TO PLANTS

The interactions between epiphytes and their hosts have been studied most frequently when the interaction is associated with disease. In a study of cultivar resistance to bacterial spot of tomato caused by *Xanthomonas campestris* pv. *vesicatoria* [25] it appeared that one component of the various forms of resistance was the reduction of epiphytic populations of the pathogen, one cultivar having the potential to reduce pathogen populations below the threshold at which infection was likely to occur. This not only reduced the likelihood of infection of individual plants but also influenced the reservoir of inoculum within the crop, reducing the rate of disease spread through the crop.

The role of plant pathogenic toxin production by epiphytic bacteria was considered by O'Brien and Lindow [19], on the assumption that nutrient limitation on plant surfaces may be mitigated by the action of plant toxins in causing cellular leakage of nutrients. However, no such role was evident.

The role that epiphytic bacteria can play in frost damage was first reported in *Pseudomonas syringae* by Maki *et al.* in 1974 [26]. The significance of ice-nucleating-active (INA+) bacteria is now well established [27]. Ice-nucleating strains have been shown to occur in only six species of bacteria: *Erwinia herbicola, E. ananas, Pseudomonas syringae, P. viridiflava, P. fluorescens* and *Xanthomonas campestris*. It appears that in some cultivars preferential growth of such bacteria may determine frost tolerance or sensitivity and, at least in the case of *Xanthomonas campestris*, the freezing damage engendered has been related to infection processes. Marshall demonstrated that INA+ bacteria reached higher populations on a frost-sensitive cultivar of oats than on cultivars that were less prone to frost damage [28]. It was found that bactericide applications reduced both INA+ bacteria and damage caused by freezing. Oat cultivars were found to vary in their response to INA+ bacteria, in some instances resisting the development of such epiphytic populations and consequently showing greater frost resistance. While the creation of aseptic plants in culture may result in the removal of such potentially harmful epiphytes, some cultivars may favour re-establishment of INA+ epiphytic populations on return to a normal environment. However, for cultivars not showing epiphytic preferences it may be possible, on weaning, to establish INA− populations, shown by Lindow [27] to be capable of providing biological control of frost damage by preventing ice nucleation under field conditions.

A similar approach may be taken by the use of epiphytic biocontrol agents of bacterial and fungal plant pathogens, providing cultivars with an established antagonistic epiphytic bacterial flora. Many bacterial epiphytes have been demonstrated to have good biocontrol activity [29–31] and could provide both weaning and established plants with some degree of protection from microbial infection.

7. RESISTANCE OF EPIPHYTIC BACTERIA TO ANTI-MICROBIAL COMPOUNDS

Luckily, in Europe, the use of antibiotics for the control of plant pathogens is normally prohibited, and the dangers inherent in their use are well illustrated by observation of their effects in the USA. Burr *et al.* [32] examined the distribution of streptomycin resistance among epiphytic bacteria in apple and pear orchards in the USA where it is used, *inter alia*, to control bacterial Fire Blight of apples and pears and Blister Spot of apples. While in some areas they found no streptomycin-resistant strains among populations of *Erwinia amylovora*, other epiphytic bacteria were found to be streptomycin resistant, including *E. herbicola* (syn. *Pantoea agglomerans*), *Pseudomonas syringae* and non-pathogenic *Pseudomonas* spp. Antibiotic resistance was shown to be plasmid borne in these species, and there was considerable homology between the plasmids, indicating the likelihood of transmission of this resistance between species by a conjugative plasmid. A similar study of epiphytes in apple orchards in Michigan [33], examining 152 streptomycin-resistant strains, concluded that it appeared likely that resistance initially arose in the non-pathogenic epiphytic bacterial population before developing in plant pathogenic bacteria.

Perhaps of greater relevance to bacterial disease control in Europe is evidence in the USA of the development of resistance to copper bactericides among epiphytic bacterial populations [34,35].

REFERENCES

1. Ercolani, G.K. (1991) Microb. Ecol. 21, 35–48.

2. Wilson, M. and Lindow, S.E. (1992) Appl. Environ. Microbiol. 58, 3908–3913.

3. Byrd, J.J., Xu, H.-H. and Colwell, R.R. (1991) Appl. Environ. Microbiol. 57, 875–878.

4. van Overbeek, L.S., van Elsas, J.D., Trevors, J.T. and Starodub, M.H. (1990) Microb. Ecol. 10, 239–249.

5. Kinkel, L.L., Wilson, M. and Lindow, S.E. (1995) Microb. Ecol. 29, 283–297.

6. Hirano, S.S., Rouse, D.I. and Upper, C.D. (1984) in Proceedings of the Second Working Group of *Pseudomonas syringae* pathovars (Panagopoulos C.G., Psallidas P.G. and Alivizatos A.S., eds) pp. 24–27, Hellenic Phytopathological Society, Athens, Greece.

7. Brown, M. (1996) Ph.D. Thesis, University of Manchester.

8. Mariano, R.L.R. and McCarter, S.M. (1993) Microb. Ecol. 26, 47–58.

9. Tahn, T.T. and Wang, Y.X. (1995) Seventh International Symposium on Microbial Ecology, Santos-Sao Paulo, Brazil. Abstracts, p. 111.

10. DeFlaun, M.F., Tanzer, A.S., McAteer A.L., Marshall, B., and Levy, S.B. (1990) Appl. Environ. Microbiol. 56, 112–119.

308

11. Pruvost, O. and Gardan, L. (1988) Agronomie 8, 925–932.

12. McGuire, R.G., Jones, J.B., Stanley, C.D. and Csizinszky, A.A. (1991) Phytopathology 81, 656–660.

13. McCracken, A.R. and Swinburne, T.R. (1979) Physiol. Plant Pathol. 15, 331–340.

14. Neilands, J.B. and Leong, S.A. (1986) Annu. Rev. Plant Physiol. 37, 187–208.

15. Bakker, P., Van Peer, R. and Schippers, B. (1991) in Biotic Interactions and Soil-Borne Diseases (Beemster A, Bollen G.J., Gerlach M., Ruissen M.A., Schippers B. and Tempel A., eds) Elsevier, Amsterdam, The Netherlands.

16. Laine, M.H., Karwoski, M.T., Raaska, L.B. and Mattila-Sandholm, T.M. (1996) Lett. Appl. Microbiol. 22, 214–218.

17. Blakeman, J.P. (1991) J. Appl. Bacteriol. 70 (Suppl.), 49S–59S.

18. Leifert, C., Sigee, D.C., Stanley, R., Knight, C. and Epton, H.A.S. (1993) Plant Pathol. 42, 270–279.

19. O'Brien, R.D. and Lindow, S.E. (1989) Phytopathology 79, 619–627.

20. Ercolani, G.L. (1969) Phytopathol. Mediterr. 8, 197–206.

21. Mew, T.W and Kennedy, B.W. (1982) Phytopathology 72, 103–105.

22. Stadt, S.J. and Saettler, A.W. (1981) Phytopathology 71, 1307–1310.

23. Georgakopoulos, D.G. and Sands, D.C. (1992) Can. J. Microbiol. 38, 111–114.

24. O'Brien, R.D., Jochinsen, K.N. and van Bruggen, A.H.S. (1991) Plant Dis. 75, 954–957.

25. McGuire, R.G., Jones, J.B. and Scott, J.W. (1991) Plant Dis. 75, 606–609.

26. Maki, L.R., Galyan, E.L., Chang-Chien, M. and Caldwell, D.R. (1974) Appl. Microbiol. 28, 456–459.

27. Lindow, S.E. (1986) in Microbiology of the Phyllosphere (Fokkema N.J. and van den Heuvel J., eds) pp. 293–311, Cambridge University Press, Cambridge, UK.

28. Marshall, D. (1988) Phytopathology 78, 952–957.

29. Wilson, M. and Lindow, S.E. (1993) Phytopathology, 83, 117–123.

30. Wilson, M., Epton, H.A.S. and Sigee, D.C. (1990) Plant Pathol. 39, 301–308.

31. Sakthivel, N. and Mew, T.W. (1991) Can. J. Microbiol. 37, 764–768.

32. Burr, T.J., Norelli, J.L., Reid, C.L., Capron, L.K., Nelson, L.S., Aldwincle, H.S. and Wilcox, W.F. (1993) Plant Dis. 77, 63–66.

33. Sobiczewski, P., Chiou, C.-S. and Jones, A.L. (1991) Plant Dis. 75, 1110–1113.

34. Andersen, G.L., Menkissoglou, O. and Lindow, S.E. (1991) Phytopathology 81, 648–656.

35. Jones, J.B., Woltz, S.S., Jones, J.P. and Portier, K.L. (1991) Phytopathology 81, 714–719.

FUNGAL INFECTIONS OF MICROPROPAGATED PLANTS AT WEANING: A PROBLEM EXEMPLIFIED BY DOWNY MILDEWS IN *RUBUS* AND *ROSA*

B. WILLIAMSON[1], D.E.L. COOKE[1], J.M. DUNCAN[1], C. LEIFERT[2], W.A. BREESE[3] and R.C. SHATTOCK[3]

[1]*Scottish Crop Research Institute, Invergowrie, Dundee DD2 SDA, UK*
[2]*Department of Plant and Soil Science, Cruickshank Building, University of Aberdeen, St. Machar Drive, Aberdeen AB9 3UU, UK*
[3]*School of Biological Sciences, University of Wales, Bangor, Gwynedd LL57 2UW, UK*

1. INTRODUCTION

Tissue culture has become an important technique for rapid clonal propagation of plants to provide stocks of high-health status, especially in woody perennials and fruit crops in which freedom from virus diseases and *Phytophthora* root rots is of paramount importance before establishment in the field, because these diseases cannot be eradicated once introduced with planting stocks [1]. Further dependency of the industries of agriculture, horticulture and forestry on tissue culture methods is to be expected as novel genes are isolated, characterised and made available for testing in different genetic backgrounds by *Agrobacterium*-mediated transformation techniques [2]. By these methods, as genes become available for disease resistance, salt and drought tolerance, high sugar levels or extended shelf-life in fruits, or any other traits of commercial value, there will be a demand for gene transfer technology which ultimately depends for its success on tissue culture. Rapid multiplication *in vitro* is necessary for evaluation of gene constructs in a wide range of transformants, rapid propagation for glasshouse and field evaluation under strict controls and eventually for sale to recoup the costs incurred in this long process. Conventional plant breeding may then benefit as genes from diverse species can be introduced to the gene pool in a crossing programme and followed in progenies by rapid molecular methods.

At the weaning stage, however, plants are vulnerable to environmental stresses and highly susceptible to some pathogens which may enter the propagation system through air currents, in watering regimes or with insects or micro-arthropods. The protection of this vulnerable germplasm from fungal pathogens at the weaning stage is, therefore, important but, initially, it is essential to prove that *in vitro* mother stocks are entirely free of pathogens. Many fungicides used on established crop plants are phytotoxic on micropropagated plants and use of certain chemical groups is ill-advised because fungicide-resistant strains of the pathogen may arise and be disseminated widely with the planting stock. For these reasons

A.C. Cassells (ed.), Pathogen and Microbial Contamination Management in Micropropagation, 309–320.
© *1997 Kluwer Academic Publishers. Printed in the Netherlands.*

rapid, sensitive and specific diagnostic methods are needed to confirm freedom of stocks from key pathogens before release of stocks.

2. FUNGAL CONTAMINANTS OF TISSUE CULTURES

Microbial contamination is difficult to eradicate from *in vitro* stocks. Recent reviews dealt with the microbial ecology of tissue-cultured plants and suggested procedures which reduce the risk of contamination [3,4]. This paper concerns the fungal saprophytes and pathogens encountered primarily during acclimatisation and weaning stages. The full extent of the problem of fungal contamination of tissue cultures is difficult to determine in commercial production because of reticence in making declarations of this nature. However, studies which compared the air spora in tissue culture laboratories with the external air at different times of the year suggest that inadequate filtration of incoming air is a major source of contamination, whilst the systemic symptomless infection of tissues by fungi is less common [5]. Species in some of the fungal genera found in this sampling exercise, such as *Alternaria*, *Botrytis*, *Phoma* and *Cladosporium*, included those which are plant pathogens. *Botrytis cinerea* is especially known to cause significant losses during acclimatisation of tissue-cultured plants [5]. A recent questionnaire sent out to 19 tissue culture laboratories in the UK and other European countries showed that 16 of them record losses due to *B. cinerea*, and six recorded *Pythium* and *Rhizoctonia* as problems (Table 1).

Table 1. Confidential survey of 19 propagators in 1996

Number of propagators reporting losses due to pathogens (out of 19)			
Botrytis	16	*Pythium*	6
Rhizoctonia	6	Downy mildews	3
Phytophthora	3	Powdery mildews	1
Fusarium	1	Brown mould	1
Penicillium	1		

Design of the layout of micropropagation facilities can substantially reduce the risk of microbial contamination. For example, refrigeration units often have areas, frequently concealed, where condensation forms on painted or wooden surfaces and fungi such as *Alternaria* will sporulate freely allowing spores to enter the air currents of the laboratory. By careful design of the laboratories it is usually possible to place the growth rooms and refrigeration equipment on the perimeter of the facility with external venting of air from this area.

2.1. Role of relative humidity in fungal pathogenesis

The prominence of *B. cinerea* in the micropropagation industry raises some interesting questions about plant defence responses and tissue maturity. The pathogen is generally recognised as one which attacks mature-to-senescent leaves or fruits to produce grey mould, a disease often of most severity on stored fruits, vegetables or cut flowers [6]. Even when inoculated as mycelium into deep wounds made on primocanes of red raspberry (*Rubus idaeus*), the pathogen spreads only in the primary cortex when the deep-seated cork layers (polyderm) begin to suberise [7]. An exception to this general pattern is the infection by *B. cinerea* of newly opened flowers of raspberry, strawberry, blackberry and blackcurrant when the conidia germinate in the stigmatic fluid and establish symptomless stylar infections [8–11]. Therefore, the reasons why *B. cinerea* can infect juvenile plant tissues *in vitro*, and at the weaning stage, are unclear.

One factor which may be significant in this respect is that, during acclimatisation, the plants must be held at high humidity to prevent severe desiccation [12]. The juvenile leaves have much thinner cuticles and reduced thickness of epicuticular wax, compared to more mature leaves grown in full light and in most cases the plantlets *in vitro* do not have functional stomata [12–15]. Conidia of *B. cinerea* have been shown to be capable of germination and infection of glasshouse-grown rose petals at relative humidities at, or above, 94% [16]. In this humidity range, airborne conidia entering culture vessels and settling on sensitive plantlets may find the thin cuticle an easy barrier to overcome, since cutinases are known to be produced early during infection [17]. This pathogen and many other recorded fungal contaminants will grow satisfactorily on the culture media developed for tissue culture of plants and overgrow plantlets rapidly. Dry conidia of *B. cinerea* can survive for several months and retain the ability to germinate and infect plant surfaces [18]. Much of the earlier work on *B. cinerea* placed emphasis on the behaviour of conidia in water droplets on plant surfaces. It was found that in water the conidia lose endogenous nutrients and germination inhibitors by leaching [19,20]; these nutrients can stimulate bacteria in the phylloplane to multiply rapidly [21]. Plant epidermal cells also release nutrients and antimicrobial substances into water droplets placed on the surface and the complexity of these interactions on plant surfaces have been studied extensively. However, to our knowledge, there have not been comparable studies with micropropagated plantlets. The thin cuticle produced as a result of culture at very high relative humidity fails to control passive water loss to the atmosphere [15] and the simultaneous loss of nutrients on to the surface may be greatly enhanced compared with mature leaves. This is an area that perhaps deserves further examination.

However, a number of publications in the last 10 years have shown that the behaviour of *B. cinerea* on plant surfaces when inoculated dry differs markedly from when conidia are sprayed in water or nutrient solutions. Dry inoculated conidia produce one to five germ tubes; the germ tubes are usually short, lack appressoria and effect penetration of the epidermis soon after germination [16,22,23] whereas conidia applied in water produce single long germ tubes and, particularly in nutrient solutions or when inoculated as mycelium, will form appressoria or dome-shaped infection cushions [24–26]. Other fungi probably have such versatility, but it is only by careful studies of their behaviour on plantlets in experiments with rigorous control of the environment that data may be derived that could form a basis for control of the contamination problem. Conidia may arrive on leaflets and remain ungerminated for long periods until the relative humidity exceeds a critical level, condensation forms on the surface, or overhead watering stimulates germination and penetration of the plant surface. *Botrytis* infections of cut flowers cause heavy losses during transport and cool storage because blossoms carrying surface conidia are sold during the 'incubation period' of the pathogen, the time from germination to appearance of symptoms [27], and this may be the problem with many micropropagated plants.

2.2. Problems of measurement and control of relative humidity

It is difficult to measure and regulate relative humidity levels precisely and experiments devised to study this environmental parameter in relation to microbial growth on living plant surfaces rarely take full account of the physics involved in water vapour dynamics occurring within thin boundary layers and in the microhabitat concerning fungal conidia on a leaf surface [28–31]. Before installing expensive air conditioning and environmental control equipment in commercial micropropagation chambers it would be helpful to acquire precision sensors and to have a knowledge of the constraints in a system in which air is being circulated over agar media containing plantlets or moist compost and transpiring rooted plants. The most important factor in the regulation of relative humidity is temperature control to within 0.1°C and this is unlikely to be achieved in commercial growth chambers. Capacitance sensors, now widely used in humidity controlled areas, are generally inadequate and inaccurate at, or above, 90% relative humidity — the most critical humidity range for many fungi [30]. In large cabinets, an aspirated wet–dry bulb hygrometer is still probably the sensor of choice, despite the large air volumes being sampled. For absolute measurements of humidity in a chamber from which aspirated air is sampled, a dew-point hygrometer would be essential, but this equipment is expensive (R. Lowe, personal communication).

3. DOWNY MILDEW IN MICROPROPAGATED *RUBUS*

Downy mildew of *Rubus* cane fruits first became a problem in the propagation industry in the UK in the mid-1980s [32] when *Peronospora rubi* was described on 'Tummelberry', a blackberry × red raspberry hybrid, and other similar hybrids and blackberries. Although some of these outbreaks affected conventionally propagated stocks derived from leaf cuttings, it became clear that micropropagated stocks of this germplasm were especially vulnerable to downy mildew shortly after rooted plantlets were received from weaning [33]. *Peronospora rubi* had been recorded rarely before this time in the UK [34,35] and a source of inoculum for these outbreaks was difficult to identify. However, in New Zealand a disease of Boysenberry (blackberry × red raspberry) known as 'dryberry' was attributed to *Peronospora sparsa*, the rose downy mildew [36].

Experiments in which downy mildew isolates taken from *Rosa* cultivar 'Can Can' were placed on the abaxial surface of leaf disks cut from young fully expanded leaves of 'Tummelberry' and, conversely, isolates from 'Tummelberry' inoculated on to *Rosa*, showed that downy mildews from both hosts can cross-infect, at least under these test conditions [37]. The fungi described in the literature as *P. rubi* and *P. sparsa* are morphologically identical, and with the results of cross-inoculations the view has been expressed that they should be regarded as conspecific, with this fungus referred to as *Peronospora sparsa* Berkeley (*sensu stricto*), which is the earliest name available [38]. Other downy mildews reported on plants in the Rosaceae, such as the pathogen found on *Prunus laurocerasus*, have been named *P. sparsa* [38]. With fungal biotrophs normally regarded as highly host-specific, these taxonomic issues have practical importance. For example, in some micropropagation companies, it is not unusual to have several species from one plant family in production within the facility and this can pose high risks for perennial plants. Miniature pot roses are prone to downy mildew in glasshouses (T. Brokenshire, personal communication) and other members of the Rosaceae may also be in production in the same glasshouse complex. In these circumstances, without due care, it is likely that *Peronospora sparsa* could enter the propagation area with devastating consequences.

Although few outbreaks of downy mildew in micropropagated *Rubus* plants have been described in the literature, we know of several occurrences involving blackberries and related hybrid berries in N. America, South Africa and the UK. In all the cases we have examined so far, no instance involved the systemic infection of the *in vitro* mother stocks. This suggests that the stocks are becoming infected by air-borne conidia within the acclimatisation period and the infection is not recognised until after the plants are released for growing-on.

3.1. Observation of fungal spores and hyphae by fluorescence microscopy

A simple and convenient method for the examination of downy mildew on leaf surfaces and inside plant tissues, originally developed for the study of symptomless infection of flowers by *B. cinerea* [11], has proved useful in detailed studies of *Rubus* germplasm [16]. To observe conidiophores, conidia and germ tubes on the underside of fresh specimens, small pieces of leaf were immersed in 0.1% aniline blue (B.D.H. C.I. 42755) prepared in 0.1 N $K_3PO_4.H_2O$ and examined immediately by fluorescence microscopy with an Olympus BH-2 microscope fitted with a UV light source using the U + V Excitation Dichroic Mirror Assembly (peak transmission for excitation, 405 nm) with an L435 supplementary barrier filter (excludes transmission above 435 nm), or other equivalent lens and filter combinations. This rapid method allows surveys of leaf surfaces, or the surfaces of filters, for fungal spores and hyphae to be made. To examine the internal tissues it is essential to fix the specimen in Carnoy solution (ethanol/chloroform/glacial acetic acid; 6:3:1, v/v/v), soften and clear in 1 N NaOH at 60°C for 1 h, blot to remove NaOH and mount in aniline blue for fluorescence microscopy. Gentle tapping of the coverslip will produce a squashed preparation which gives optimal resolution of the fungal hyphae after storing 24 h at 4°C.

The infection of *in vitro*-grown leaves by *P. rubi* (= *P. sparsa*) was studied by use of plants produced on a modified Murashige and Skoog medium containing activated charcoal [39] and included blackberry cultivar 'Loch Ness', the blackberry × red raspberry hybrids 'Tummelberry', 'Tayberry', 'Sunberry' and the red raspberry genotypes 'SCRI 8042E6' and 'Autumn Bliss'. Whole detached leaflets were inoculated on the abaxial surface with a suspension of 10–20,000 conidia/ml and incubated at 15°C under a 16 h photoperiod at 20 mmol/m^2/s and *c*. 100% relative humidity.

Sporulation of the pathogen in the detached leaflets occurred 5–7 days after inoculation. Infection occurred by direct penetration of abaxial epidermal cell walls and hyphae caused a callose plug to form in the epidermal wall around the infecting hypha, visible by fluorescence microscopy as a bright ring. In similar studies with leaf disks cut from more mature leaves produced in glasshouses and polytunnels and inoculated by dipping in conidial suspensions, penetration occurred through both the adaxial and abaxial surface, though occasionally penetration occurred via the stomata.

Intercellular hyphae grew deep within the mesophyll (Fig. 1) in the interveinal regions of the lamina inserting simple dichotomously branched haustoria into each cell, again causing callose rings to develop at the point of entry (Fig. 2).

Intercellular hyphae also spread systemically from the laminae to petioles within the cortex, where the frequency of haustoria was lower than in mesophyll tissues (Fig. 3). In *in vitro*-grown leaflets, the pathogen formed oogonia with paragynous antheridia freely in the mesophyll (Fig. 4) between 11 and 21 days after inoculation and produced thick-walled oospores (Figs. 5 and 6) in leaves of blackberry cultivars 'Loch Ness', 'Tummelberry' and 'Tayberry' and in the red raspberry genotypes 'SCRI 8042E6' and 'Autumn Bliss', but not in 'Sunberry' [16,33,37].

3.2. Longevity of oospores and persistence of the downy mildew problem

The thick-walled oospores of fungi in the Peronosporaceae are long-term survival structures which once formed are difficult to eradicate. It should be viewed with concern that plant debris could harbour oospores and persist in the vicinity of a micropropagation facility. Sexual reproduction in the fungus also increases the risk of fungicide-resistant strains arising if certain chemical groups, such as the acylalanines, are used repeatedly at the propagation stage. Already there are cases in which routine treatment of strawberries with this class of chemical at the propagation stage has suppressed symptoms of red core disease

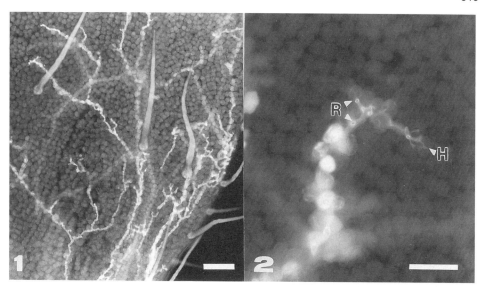

Figure 1. Intercellular hyphae of *Peronospora rubi* spreading through the mesophyll of 'Tummelberry' leaf grown *in vitro*. Specimen cleared, stained in aniline blue and viewed by fluorescence microscopy [16] (Bar = 100 mm).

Figure 2. Hyphae of *P. rubi* in palisade mesophyll of 'Tummelberry' leaf 96 h after inoculation. Dichotomously branched haustoria (H) are visible in faintly stained host cells and a strongly stained ring (R) at the point of penetration of the mesophyll cell indicates callose deposition. Host cells penetrated by haustoria more distant from the hyphal apex show strong fluorescence (Bar = 50 mm).

(*Phytophthora fragariae* var. *fragariae*) and led to fungicide resistance in the pathogen and spread of the new strains with planting stocks. With few chemicals available to the horticultural industry, and the number likely to decrease in the future, the use of chemicals in this way should be avoided and alternative control strategies are required. These might include precision control of humidity and avoidance of condensation, reduction of humidity after strengthening shoots and roots by inclusion of the growth retardant paclobutrazol (which also improves stomatal physiology, increases deposition of epicuticular wax) [40], or biological control [41] .

4. TOWARDS SPECIFIC DETECTION OF DOWNY MILDEWS IN ROSACEAE

Abundant interspecific DNA sequence variation in the internal transcribed spacer (ITS) regions of the ribosomal RNA gene has resulted in these regions being used widely in rapid species identification and molecular detection of fungi [42,43]. They have been used for the characterisation of Oomycete fungi [44] but there are few reports on their use in the downy mildews. The reason for this may be because the primers ITS5 and ITS4 [45] are

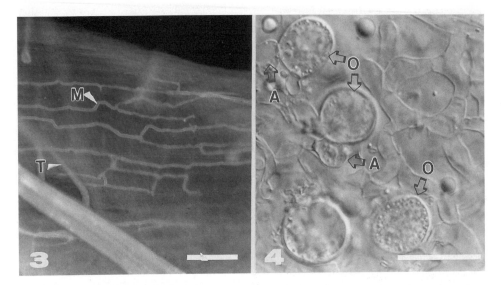

Figure 3. Intercellular mycelium (M) of *P. rubi* spreading systemically in a petiole beneath trichome hairs (T) 120 h after inoculation of a 'Tayberry' leaf lamina *in vitro*. Prepared as in Fig. I above (Bar = 100 mm).

Figure 4. Oogonia (O) and antheridia (A) of *P. rubi* present in the spongy mesophyll of leaf of red raspberry selection 'SCRI 8042E6' within 11 days of inoculation *in vitro*. Specimen fixed, cleared and viewed by differential interference contrast (DIC) microscopy [16] (Bar = 50 mm).

based in a highly conserved region of the large and small subunits and amplify the ITS regions of all fungi and plants present in a given sample. After PCR, the reaction mix will therefore contain many different amplification products from the mixture of DNA present in the initial sample thus confounding the identification of the downy mildew fragment.

In this study, DNA was extracted from infected leaf material and a semi-nested PCR approach used to overcome the problem of a mixed DNA sample. By comparing 18S DNA sequences of *Phytophthora megasperma* [46] with those of other Oomycetes, true fungi and plants, a PCR primer was selected which was subsequently shown to be specific to members of the order Peronosporales. In the first round PCR reaction, this forward primer (DC6) was combined with the universal reverse primer ITS4 and a 1.2 kb fragment was amplified. A 1 ml aliquot of the first reaction was then re-amplified using a modified universal primer ITS6 (similar to ITS1) and the universal reverse primer ITS4. The resultant 900-bp PCR products were similar in size to those amplified from *Phytophthora* [44]. After purification using a Wizard kit (Promega), the fragments were sequenced on an automated sequencer (Applied Biosystems, Model 373) using the Perkin Elmer sequencing kit.

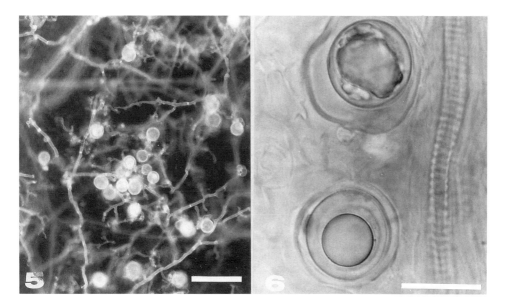

Figure 5. *P. rubi* oospores formed abundantly in mesophyll of red raspberry 'SCRI 8042E6' 18 days after inoculation. Stained with aniline blue and viewed by fluorescence microscopy (Bar = 100 mm).

Figure 6. *P. rubi* oospores in 'Tummelberry' leaf 27 days after inoculation. Prepared and viewed as in Fig. 4 (Bar = 20 mm).

Sequencing proved the PCR products were pure. Using this method, fresh leaf material from the named hosts (in parentheses) bearing the following pathogens was extracted and tested: *Peronospora rubi* (blackberry, two isolates), *Peronospora sparsa* (rose, two isolates), *P. viciae* (pea), *P. parasitica* (oilseed rape), *P. cristata* (poppy), *Albugo candida* (shepherd's purse), *Phytophthora infestans* (potato). Subsequent alignment of the ITS regions from isolates of *P. sparsa* and *P. rubi* showed complete identity, adding weight to the argument that they are conspecific. By comparing these *Rubus* downy mildew ITS sequences with the other downy mildews, regions have been chosen for the design of PCR primers specific to *P. sparsa* (= *P. rubi*) which can be used in a sensitive and specific detection protocol. Such methods have already been successfully applied to *P. fragariae* in strawberry and raspberry roots [47] and *Colletotricum acutatum* in strawberry [48].

In order to minimise the threat of pathogen attack on micropropagation planting systems, protocols for the efficient use of the sensitive and selective PCR approaches will need to be developed. Routine sampling of planting material, irrigation water and air supplies may be necessary. It is feasible, for example, to carry out PCR detection on water filters or spore traps sited near air vents. These approaches are amenable to the detection of many other fungal (*Pythium* or *Botrytis* spp.) and bacterial pathogens and such rapid identification of pathogen or contamination problems would allow early treatment by modification of the environment, chemicals or biological control agents.

ACKNOWLEDGEMENTS

We thank N. Williams for DNA extraction and PCR work on downy mildews, R. Lowe for advice on measurement of relative humidity and design of equipment, and commercial companies willing to participate in the survey of fungal contamination. We also thank The Scottish Office of Agriculture, Environment and Fisheries Department for funding for BW, DELC and JMD, and The Ministry of Agriculture, Fisheries and Food (Grant No. CSA 1341) for funding the work on *Rubus* downy mildew. We are grateful to all those who have supplied downy mildew isolates.

REFERENCES

1. Williamson, B. (1993) in Proceedings of an EC Workshop on Perspectives for the European Soft Fruit Sector, Dundee, 16–17 November 1992 (Williams G.H., Szmidt R.A.K., Dixon G.R. and McNicol R.J., eds) pp. 76–86.

2. Tinland, B. (1996) Trends Plant Sci. 1, 178–184.

3. Leifert, C., Morris, C.E. and Waites, W.M. (1994) Crit. Rev. Plant Sci. 13, 139–183.

4. Leifert, C. and Waites, W.M. (1994) in Physiology, Growth and Development of Plants in Culture (Lumsden P.J., Nicholas J.R. and Davis W.J., eds) pp. 363–378, Kluwer Academic Publishers, Dordrecht, The Netherlands.

5. Danby, S., Berger, F., Howitt, D.J., Wilson, A.R., Dawson, S. and Leifert, C. (1994) in Physiology, Growth and Development of Plants in Culture (Lumsden P.J., Nicholas J.R. and Davis W.J., eds) pp. 397–403, Kluwer Academic Publishers, Dordrecht, The Netherlands.

6. Coley-Smith, J.R., Verhoeff, K. and Jarvis, W.R. (1980) The Biology of Botrytis. London, Academic Press.

7. Jennings, D.L. and Williamson, B. (1982) Ann. Appl. Biol. 100, 375–381.

8. McNicol, R.J., Williamson, B. and Dolan, A. (1985) Ann. Appl. Biol. 106, 49–53.

9. Bristow, P.R., McNicol, R.J. and Williamson, B. (1996) Ann. Appl. Biol. 109, 545–554.

10. Williamson, B., McNicol, R.J. and Dolan, A. (1987) Ann. Appl. Biol. 111, 285–294.

11. McNicol, R.J. and Williamson, B. (1989) Ann. Appl. Biol. 114, 243–254.

12. Kerstiens, G. (1994) in Physiology, Growth and Development of Plants in Culture (Lumsden P.J., Nicholas J.R. and Davies W.J., eds) pp. 132–142, Kluwer Academic Publishers, Dordrecht, The Netherlands.

13. Sutter, E. and Langhans, R.W. (1979) J. Am. Soc. Hortic. Sci. 104, 493–496.

14. Sutter, E. and Langhans, R.W. (1982) Can. J. Bot. 60, 2896–2902.

l5. Johansson, M., Kronestedt-Robards, E.C. and Robards, A.W. (1992) Protoplasma 166, 165–176.

16. Williamson, B., Breese, W.A. and Shattock, R.C. (1995) Mycol. Res. 99, 1311–1316.

17. Salinas, J. (1992) Function of cutinolytic enzymes in the infection of gerbera flowers by *Botrytis cinerea*. PhD Thesis, Wageningen University, The Netherlands.

18. Salinas, J., Glandorf, D.C.M., Picavet, F.D. and Verhoeff, K. (1989) Neth. J. Plant Pathol. 95, 51–64.

l9. Blakeman, J.P. (1975) Trans. Br. Mycol. Soc. 65, 239–247.

20. Brodie, I.D.S. and Blakeman, J.P. (1977) Trans. Br. Mycol. Soc. 68, 445–447.

21. Brodie, I.D.S. and Blakeman, J.P. (1976) Physiol. Plant. Pathol. 9, 227–239.

22. Cole, L., Dewey, F.M. and Hawes, C.R. (1996) Mycol. Res. 100, 277–286.

23. Salinas, J. and Verhoeff, K. (1995) Eur. J. Plant. Pathol. 101, 377–386.

24. Garcia-Arenal, F. and Sagasta, E.M. (1980) Phytopathol. Z. 99, 37–42.

25. Van der Heuvel, J. and Waterreus, L.P. (1983) Plant Pathol. 32, 263–272.

26. Fourie, J.F. and Holz, G. (1994) Plant Pathol. 43, 309–315.

27. Williamson, B. (1994) in Ecology of Plant Pathogens (Blakeman J.P. and Williamson B., eds) pp. 187–207, CAB International, Oxford.

28. Hartmann, H., Sutton, J.C. and Thurtell, G.W. (1982) Phytopathology 72, 914–916.

29. Harrison, J.G. and Lowe, R. (1989) Plant Pathol. 38, 585–591.

30. Harrison, J.G., Lowe, R. and Williams, N.A. (1994) in Ecology of Plant Pathogens (Blakeman J.P. and Williamson, B., eds) pp. 79–97, CAB International, Oxford.

31. Butler, D.R., Reddy, R.K. and Wadia, K.D.R. (1995) Plant Pathol. 44, 1–9.

32. McKeown, B. (1988) Plant Pathol. 37, 281–284.

33. Wallis, W.A., Shattock, R.C. and Williamson, B. (1989) Acta Hortic. 262, 227–230.

34. Francis, S.M. and Waterhouse, G.M. (1988) Trans. Br. Mycol. Soc. 91, 1–62.

35. Hall, G. (1989) CMI Descriptions of Pathogenic Fungi and Bacteria, No. 976. Mycopathologia 106, 195–197.

36. Tate, K.G. (1981) N.Z. J. Exp. Agric. 9, 371–376.

37. Breese, W.A., Shattock, R.C., Williamson, B., and Hackett, C. (1994) Ann. Appl. Biol. 125, 73–85.

38. Hall, G., Cook, R.T.A. and Bradshaw, N.J. (1992) Plant Pathol. 37, 224–227.

39. Graham, J., McNicol, R.J. and Kumar, A. (1990) Plant Cell Tissue Org. Cult. 20, 35–39.

40. Smith, E.F., Roberts, A.V. and Mottley, J. (1990) Plant Cell Tissue Org. Cult. 21, 141–145.

41. Berger, F., Li, H., White, D., Frazer, R. and Leifert, C. (1996) Phytopathology 86, 428–433.

42. Sherriff, C., Whelan, M.J., Arnold, G.M., Lafay, J-F., Brygoo, Y. and Bailey, J.A. (1994) Exp. Mycol. 18, 121–138.

43. Cooke, D.E.L., Duncan, J.M. and Unkles, S.E. (1995) OEPP/EPPO Bull. 25, 95–98.

44. Cooke, D.E.L., Kennedy, D.M., Guy, D.C., Russell, J., Unkles, S.E. and Duncan, J. M. (1996) Mycol. Res. 100, 297–303.

45. White, T.J., Bruns, T., Lee, S. and Taylor, J. (1990) in PCR Protocols, A guide to Methods and Applications (Innis M.A., Gelfand D.H., Sninsky J.J. and White T.J., eds) pp. 315–322, Academic Press, San Diego.

46. Auwera, G., Chapelle, S. and Wachter, R. (1994) FEBS Lett. 338, 133–136.

47. Lacourt, I., Cooke, D.E.L. and Duncan, J.M. (1996) in Diagnostics in Crop Production, BCPC Symposium Proceedings No. 65, pp. 145–149.

48. Sreenivasaprasad, S., Sharada, K., Brown, A.E. and Mills, P.R. (1996) Plant Pathol. 45, 650–655.

FROM LABORATORY TO APPLICATIONS: CHALLENGES AND PROGRESS WITH *IN VITRO* DUAL CULTURES OF POTATO AND BENEFICIAL BACTERIA

J. NOWAK[1], S.K. ASIEDU[1], S. BENSALIM[1], J. RICHARDS[1], A. STEWART[1], C. SMITH[1], D. STEVENS[1] and A.V. STURZ[2]

[1]*Department of Plant Science, Nova Scotia Agricultural College, P.O. Box 550, Truro, N.S., Canada B2N 5E3*

[2]*PEI Department of Agriculture, Fisheries & Forestry, Charlottetown, PEI, Canada C1A 7N3*

1. INTRODUCTION

In natural habitats, free-living micro-organisms play an important role in plant growth, development and adaptation to extreme environments [1-4]. They form close associations with plant roots and colonize internal and external tissues [2–5]. Production of axenic plantlets under tissue culture conditions devoids plants of these natural allies. Combination of the culture conditions, high humidity and high sugar in particular, and lack of microbial elicitors triggering or enhancing certain metabolic pathways, make tissue culture transplants vulnerable to pathogens and other environmental stresses [6]. It is well documented that plantlets have reduced photosynthetic capacity [7], lower wax deposits [8], poorly functioning stomata [9,10], underdeveloped root system, and very few leaf and root hairs [6,10]. There is a great possibility that reintroduction of certain micro-organisms or their combinations to tissue culture propagules, bacteria and vesicular-arbuscular mycorrhiza in particular, can be utilized in agricultural and horticultural practices for the purpose of transplants protection against diseases, improvement of establishment and overall performance ([3,4,6,7] and papers by A.C. Cassells and E. Wilhelm in this Proceedings).

Despite the progress in general understanding of plant–microbial interactions [1–5], application of microbial inoculants to tissue culture propagules (bionization) is limited to a few cases [3,6,7,10–12]. Better knowledge of these interactions may lead to the development of new culture management systems in plant micropropagation. Our laboratory has adapted an *in vitro* dual culture system of beneficial bacteria and potato for the purpose of the improvement of plantlet stress tolerance [6,11]. Plantlets co-cultured with our most effective isolate (*Pseudomonas* sp., strain PsJN) grow faster, have significantly more secondary roots, root and leaf hairs, better functioning stomata and contain more lignin [10]. They also survive direct transplanting from culture vessels to the field significantly better than non-inoculated controls [3,6]. Moreover, the bacterium does not grow on potato tissue culture medium by itself and is capable of establishing endophytic

A.C. Cassells (ed.), Pathogen and Microbial Contamination Management in Micropropagation, 321–329.
© *1997 Kluwer Academic Publishers. Printed in the Netherlands.*

(in xylem vessels) and epiphytic populations, allowing clonal multiplication of plantlets by nodal explants *in perpetuum*, without the need for re-inoculation [10]. The paper outlines recent data obtained with this system and discusses benefits of the bacterization and challenges related to its utilization.

2. PROCEDURE

2.1. Material and culture condition

Disease indexed plantlets of cultivars 'Desirée', 'Kennebec' and 'Russet Burbank' were from the Plant Propagation Center, New Brunswick Department of Agriculture, Fredericton, NB, Canada. Tubers of an abscissic acid deficient mutant 11401-01 were kindly supplied by Dr. Henry DeJong of Agriculture and Agri-Food Canada Research Station, Fredericton, NB, Canada; the culture was initiated in our laboratory. All the other clones (heat resistant) were from the International Potato Center (CIP), Lima, Peru. Cultures were maintained and multiplied on a hormone-free, MS-based medium using single node explants [13]. Culture conditions and plantlet bacterization with *Pseudomonas* sp., strain PsJN, were as described earlier [10].

2.2. Clonal responses to bacterization

Explants taken from 6-week-old non-bacterized control and bacterized plantlets were cultured on 10 ml medium in 25 × 200 mm test tubes, one node per tube, 12 replicates per clone/treatment combination. The cultures were grown under 240 $\mu E/m^2/s$ fluorescent/incandescent light, 12 h photoperiod at either 20:15 or 33:25°C day/night temperature. Growth responses were evaluated after 6 weeks in culture. Dry weights were determined after forced air drying for 48 h at 60°C.

2.3. Evaluation of induced resistance

Four-week-old plantlets cultivar 'Kennebec', non-bacterized, bacterized and jasmonate conditioned (grown on the medium containing 2.5 μM filter sterilized jasmonic acid; Apex Organics Ltd., Devon, UK), were inoculated with *Verticillium albo-atrum* by 1 min root dip in 10^3 conidia/ml suspended in sterile, distilled water. Control plantlets were dipped in water. The plantlets were then planted in a peat-based growing medium (ProMix), nine plants per tray (30 × 25 × 7 cm^3, Kord fiber pack, Halifax Seed, Halifax, NS, Canada), four trays per treatment, and grown in a growth chamber as above, using 16 h photoperiod and 21:18°C day/night temperature, for 8 weeks. A similar procedure was applied to evaluate transplants response to *Rhizoctonia solani*. Bacterized and non-bacterized plantlets cultivars 'Desirée' and 'Kennebec' were inoculated with 3.5 × 10^4 mycelial fragments/ml. Other conditions were as above.

2.4. Phenylalanine ammonia lyase and phenolics

Phenylalanine ammonia lyase (PAL) and soluble and bound phenolics were extracted from non-bacterized and bacterized potato plantlets, cultivar 'Kennebec', grown as described earlier [10]. PAL activity was determined in leaves, stems and roots of 2-, 4- and 6-week-old plantlets, using three composite samples of four plantlets each. Enzyme extraction and activity determination were as in Knogge and Weissenbock [14]. Phenolics analysis was done on two or three pooled samples of 12 plantlets per treatment. The samples were extracted under nitrogen gas and prepared for analysis according to Van Sumere [15]. HPLC separation was as in Wulf and Nagel [16]. Compound identification was confirmed by mass spectroscopy using Sciex mass spectrometer (Perkin Elmer, Irvine, CA, USA). An internal standard, 3,4-dimethoxycinnamic acid, and standard phenolics were from Sigma Chemicals (St. Louis, MO, USA).

3. RESULTS AND DISCUSSION

The bacterium was a contaminant isolated by J. Nowak from surface sterilized onion roots treated with *Glomus vesiculiferum* inoculant provided by Dr. W. Robertson, NSRF, Dartmouth, NS, Canada (Fig. 1A). The isolate was identified as a non-fluorescent *Pseudomonas* sp. and designated as strain PsJN [10]. The bacterium is capable of promoting in vitro growth of potato [3,6,10], tomato [6] and other crops [6]. It does not lose its growth promoting abilities when maintained in planta, freeze-dried with casein or stored frozen at -20°C on Protect™ beads (Technical Service Cons., Heywood, Lancashire, UK). Both, shoot and root growth promotion were recorded when potato plantlets were grown on sucrose or lactose as carbon source (Fig. 1B). Differential responses of potato cultivars to bacterization with the strain PsJN were recorded when cultures were grown at 20:15°C and 33:25°C (Table 1). The benefits of bacterization, root growth in particular, were more pronounced at 33:25 than at 20:15°C day/night temperatures. There was a strong clone × temperature interaction ($P<0.01$); however, no clear pattern between the growth responses to bacterization and heat stress tolerance was found. Heat-sensitive clone 11401-01 and heat-resistant clone Maine-47 reacted the strongest; heat-sensitive 'Russet Burbank' and heat-resistant DTO-33 were medium responders. Plantlets co-cultured with the bacterium could also withstand low levels of *Verticillium albo-atrum* [6] and *Rhizoctonia solani* (Table 2) much better than the non-bacterized controls. Bacterized transplants challenged with *R. solani* had less and smaller root lesions and gave higher tuber yields than non-bacterized controls. 'Kennebec', which is more susceptible to the disease, benefited from bacterization more than the less susceptible 'Desirée' (Table 2). Searching for chemical signals of growth promotion and tuberization we found that addition of 1–5 μM jasmonate to the culture medium stimulates plantlet growth (Nowak *et al.*, in preparation) and *in vitro* tuberization [17] similar to the bacterium. To clarify whether or not jasmonate is also involved in the induced resistance we tested plantlet response to *V. albo-atrum*. In an earlier study with Tn5 transposon mutants of the strain PsJN (supplied by Drs. K. Conn and G. Lazarovits, Agriculture and Agri-Food Canada, London, ON, Canada), we found

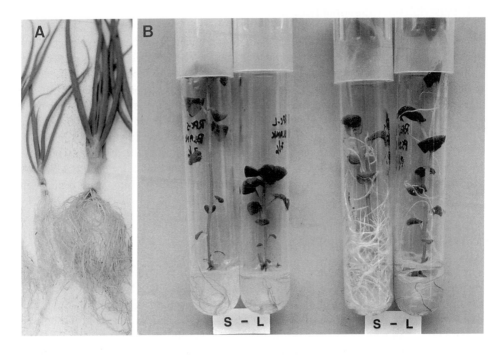

Figure 1. Responses of onion and potato to bacterization with *Pseudomonas* sp., strain PsJN. A and B – non-inoculated controls (left) and bacterized (right) 8-week-old greenhouse grown sweet Spanish onions, cultivar 'Riverside' (A) and 6-week-old potato plantlets cultivar 'Red Pontiac' (B) on sucrose (S) and lactose (L) media. C – yield of potato minitubers cultivar 'Kennebec' before (top) and after (bottom) infection with *Verticillium albo-atrum* (2.4×10^3 conidia/ml).

that the mutants promoting plantlets growth similar to the wild type were ineffective as triggers of the induced resistance responses to *V. dahliae* in tomato (Pillay and Nowak, submitted) and *V. albo-atrum* in potato (Stewart *et al.*, in preparation). As demonstrated in Fig. 1C, jasmonate conditioned plantlets behaved similar to controls, exhibiting severe yield reduction when challenged with V. albo-atrum. Verticillium infected bacterized plantlets gave 30% more tubers (3.2 vs. 2.2/plant) and 56% higher yield (80.2 vs. 35.2 g/plant) than those conditioned with jasmonate. At week 3 after transplanting, the disease rating was 0.83, 2.39 and 2.69 (on the scale 0 = no symptoms, 5 = dead), for the bacterized, control, and jasmonate treatments, respectively.

Table 1. Bacterization and temperature effects on *in vitro* growth of selected clones of potato[a]

Clone and treatment[b]	20:15°C[c]				33:25°C			
	Shoot			Root	Shoot			Root
	Nodes[d]	Length[e]	DW[f]	DW	Nodes	Length	DW	DW
'Desiree'								
B	11.0	10.2	40.2	18.8	10.3	11.3	48.8	23.2
C	9.7	4.8	16.1	5.8	6.6	5.4	17.7	11.7
'Kennebec'								
B	12.6	10.2	42.0	7.7	10.3	12.8	52.1	57.0
C	12.2	6.2	33.3	7.2	6.2	5.5	25.9	21.6
'R. Burbank'								
B	12.3	8.4	31.6	5.7	10.1	13.4	54.6	39.3
C	11.3	5.0	25.5	5.7	7.8	5.3	22.0	14.1
Maine-47								
B	11.6	10.0	57.6	27.9	13.5	15.4	77.6	63.1
C	11.9	6.5	40.1	9.6	11.2	8.5	47.4	28.4
DT0-33								
B	11.1	6.3	24.3	11.0	11.9	11.9	53.2	32.8
C	9.9	4.5	18.5	7.4	7.7	4.9	24.3	17.1
LT-1								
B	11.3	11.2	52.9	25.4	8.9	9.8	40.1	37.9
C	12.1	6.3	30.9	10.8	7.5	6.2	34.5	31.9
LT-2								
B	13.2	8.6	32.1	8.1	10.2	9.9	47.5	33.9
C	12.3	6.5	24.3	8.4	9.9	7.6	30.1	17.7
11401-01								
B	17.2	13.8	36.3	5.8	13.4	13.7	78.7	50.6
C	7.0	3.3	8.0	2.0	11.8	7.6	14.9	2.8

[a]Six-week-old plantlets, *n*=12; [b]Bacterized (B) and non-bacterized (C) plantlets; [c]Day/night temperatures; [d]Number per plantlet; [e]cm; [f]Dry weight, mg/plantlet (shoots with leaves); LSD (*P*=0.05) values: node number = 3.0, shoot length = 2.9, shoot DW = 12.7, root DW = 10.2.

Table 2. Disease spread and tuber yield of bacterized (B) and non-bacterized (C) potato plantlets challenged with *Rhizoctonia solani*

Cultivar and treatment	Root infection			Tuber yield	
	Lesion size (mm)	Lesion number	% plants infected	Per plant (g)	Number
'Kennebec'					
B	6.3±2.3	2.0±0.8	50	33.5±3.9	3.8±1.2
C	8.5±1.0	5.9±0.7	100	26.7± 7.9	2.8±1.1
'Desiree'					
B	8.0±2.5	1.4±0.5	60	29.5±4.0	2.6±1.3
C	14.4±3.5	2.1±0.5	80	27.8±4.7	3.6±1.7

The above phenomena were due to the induced resistance response of the inoculated plantlets rather than to cross protection. In an *in vitro* plate bioassay, the bacterium did not inhibit growth of the fungi. Two- and 4-week-old plantlets grown from nodal explants taken from bacterized stock plants always exhibited significantly higher levels of PAL activity than the plantlets derived from non-bacterized controls (Table 3). Four-week-old bacterized plantlets (usual transplanting age) had also much more lignin deposits around vascular bundles (data not shown). In 6-week-old cultures the differences in PAL activity between the bacterized and controls became less pronounced (Table 3), probably due to a more advanced developmental stage of the bacterized plantlets. At week 4 bacterized plantlets also contained more free phenolics, chlorogenic and caffeic acids in particular (Table 4). The elevated levels of chlorogenic and caffeic acids in leaves indicate that bacterized transplants can have improved resistance to foliar diseases. It is well recognized that phenolics function as signals in plant developmental processes and in plant-microbial interactions and can also inhibit some extracellular enzymes secreted by fungal pathogens [18]. Higher activity of PAL, better lignification and elevated free phenolics content at early stages of growth demonstrate high speed of the induced resistance response in the bacterized plantlets. The speed of plant response to stress signal(s), and the degree of this response, are critical in plant–pathogen interactions (reviewed by Benhamou [19]). Our studies indicate that plantlets grown in the presence of the bacterium are developmentally and metabolically 'sensitized' to respond faster to environmental signals than the axenic controls. The degree of plantlet 'sensitization' and consequently its response to transplant stress may depend on the level of expression of several genes which in turn can depend on plant genotype and culture conditions (e.g. temperature, sugar, CO_2, culture ventilation, presence of other organisms, etc.).

Table 3. Phenylalanine ammonia lyase (PAL) activity in bacterized (B) and non-bacterized (C) potato plantlets cultivar 'Kennebec' of different age

Plantlet age (weeks)	Treatment	PAL activity[a]		
		Leaves	Stem	Roots
2	B	43.6	41.7	19.3
	C	36.1	15.8	9.0
	$LSD_{0.05}$	0.9	3.6	0.6
4	B	36.8	17.9	7.6
	C	23.3	14.5	5.6
	$LSD_{0.05}$	0.4	0.2	0.2
6	B	32.5	12.0	11.5
	C	33.1	14.1	8.9
	$LSD_{0.05}$	0.5	0.2	2.4

[a] μg cinnamic acid/g per hour.

Table 4. Soluble (Sl) and bound (Bd) phenolics content in 4-week-old bacterized (B) and non-bacterized (C) potato plantlets cultivar 'Kennebec'

Tissue	Trt	Form	Phenolics (μg/g fresh weight)				
			Chl ac[a]	Caf ac[b]	Con al[c]	Fer ac[d]	Cin ac[e]
Leaf	B	Sl	279.1	33.7	5.3	3.4	0.6
		Bd	t[f]	1.5	0.9	1.0	0.7
	C	Sl	52.1	10.7	3.0	2.4	0.6
		Bd	t	t	0.5	1.0	0.8
Stem	B	Sl	30.8	7.9	1.0	2.0	t
		Bd	0.6	1.1	1.2	2.8	0.6
	C	Sl	8.8	3.3	1.0	1.9	t
		Bd	1.1	t	0.6	1.8	0.6
Root	B	Sl	29.2	9.0	1.1	0.4	0.6
		Bd	t	0.9	2.8	10.3	0.7
	C	Sl	27.8	8.9	0.8	0.5	0.4
		Bd	0.6	0.5	1.8	8.5	0.6

[a] Chlorogenic acid, [b] Caffeic acid, [c] Coniferyl alcohol, [d] Ferulic acid, [e] Cinnamic acid, [f] Traces.

In tissue culture the bacterium is translocated to tubers; populations of 10^5–10^6 and 10^6–10^7 CFU/g tuber fresh weight of endophytic bacteria were recorded in microtubers cultivars 'Shepody' and 'Red Pontiac', respectively (Nowak, J. and Sturz, T., in preparation). Much lower levels were found, however, probably due to the poor competitiveness of this strain, in greenhouse and field experiments ([20] and Lazarovits, G., personal communication). Improvement of competitive ability of the strain PsJN is one of the challenges of the system. Populations of endophytic bacteria in seed tubers are enormous [21]. Combination of certain strains can have synergistic effects on plants [22]. We believe that explant inoculation (bionization per analogy to immunization) with combination of strains capable of establishment in vascular system and translocation to tubers, rather than with single strains, could enhance their stability *in planta* and guarantee predicted responses of potato clones to environmental stresses. Tissue culture systems, heterotrophic and autotrophic, allow testing of different combinations of such strains. Extensive research support from industrial partners is, however, difficult to solicit as current regulations for nuclear seed potato production do not allow co-culturing of plantlets with other organisms. In view of the fact that we cannot culture over 99% naturally occurring micro-organisms using standard media [23], these regulations should be re-visited and modified.

REFERENCES

1. Rovira, A.D. (1991) in The Rhizosphere and Plant Growth (Keister D.L. and Gegan P.B., eds) pp. 3–13, Kluwer Academic Publishers, Dordrecht, NL.

2. Glick, B.R. (1995) Can. J. Microbiol. 41, 109–117.

3. Lazarovits, G. and Nowak, J. (1997) HortSci. 32(2), 188–192.

4. Kloepper, J.W. (1996) BioScience 46(6), 406–409.

5. Lynch, L.M. (1990) in The Rhizosphere (Lynch L.M., ed.) pp. 1–10, John Wiley, Chichester, UK.

6. Nowak, J., Asiedu, S.K., Lazarovits, G., Pillay, V., Stewart, A., Smith, C. and Liu, Z. (1995) in Ecophysiology and Photosynthetic In Vitro Cultures (Carre F. and Chagvardieff P., eds) pp. 173–180, CEA, Aix-en-Provence, France.

7. Desjardins, Y. (1995) in Ecophysiology and Photosynthetic In Vitro Cultures (Carre F. and Chagvardieff P., eds) pp. 145–160, CEA, Aix-en-Provence, France.

8. Pierik, R.L.M. (1987) In Vitro Culture of Higher Plants. Martinus Nijhoff Publ., Dordrecht, NL.

9. Ziv, M., Schwartz, A. and Fleminger, D. (1987) Plant Sci. 52, 127–134.

10. Frommel, M.I., Nowak, J. and Lazarovits, G. (1991) Plant Physiol. 96, 928–936.

11. Agricell Report (1987) 9, 38.

12. Varga, Sz.S., Kornanyi, P., Pereininger, E. and Gyurjan, I. (1994) Phys. Plant. 90, 786–790.

13. Sipos, J., Nowak, J. and Hicks, G. (1988) Am. Potato J. 65, 353–364.

14. Knogge, W. and Weissenbock, G. (1986) Planta 167, 196–205.

15. Van Sumere, C.F. (1989) in Methods in Plant Biochemistry (Dey D.M. and Harborne J.B., eds) Vol. 1, pp. 29–73, Academic Press, London, UK.

16. Wulf, L.W. and Nagel, C.W. (1976) J. Chromatogr. 116, 271–279.

17. Pruski, K., Nowak, J. and Lewis, T. (1993) In Vitro Cell. Dev. Biol. 29A (3), 60.

18. Metraux, J.P. and Raskin, I. (1993) in Biotechnology in Plant Disease Control (Chet I., ed.) pp. 191–209, Wiley-Liss, Inc., New York, NY, USA.

19. Benhamou, N. (1996) Trends Plant Sci. 1(7), 233–240.

20. Frommel, M.I., Nowak, J. and Lazarovits, G. (1993) Plant Soil 150, 51–60.

21. Sturz, A.V. (1995) Plant Soil 175, 257–263.

22. Frommel, M.I., Pazos, G.S. and Nowak, J. (1991) Fitopatologia 26, 66–73.

23. Hugenholtz, P. and Pace, N.R. (1996) Tibtech 14(6), 190–197.

BACILLUS SUBTILIS AN ENDOPHYTE OF CHESTNUT (*CASTANEA SATIVA*) AS ANTAGONIST AGAINST CHESTNUT BLIGHT (*CRYPHONECTRIA PARASITICA*)

EVA WILHELM[1], WOLFGANG ARTHOFER[1], ROLAND SCHAFLEITNER[1] and BIRGIT KREBS[2]

[1] *Austrian Research Centre Seibersdorf, A-2444 Seibersdorf, Austria*
[2] *FZB Biotechnik GmbH, Berlin, Germany*

1. INTRODUCTION

Chestnut blight (*Cryphonectria parasitica*) is a severe disease of European (*Castanea sativa*) and American chestnut (*Castanea sativa*). Antagonistic bacteria have been shown to have a potential use in the control of plant diseases [1]. This report describes our approach for screening and testing bacterial isolates for antagonistic effects on the growth of *Cryphonectria parasitica in vitro*.

2. PROCEDURE

2.1. Screening and testing of antagonistic bacteria

Surface bacteria from leaves of lilac, birch, roses and chestnut as well as endophytic bacteria of the xylem sap of healthy chestnut stems were isolated on selective media [2]. For further cultivation of the bacterial isolates PD-broth (Difco) or PD-agar was used. For a general *in vitro* screening of antagonistic bacteria multi-plates containing 200 µl PD broth were inoculated with a 1µl loopful of bacteria colonies and cultivated on a shaker for 24 h at 30°C. Twenty microlitres of each bacterial solution were spotted three times in a petri dish on PD agar.

After 7 days growth (30°C in the dark) the cultures were overlaid with a conidial suspension (2×10^6 conidia/ml) of a reference strain of *Cryphonectria parasitica* from the German culture type collection (DSM). Seven days later the petri dishes were evaluated for antagonistic effects.

The most efficient endophytic bacterial isolates were selected and used for further screening with the Austrian *Cryphonectria parasitica* isolates. The most efficient isolate was named L25 and was used for all further experiments.

A.C. Cassells (ed.), Pathogen and Microbial Contamination Management in Micropropagation, 331–337.
© *1997 Kluwer Academic Publishers. Printed in the Netherlands.*

2.2. Screening on other phytopathogenic fungi

The effectiveness of isolate L25 was tested against ten other phytopathogenic fungi (*Alternaria radicina, Botrytis cinerea, Fusarium culmorum, Fusarium oxysporum* f. sp. *cucumerinum, Guamannomyces graminis, Gerlachia niveale, Rhizoctonia solani, Phomopsis sclerotioides, Sclerotinia sclerotiorum* and *Stromatinia fresiae*) at temperatures of 15°C and 25°C. The results were compared to the reference *Bacillus subtilis* isolate ATCC 6051.

The bacteria were cultivated for 2 days on nutrient media I (Merck 7881) at 30°C. The fungal strains were incubated for 10 days prior testing on maltextract agar (Merck 5398) at 25°C. Screening for antagonistic effects was performed in a co-culture of bacteria and fungi measuring the fungal growth inhibition zone. The petri dishes were incubated for 5–10 days at temperatures of 25°C and 15°C. Testing for each fungal species was repeated four times.

2.3. Coculture of L25 and L26 with *Cryphonectria parasitica* on dormant chestnut stems

Dormant chestnut stem sections (diameter approx. 20 mm, length 30 cm) were surface sterilised by 10 min treatment with NaOCl solution (approx. 3%) followed by two rinses with sterile water. The ends of the stems were closed with paraffin. The bark was removed with a cork borer (diameter: 5 mm); 10 ml of the bacterial cultures or PD-broth were inoculated into the cavities. Four groupings were made: A, control 1 (only wounding); B, control 2 (10 µl PD broth); C, isolate L25; D, isolate L26. These groups were further divided into two classes: 0, fungal challenge simultaneously after bacterization; 3, fungal challenge 3 days after bacterization. Six replications were made for each treatment.

Agar plugs carrying mycelium of an Austrian virulent *Cryphonectria parasitica* isolate SR1 measuring approx. 25 mm^2 were put in the wounds. The sticks were placed in plastic containers at room temperature in the dark. After 18 days the lesion areas were measured using a planimeter. Multifactorial variance analysis was calculated after the SAS programme version 6.01, procedures ANOVA and GLM.

2.4. Inoculations on *in vitro* chestnut shoots

The Austrian chestnut clone GG 10 (Gross Gelb) and the Spanish hybrid (*C. sativa* × *C. crenata*) clone C 125 (provided from Antonio Ballester, Santiago de Compostela, CSIC) were micropropagated via shoot tip and nodal cuttings on P24 medium (Patent no. 92 90 25 31.0) with 0.2 mg/l BAP. Plant material was cultivated in glass tubes with 12 ml media supplemented with 3% sucrose and 0.8% agar at a temperature of 25°C (Philips TLD lamps) with a 16 h photoperiod at 65 µE/m^2/s. *In vitro* shoots with 5 to 6 cm length were inoculated by applying a small piece of agar with isolate L25 to the wound.

After 1 and 2 weeks, respectively, the inoculated shoots and control shoots were challenged with 1 µl conidial solution (approx. 40×10^3 conidia/ml) of a virulent fungus strain. Twelve replications per treatment were made. Disease symptoms were scored for each shoot in weekly intervals for up to 6 weeks, using the following rating scheme:

0, no symptoms;
1, hyphae at the wound;
2, hyphae along the stem;
3, massive growth of hyphae;
4, massive necrosis;
5, death of explant.

Shoots recovered after fungal infection were subcultured again as described above.

2.5. Preparation of protein extracts and intercellular fluid (IF)

Shoots were harvested in time intervals from 0 to 9 days, separated into leaves and stems, rapidly frozen in liquid nitrogen and then stored at -80°C. Six shoots were pooled for extraction. Two hundred milligrams of frozen leaves or stems were ground in liquid nitrogen in a mortar. The powder was transferred to 1.5-ml microtubes and 100 mM Tris–HCl buffer, pH 8.8, containing 5% PVP, 5% glycerol and 5 mM PMSF were added in a ratio of 1:2 (w/v). The powder was resuspended by vortexing for 10 s. The suspensions were centrifuged at 17,500 × g at 4°C for 10 min; the supernatant was transferred to new tubes and centrifuged again at 17,500 × g at 4°C for 15 min. The resultant supernatant was aliquoted and rapidly frozen in liquid nitrogen and stored at -80°C until analysis.

IF from 500 mg leaves or stems was recovered by submersing the fresh plant material under 20 ml 100 mM Tris–HCl, pH 8.8, supplemented with 5% PVP, 5% glycerol and 5 mM PMSF in 40-ml beakers. The plant material was kept under the buffer surface by an air-permeable plug formed from aluminium foil. The beakers were placed in an exsiccator and a vacuum of 40 mbar air pressure was established for 10 min; then the exsiccator was slowly aerated. The plant material was blotted dry and put into syringes. The syringes containing the plant material were placed in SS34 centrifuge tubes and centrifuged at 2000 × g and 4°C for 10 min. The IF passing through the tip of the syringe into the centrifuge tube was recovered, aliquoted, rapidly frozen in liquid nitrogen and stored till analysis at -80°C.

The amount of contamination of the IF by intracellular compounds was determined after Roulin [3].

Protein content measurements in the crude extracts and the IF were made according to Bradford [4] using bovine serum albumin (Biorad Inc.) as standard.

2.6. Quantitative enzyme assays

The measurements in the IF were repeated three times. ß-1,3-Glucanase activity was measured as described [6] with slight modifications. IF (0.2 μg protein/sample) was added to 12-ml plastic tubes containing reaction mixtures and blanks as described [6]. After 30 min of incubation at 37°C on a shaker, the test tubes were heated to 100°C for 1 min to stop the reaction. The amount of reducing sugar moieties, released from laminarin by plant ß-1,3-glucanases were determined as described previously [3] using the formula described in Boller [6].

Chitinase activity measurements in IF, containing 0.2 μg protein/sample were set up and incubated as described [8]. The chitinase reactions were stopped by sedimenting undigested colloidal chitin at $17,500 \times g$ for 3 min; 300 μl of the supernatant were transferred to 15-ml glass reaction tubes and incubated for 2 h with 5 μl Cytohelicase solution (Sigma) (30 mg/ml), previously desalted by gel filtration on a Sephadex G-25-column. The amount of N-acetylglucosamine released was determined as described by Boller [8].

3. RESULTS AND DISCUSSION

In total, 177 bacterial isolates were screened for their antagonistic effects on the growth of the chestnut blight fungus. Chestnut yielded 82 bacterial isolates, both from epiphytic and endophytic origin. Forty isolates (23%) exerted growth suppression on *Cryphonectria parasitica*. Twelve isolates originated from the xylem fluid of healthy chestnuts (Fig. 1). Growth suppression ranged from small inhibition zones to nearly total growth inhibition. L25, the most efficient endophytic isolate, was characterized as *Bacillus subtilis* and used for further tests.

The screening for antagonistic effects on other phytopathogenic fungi revealed that L25 is able to suppress the growth of quite a broad range of different fungal species (Fig. 2). L25 is more effective at the higher temperature of 25°C than at 15°C. The reference isolate ATCC 6051 showed only a slight inhibition effect at 25°C, whereas at 15°C no effect on any fungal growth suppression could be observed.

The experiments on co-cultivation of bacteria with the chestnut blight fungus on dormant chestnut stems proved the effectiveness of the antagonistic capacity of *Bacillus subtilis* (Fig. 3). A highly significant antagonistic effect expressed as reduced lesion area could be observed on bacterized chestnut stems when bacteria were applied 3 days prior to fungal challenge. Simultaneous application of bacteria and fungal mycelium resulted in no significant symptom reduction compared to the control. No differences in the lesion areas of either control group could be observed. The isolate L25 caused a 71% reduction in the lesion area in relation to the control. Isolate L26 also exhibited antagonistic effects (44%), whereas the suppression effect was less severe compared to isolate L25.

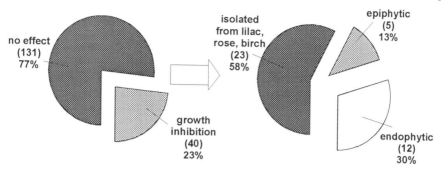

Figure 1. *In vitro* effects of isolated bacteria on chestnut blight.

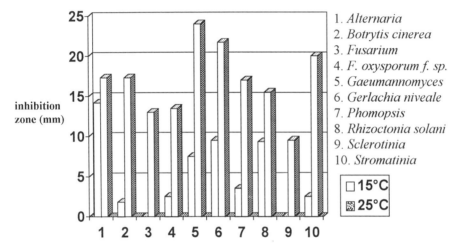

1. *Alternaria*
2. *Botrytis cinerea*
3. *Fusarium*
4. *F. oxysporum f. sp.*
5. *Gaeumannomyces*
6. *Gerlachia niveale*
7. *Phomopsis*
8. *Rhizoctonia solani*
9. *Sclerotinia*
10. *Stromatinia*

Figure 2. Screening for antagonistic effects on other phytopathogenic fungus species.

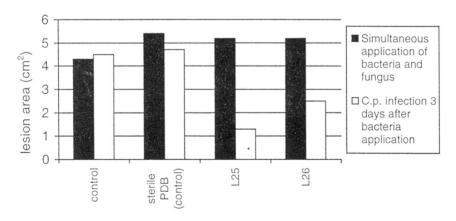

Figure 3. Co-culture of bacteria and chestnut blight on dormant chestnut stems.

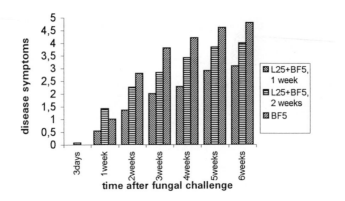

Figure 4. Effect of bacterization on disease symptom development on *in vitro* shoots.

Inoculation experiments with *in vitro* chestnut shoots exhibited a distinct growth retardation on the fungus after inoculation with bacteria. No curative effect of the bacteria on diseased *in vitro* chestnuts could be observed. Responses of chestnut clones were different. Clone 125 was more susceptible to the pathogen than clone GG 10. Preinoculation with bacteria 1 week before fungal infection suppressed mycelium development considerably more than at a 2-week interval (Fig. 4). Multiplicated shoots of clone GG 10 still exhibited a reduced susceptibilty against chestnut blight. It could be shown that, although weaker, the protection effect by bacterization with *Bacillus subtilis* isolate L25 can be maintained through the second multiplication phase. Inoculation with *Bacillus subtilis* influenced the multiplication rate negatively. The micropropagation rate expressed in newly formed shoots and nodal segments was significantly reduced versus uninoculated control shoots of clone GG 10.

Investigations on the biochemical level indicated that inoculations with *Bacillus subtilis* induced the production of PR (pathogen related) proteins, such as an acidic extracellular chitinase and an acidic ß-1,3 glucanase isoenzyme, which did not appear in wounded control plants. Quantitative enzyme assays revealed that the highest amount of PR proteins could be detected 3–7 days after inoculation (Fig. 5), which coincides also with the bacterization interval shown in dormant chestnut stems and also *in vitro* chestnut shoots.

The mechanisms for the plant protection effect are still unclear. *Bacillus subtilis* is a known producer of cyclic peptide antibiotics such as Iturin, Bacillomycin, etc. Preliminary HPLC analysis showed that the active fungistatic compound is related to the fengomycin-type. Additionally, *Bacillus subtilis* is supposed to act as an elicitor triggering the host defences by changing the host plant metabolism, which seems to be related to systemic acquired resistance.

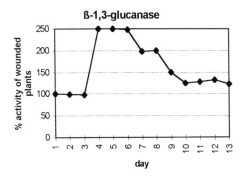

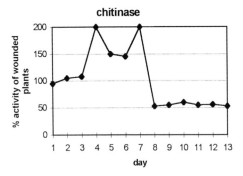

Figure 5. Chitinase and ß1-3, glucanase enzyme assays.

REFERENCES

1. Schreiber, L.R., Gregory, G.F., Krause, C.R. and Ichida, J.M. (1988) Can. J. Bot. 66, 2338–2346.

2. Sands, D.C. and Scharenen, A.L. (1978) Phytopathol. News 12, 43.

3. Roulin, S. (1992) Dissertation, Faculté des Sciences, Université Fribourg, Suisse.

4. Bradford, M.M. (1976) Anal. Biochem. 72, 248–254.

5. Boller, T. (1992) In: Molecular Plant Pathology, Vol II (Gurr J., McPherson M.J. and Bowles D.J., eds.), pp. 23–30, IRL Press, New York.

6. Boller, T. and Mauch, F. (1988) Methods Enzymol. 161, 430–434.

BIOLOGICAL CONTROL OF *BOTRYTIS, PHYTOPHTHORA* AND *PYTHIUM* BY *BACILLUS SUBTILIS* COT1 AND CL27 OF MICROPROPAGATED PLANTS IN HIGH-HUMIDITY FOGGING GLASSHOUSES

HONG LI[1], DUNCAN WHITE[2], KATHRYN A. LAMZA[2], FRANK BERGER[3] and CARLO LEIFERT[2]

[1]*Department of Biology, Yunnan Normal University, Kunming, Yunnan 650092, PR China*
[2]*Department of Plant and Soil Science, University of Aberdeen, Cruickshank Building, Aberdeen, AB9 2UD, UK*
[3]*Institute für Pflanzenbau und Tierhygiene in den Tropen und Subtropen, Universität Göttingen 37000, Göttingen, Germany*

1. INTRODUCTION

Fogging glasshouses are used to acclimatise micropropagated plants for up to 6 weeks after planting [1]. However, plants grown in these high humidity environments are very susceptible to soil-borne damping-off by *Phytophthora* and *Pythium* and attack by air-borne grey mould *Botrytis* and need to be sprayed frequently with fungicides. This has resulted in the development of resistance within the pathogen population.

Biocontrol of fungal diseases by an applied bacterial inoculum is now seen as a viable method of controlling some diseases and several commercial biocontrol agents are available [2]. The often unpredictable/variable activity of the biocontrol agent compared with the fungicide is one of the reasons why biocontrol has not gained a large market share in comparison to fungicides.

There are several mechanisms by which biocontrol is thought to work including the production of antifungal antibiotics, competition for nutrients and rhizosphere colonisation [3]. Antibiotic activity by potential biocontrol strains is often determined by *in vitro* plate assays on artificial media. However, there is as yet no direct evidence that *in vitro* activity corresponds with *in vivo* activity. The aim of this paper is to compare the biocontrol activity of several *Bacillus* strains with microplants commonly grown under fogging glasshouse conditions and to determine possible constraints to the biocontrol.

A.C. Cassells (ed.), Pathogen and Microbial Contamination Management in Micropropagation, 339–344.
© *1997 Kluwer Academic Publishers. Printed in the Netherlands.*

2. MATERIALS AND METHODS

2.1. Plant material and infection of plants

Astilbe hybrida L., *Aster hybrida* L. 'Pearl Star', Hermocallis 'Stella d'ora', *Daphne blayana* Freyer and *Photinia fraseri* Lindl. were micropropagated by shoot tissue culture and subsequently grown in peat compost in polystyrene trays [4]. Plants were infected with *Pythium ultimum* and *Phytophthora* spp. by introducing oospores into the water used to adjust substrate moisture levels. *Botrytis fuckeliana* was isolated from a previously infected crop of *Astilbe* and grown on malt extract agar [5]. Conidia formation was induced by UV exposure and the spore suspension sprayed onto new *Astilbe* plants using an atomiser at 10^6 conidia/ml every 5 days. *Botrytis cinerea* was obtained from infected *Astilbe* plants and grown on cabbage agar or on malt extract agar to induce sporulation and sprayed onto plants at 2×10^3 conidia/ml every 7 days [6].

2.2. Preparation, inoculation and enumeration of biocontrol agents

Bacillus subtilis Cot1 was initially isolated from *Cotinus* tissue cultures and was active against oomycete fungi but not *Botrytis cinerea* [7]. Cultures were initially prepared by growing in nutrient broth containing 3% sucrose at 20°C for 30 h. Microplants and seedlings were inoculated by dipping them into cells suspensions containing 10^5–10^9 cfu. *B. subtilis* Cot1 was isolated from plant rhizosphere by sectioning washed roots and homogenising in a blender in 10 ml one-quarter Ringer's solution. Bacteria were enumerated by serial diluting and culturing on NA containing cycloheximide (60 µg/ml) and on Murashige and Skoog's medium containing 3% sucrose. *B. pumilus* CL45 and *B. subtilis* CL27 were originally isolated from *Brassica* leaves [8] and were inoculated onto plants using a cabbage nutrient broth [6].

3. RESULTS AND DISCUSSION

Biocontrol of *Pythium* on *Photinia* was achieved using a cell suspension of *B. subtilis* Cot1 between 10^8 and 10^9 cfu/ml (equivalent to 10^5 and 10^6 cfu/g RFW) and at 10^1–10^2 oospores/g peat, at levels similar to the commercial fungicide (Table 1). No protection, however, was found at higher pathogen levels or lower Cot1 inoculum levels. The levels of *Pythium* at which biocontrol was effective is consistent with levels found in horticultural composts. Biocontrol activity against *Phytophthora* and *Pythium* was also observed in *Aster* and *Hemerocallis* using high levels of *B. subtilis* Cot1 (Table 2). Biocontrol in *Daphne* was, however, either poor or no activity was observed in comparison with the other plants. To determine why biocontrol was ineffective with *Daphne*, rhizosphere colonisation of Cot1 was examined. High levels of Cot1 were present in the rhizosphere of *Photinia* after 7 and 28 days of plant growth. However, significantly lower levels of Cot1 were present on the *Daphne rhizosphere* (Table 3). Rhizosphere colonisation is often seen as a prerequisite for effective biocontrol by ensuring that the biocontrol agent is present at the target site of pathogen activity. Competition between rhizosphere microbial flora may limit the extent

Table 1. Effect of inoculum densities of B. subtilis Cot1 and Pythium ultimum on damping-off of Photinia. Different letters indicate significant differences ($P<0.01$) according to Tukey's Honestly Significant Difference test after arcsine transformation. Lowercase letters indicate differences between fungal inocula in the same antifungal treatment; uppercase letters are between antifungal treatment at the same dilution

Treatment	Dose (cfu/ml)	Pythium inoculum (oospores/g peat)			
		10^1	10^2	10^3	10^4
Metalaxyl		6 aA	2 aA	6 aA	3 aA
B. subtilis	10^9	6 aA	7 aA	14 bB	61 cB
	10^8	2 aA	6 aA	54 bC	69 cB
	10^7	30 aB	74 bB	97 cD	95 cC
	10^5	95 aC	94 aC	96 aD	98 aC
Untreated		94 aC	99 aC	96 aD	98 aC

Table 2. Percent damping-off in microplants. Plants were grown in peat compost infected with 10^2 oospores/g peat and treated using 10^9 cfu/ml B. subtilis Cot1 or metalaxyl fungicide. For treatments of the same plant, different letters indicate significant differences ($P<0.05$) according to Tukey's Honestly Significant Difference test after arcsine transformation

Plant	Pythium			Phytophthora		
	Fungicide	Cot1	Untreated	Fungicide	Cot1	Untreated
Aster	6a	2a	96b	2a	4a	100b
Daphne	4a	62a	86cd	5a	94bc	94b
Photinia	4a	7a	99b	6a	3a	96c
Hemerocallis	7a	1a	99b	9a	3a	74c

of colonisation by an introduced inoculum [9]. However, for these experiments, aseptic tissue-culture plants were used indicating no prior established microbial flora. It is likely, therefore, that the poor biocontrol activity observed on Daphne could be attributed to rhizosphere conditions. This was further supported by using spent tissue culture

Table 3. Rhizosphere colonisation of *B. subtilis* Cot1 on micropropagated *Photinia* or *Daphne* plants. Plants were inoculated with 4×10^6 cfu/g root fresh weight (RFW) and results are the means of five replicates. Counts were log transformed prior to analysis by Tukey's Honestly Significant Difference test and there were significant interactions ($P<0.001$) between plant species and between root sections but not between assessment dates

| | $\log_{10}$ spores/g RFW | | | | | |
| | *Photinia* | | | *Daphne* | | |
Assessment date	Upper	Centre	Tip	Upper	Centre	Tip
7	6	5.52	5	2.85	3.5	2.75
28	6.24	5	5.25	3.7	2.55	2.15

experiments in which spent tissue culture media of *Daphne* inhibited the growth of Cot1 [4].

Biocontrol of *Botrytis cinerea* was determined using *B. subtilis* CL27 and *B. pumilus* CL45. *B. subtilis* CL27 also shows biocontrol activity against *Botrytis fuckeliana*. In *in vitro* plate assays both CL27 and CL45 inhibited *Botrytis cinerea* (results not shown) [6]. However, using a seedling-based assay, biocontrol of *B. cinerea* on *Astilbe* was only found with CL27 (Table 4). Studies using UV-induced biocontrol negative strains and thin-layer chromatography [6] have indicated that it is likely that these two strains produce antifungal antibiotics. *In vitro* pH and nutrient assays indicated a different pH range of activity of the antibiotics from CL27 and CL45 [6]. This may suggest a possible reason why *in vivo* biocontrol activity was not found with CL45.

Table 4. Amount of *Botrytis* infection in micropropagated *Astilbe* using two biocontrol strains *Botrytis* score is: 6, all plants showing mycelial growth; 5, >80%; 4, >60%; 3, >40%; 2, >20%; 1, 1–20%

| | | | *B. subtilis* CL27 | | *B. pumilus* CL45 | |
	Untreated control	Fungicide (first rotation)	Cell-free filtrate	Washed cells	Cell-free filtrate	Washed cells
Botrytis score	6.0	0.75	0.7	1.1	6.0	6.0

To determine the long-term effectiveness of a biocontrol agent when repeatedly applied to a pathogen-infected plant, *B. subtilis* CL27 was applied to *Botrytis fuckeliana* infected *Astilbe* every 5 days. After 4 weeks, an increase in the *Botrytis* growth score was measured, similar to the untreated plants, indicating that biocontrol did not remain effective (Table 5). The lack of biocontrol is likely to be due to pathogen development of resistance to the antifungal antibiotic [5].

Table 5. Grey mould *(Botrytis fuckeliana)* development on treated *Astilbe* plants 4 weeks after transfer into compost. Disease development was assessed by coring leaves for visible signs of fungal growth on the leaves. The key for the disease score rating is as given in Table 4

	Botrytis disease score									
	Glasshouse			Growth room				Glasshouse		
	Crop number									
	1	2	3	4	5	6	7	8	9	10
Untreated	6	6	5	6	6	6	6	6	6	6
CL27	0	1	0	0.5	1.5	0	0	1	4	6
Fungicide	0	0	0.6	0	0	0	0	1	0	0.5

In conclusion, it is now apparent that biocontrol methods can replace fungicides during weaning of tissue culture plants. However, the same problems (e.g. development of resistance, failure of treatments in certain crops) observed with fungicides may also occur with biocontrol agents.

ACKNOWLEDGEMENT

We would like to thank the British Ministry of Agriculture, Fisheries and Food (MAFF) for their support (MAFF Open Contract Grants CSA 1710 and CSA 2767).

REFERENCES

1. Leifert, C., Clark, E. and Rothery, C. (1993) Biol. Sci. Rev. 5, 31–35.
2. Leifert, C., White, D., Killham, K., Malathracis, N.E., Wolf, G.A. and Li, H. (1996) in Advances in Biological Control of Plant Diseases. (Wenhua T., Cook R.J. and Rovira A., eds)

pp. 72–78. Proceedings of the International Workshop on Biological Control of Plant Diseases, 22–27 May, China Agricultural University Press, Beijing, China.

3. Baker, K.F. (1987) Annu. Rev. Phytopathol. 25, 67–85.

4. Berger, F., Li, H., White, D., Frazer, R. and Leifert, C. (1996) Phytopathology 86, 428–433.

5. Li, H. and Leifert, C. (1994) J. Plant Dis. Prot. 101, 414–418.

6. Leifert, C., Li, H., Chidburee, S., Hampson, S., Workman, S., Sigee, D., Epton, H.A.S. and Harbour, A. (1995) J. Appl. Bacteriol. 78, 97–108.

7. Leifert, C., Waites, W.M. and Nicholas, J.R. (1989) J. Appl. Bacteriol. 67, 353–361.

8. Leifert. C., Sigee, D., Epton, H.A.S., Stanley, R. and Knight, C. (1992). Phytoparasitica 20, 143S–148S.

9. Kloepper, J.W. and Sehroth, M.N. (1981) Phytopathology 71, 1020–1024.

MICROBIAL CHARACTERISATION AND PREPARATION OF INOCULUM FOR *IN VITRO* MYCORRHIZATION OF STRAWBERRY IN AUTOTROPHIC CULTURE

G.L. MARK, J. MURPHY and ALAN C. CASSELLS
Department of Plant Science, University College, Cork, Ireland

1. INTRODUCTION

Strategies to control losses *in vitro* due to microbial contaminants (vitropaths [1]) have been the subject of extensive research (see Leifert and Woodward, this volume). Significant additional losses, however, may also occur at microplant establishment (Williamson *et al.*, this volume). Polygenic disease resistance develops in seedlings as the seedling develops and it is recognised that seedlings are vulnerable to pathogens if growth is checked due to adverse environmental conditions [2]. Microplants, especially if vitrified ('hyperhydrated'), may undergo a growth check at establishment [3]. Hardening of plants *in vitro* ('*in vitro* weaning') has been shown to improve plant establishment and subsequent growth [4] and should assist in reducing losses due to damping-off.

Another aspect of micropropagation related to disease susceptibility, is the absence of a microflora associated with aseptically produced microplants (Epton, this volume). The aseptic microplant is a 'biological vacuum' and represents a niche for microbial colonisation. In the absence of a biological balance provided by the plant microflora, promiscuous colonisation by environmental micro-organisms, that are facultative pathogens, may result in crop disease or disease risk to consumers where, for example, edible plants are colonised by *Escherichia coli* from organic fertiliser or *Bacillus subtilis* from contaminated water supplies ([5]; Cassells, this volume; Zenkteler *et al.*, this volume).

The beneficial effects of vesicular arbuscular mycorrhizal fungi (AMFs) on plant growth have been extensively documented in protection against biotic and abiotic stress [6]. Inoculation of microplants has been shown to have beneficial effects in relation to improved weaning *in vivo* [7,8] and in protection against disease [9]. Further, mycorrhizal colonisation, with helper organisms may, by competitive exclusion, block potential sites for pathogen colonisation.

In previous studies on mycorrhization of microplants, attempts have been made either to sterilise inoculum for inoculation in heterotrophic culture [10,11] or non-aseptic inoculum has been used *in vivo*. Considerable bacterial contamination of AMF inoculum has been

A.C. Cassells (ed.), Pathogen and Microbial Contamination Management in Micropropagation, 345–350.
© *1997 Kluwer Academic Publishers. Printed in the Netherlands.*

reported after surface-sterilisation treatment [10]. In addition to possible inhibition of AMF spore germination by components of the plant tissue culture medium, the use of non-sterile inoculum in heterotrophic culture is ruled out as the contaminants over-run the cultures. Inoculation of autotrophic cultures, however, is possible with non-sterile inoculum ([12]; see also Long, this volume). Autotrophic culture also offers the opportunity for the establishment of gnotobiotic cultures by inoculation with mycorrhizal fungi and characterised bacterial helper organisms or mycorrhizal fungi-compatible characterised micro-organisms, e.g. biocontrol bacteria. Here an autotrophic system for the establishment of mycorrhizal fungi is described [12] using as inoculum AMF spores with characterised bacterial contaminants, to evaluate the microplant–inoculant interaction *in vitro* and after establishment.

2. MATERIALS AND METHODS

2.1. Production and characterisation of AMF inoculum

The AMF isolate used was *Glomus fistulosum* BEG-31, supplied by Dr. M. Vestberg, and registered with the European bank of Glomales (INRA, Dijon, France). It was grown on the roots of onion (*Allium cepa*) and the spores were isolated as described [9]. Spores were aseptically rinsed in sterile distilled water (SDW) three times and stored in SDW at 6°C until required.

Before use, or after decontamination treatment, percentage spore germination was determined as follows. Spores were aseptically transferred to plastic petri dishes on acid water agar containing 10 g/l Oxoid Bacteriological agar no. 1 (code L11: Unipath Ltd., Basingstoke, UK) pH 6.2. The spores were incubated in the dark for 2 weeks at $24 \pm 2°C$. The plates were examined for the presence of germination tubes and hyphae with a light microscope (Magnification: $\times 100$).

Spores were surface sterilised by the following methods: (i) the spores were soaked in 30 g/l w/v chloroamine-T for 10 min followed by three washes in sterile distilled water; (ii) spores were soaked in 20 g/l chloroamine-T w/v with 200 mg/l streptomycin sulphate and 1 drop of Tween 80 for 10 min, followed by ten washes in distilled water with a 2 min interval between washes 1–9, and 1 h before the final wash (Dr. S. Edwards, personal communication); (iii) spores were soaked in 30 g/l chloroamine-T w/v with 250 mg/l streptomycin sulphate and 1 drop of Tween 20 for 10 min, the spores were washed 10 times in sterile distilled water with 2-min intervals between washings.

After surface sterilisation, spore viability was determined as above. To determine the microbiological status of the spores they were incubated on bacteriological media, i.e. YDC, NGA, King's B and D-1 [13]. Isolates were streaked and single spore colonies were isolated in pure culture. Gram stain, oxidase and catalase tests were carried out to group the isolates which, as appropriate, were further characterised using API kits [5].

2.2. Establishment of AMF in autotrophic culture

All media and instruments were sterilised by autoclaving at 105 KPa, 121°C for 15 min. The growth-room conditions were photosynthetic photon flux 30 $\mu mol/m^2/s$, 16 h day; 22±2°C; RH day 55%–night 100%. At all stages plants were grown in Magenta culture vessels (Sigma Chemical Co., Poole, Dorset, UK).

Strawberry (*Fragaria × ananassa*) microplant cultivars 'Elvira', and 'Tenira' were obtained from Reinhold Hummel GbR, Stuttgart, Germany. These were proliferated on full Murashige and Skoog [14] medium (4.41 g/l; cat. no. 26-100-24, Flow Laboratories, Irvine, UK), 21.6 g/l sucrose, 7 g/l agar, 1.0 mg/l indolebutyric acid and 1.0 mg/l benzyl-amino purine at pH 5.6 in Magenta culture vessels. After 4 weeks on proliferation medium, single crowns were transferred for a further 3 weeks to a rooting medium containing 4.411 g/l full Murashige and Skoog medium, 30.0 g/l sucrose, 7 g/l agar, 0.1 mg/l indolebutyric acid at pH 5.8. Plantlets, once rooted, had their roots aseptically washed free of any adhering medium before further use.

Heterotrophic (control) culture: rooted plantlets were transferred to medium containing 4.41 g/l full Murashige and Skoog medium, 30.0 g/l sucrose, 5 g/l agar, pH 5.8. Two rooted plantlets were placed in each Magenta tub with gas permeable polyvinylchloride lids [4]; this treatment was replicated 20 times. All replicates were incubated in the growth room in a Latin-square design.

Autotrophic cultures were grown on a carbon-free medium of half strength (2.17 g/l) Murashige and Skoog [14] basal-salt mixture (cat. no. M-5524; Sigma-Aldrich, Dorset, UK) in distilled water. Hortifoam, (Plant Biotechnology (UCC) Ltd., Cork Ireland) polyurethane foam was imbibed with 50 ml of the medium and acted as a support substrate for the plantlets instead of agar [12]. The foam had a density of 16 ± 1 kg/m^3 and has a hardness of 72.5 ± 12.5 N. Two rooted plantlets were placed in each foam in the Magenta vessels with gas permeable polyvinylchloride lids [4].

2.3. *In vitro* inoculation with AMF

Ten spores were taken up in a micro-pipette (set volume was 10 µl) in a transparent sterile tip. These were then transferred to the root zone underneath the crown of the rooted microplant in the imbibed Hortifoam.

2.4. Determination of mycorrhizal colonisation

Samples of root material were cleared and stained with 0.05% trypan blue [15] and the percentage root length colonisation was determined using the grid line intersect method of Giovanetti and Mosse [16].

2.5. Measurement of *in vitro* and *in vivo* photosynthetic rates

In vitro and *in vivo* photosynthesis rates were measured by infra-red gas-analysis (IRGA), using a CIRAS-1 system (PP Systems, Herts., UK) [12].

3. RESULTS

3.1. AMF spore germination assay

Preliminary spore germination assays were carried out to determine the influence of the salt components of the medium. The data (Table 1) show that germination was inhibited by approx. 50% in full strength Murashige and Skoog [14] mineral salts (macro and trace elements, no organics) but was not inhibited in half-strength medium.

Table 1. Influence of the mineral components of Murashige and Skoog [14] medium on the germination of AMF spores

Treatment	Median	95% confidence limits
Acid water agar (pH 6.2)	45	[30.0–60.0]
Half MS (pH 6.2)	50	[22.2–64.3]
Full MS (pH 6.2)	26.2	[11.1–44.4]

3.2. Decontamination of AMF spores

BEG-31 spores surface sterilised using procedure (i) were contaminated with a range of fungi and bacteria, whereas spores surface sterilised by procedure (iii) were least contaminated. The procedures did not reduce spore viability. Isolates from procedures (ii) and (iii) were grouped and further characterised using API 20NE kits. Two bacteria were isolated after procedure (ii); the first was identified as *Aeromonas salm. salmonicida*; the second isolate was Gram positive and tentatively identified as a *Lactobacillus* sp. One isolate was isolated following procedure (iii), identified as common environmental bacterium and tissue culture contaminant *Xanthomonas maltophilia* (98.8% i.d.).

3.3. Confirmation of mycorrhization of *in vitro* inoculated microplants

Spore germination assays (Table 1) had shown a germination rate of *ca.* 50%, consequently, 'Elvira' microplants were inoculated with four spores/microplant. Mycorrhizal establishment *in vitro* was confirmed with mean total colonisation 15.2 +/- 2.1%. Mean arbuscular colonisation was 10.6+/- 1.8%. This experiment was repeated with 'Tenira'.

3.4. Characterisation of *in vitro*-mycorrhized microplants

Percentage of 'Elvira' microplant establishment in non-mycorrhized plants was 93%, increasing to 97% in mycorrhized plants. No special facilities, e.g. misting or fogging, were provided to assist establishment.

Measurement of photosynthetic activity *in vitro* and after establishment showed no inhibition of photosynthesis *in vitro* in the gnotobiotic ('Elvira' microplant-AMF-*X. maltophilia*) culture (Table 2).

Table 2. Photosynthetic activity *in vitro* and *in vivo* in 'Elvira' microplants from agar-gelled heterotrophic cultures (control); non-mycorrhized autotrophic cultures and mycorrhized autotrophic cultures. Data were determined 15 days and 1 day before establishment

Treatment	Photosynthetic activity (mmol CO_2/s/unit leaf area)	
	Day -15	Day -1
Agar (-AMF)	0.26 ± 0.14	0.18 ± 0.05
Foam (-AMF)	0.31 ± 0.11	0.37 ± 0.04
Foam (+AMF)	0.26 ± 0.11	0.37 ± 0.11

4. DISCUSSION

Micropropagation has established a niche in the plant production industry; in Europe alone over 180 million microplants are produced each year that have an estimated annual market value in excess of 54 million ECU [17]. To be competitive, the micropropagator must produce propagules that satisfy the customers' requirements at an economic price. To increase profits the micropropagator can seek to increase efficiency, e.g. reducing losses due to disease *in vitro* and at establishment will decrease unit costs. Additionally, profits can be increased by producing propagules which will command a higher price, e.g. novel genotypes, high health status material, or microplants pre-inoculated with beneficial micro-organisms.

The results presented here demonstrate that mycorrhization is possible in commercial autotrophic culture media under standard growth conditions. In this system mycorrhization can be achieved without competition from resident substrate micro-organisms [12], including wild strains of AMF, and this increases the efficiency of inoculation by reducing the amount, i.e. cost, of inoculum. The inoculum used here was non-axenic but the microbial contaminants, while arbitrarily selected, were characterised. This inoculum was used to confirm that gnotobiotic autotrophic cultures with AMF and bacteria were possible. Further studies are under way to establish AMF cultures with characterised beneficial bacteria.

REFERENCES

1. Herman, E.B. (1990) Acta Hortic. 280, 233–238.

2. Schumann, G.L. (1991) Plant Diseases: Their Biology and Social Impact, APS Press, St. Paul, USA.

3. Preece, J.E. and Sutter, E.G. (1991) in Micropropagation: Technology and Application (Debergh P.C. and Zimmerman R.H., eds) pp. 71–94, Kluwer, Dordrecht.

4. Cassells, A.C. and Walsh, C. (1994) Plant Cell Tissue Org. Cult. 37, 171–178.

5. Cassells, A.C. and Tahmatsidou, V. (1996) Plant Cell Tissue Org. Cult. (in press).

6. Schönbeck, F., Grunewaldt-Stöcker, G. and Von Alten, H. (1994) in Epidemiology and Management of Root Diseases (Campbell C.L. and Benson D.M., eds) pp. 65–81, Springer-Verlag, Berlin, Germany.

7. Wang, H., Parent, S., Gosselin, A. and Desjardins, Y. (1983) J. Am. Soc. Hortic. Sci. 118, 896–901.

8. Uosukainen, M. and Vestberg, M. (1994) Agric. Sci. Finl. 3, 269–279.

9. Mark, G.L. and Cassells, A.C. (1996) Plant Soil 185, 233–239.

10. Schenck, N.C., Graham, S.O. and Green, N.E. (1975) Mycologia 67, 1189–1192.

11. Tommerup, I.C. (1985) Trans. Br. Mycol. Soc. 85, 267–278.

12. Cassells, A.C. and Mark, G.L. (1996) Agronomie 16, 563–571.

13. Schaad, N.W. (1980) Laboratory Guide for Identification of Plant Pathogenic Bacteria, APS Press, St. Paul, MN, USA.

14. Murashige, T. and Skoog, F. (1962) Physiol. Plant. 15, 473–497.

15. Philips, J.M. and Hayman, D.S. (1970) Trans. Br. Mycol. Soc. 55, 158–160.

16. Giovanetti, M. and Mosse, B. (1980) New Phytol. 84, 489–500.

17. Lovato, P.N., Gianinazzi-Pearson, V., Trouvelot, A. and Gianinazzi, S. (1996) Adv. Hortic. Sci., 10, 46–52.

ACCLIMATION RESULTS OF MICROPROPAGATED BLACK LOCUST (*ROBINIA PSEUDOACACIA* L.) IMPROVED BY SYMBIOTIC MICRO-ORGANISMS

I. BALLA[1], J. VÉRTESY[1], K. KÖVES-PÉCHY[2], I. VÖRÖS[2], Z. BUJTÁS[3] and B. BÍRÓ[2]

[1]*Research Institute for Fruitgrowing and Ornamentals, Budapest, XXII. Park u. 2, POB 108 H-1775, Hungary*
[2]*Inst. for Soil Sci. and Agric. Chemistry of the Hungarian Academy of Science Budapest, II. Herman O. u. 15. H-1022, Hungary*
[3]*Forest Research Institute Budapest, II. Frankel Leo u. 42-44, H-1023, Hungary*

1. INTRODUCTION

Robinia pseudoacacia is the most important of the fast-growing forest tree species in Hungary [1]. It is selected for varieties with (1) straight stem for timber production, (2) long flowering time for honey production, (3) large pods as deer-fodder, (4) decorative pink flowers as ornamental trees.

Micropropagation of several varieties has been elaborated to speed up production [2–4] and to produce healthy material in surroundings infected with *Robinia* mosaic virus and Tomato black ring virus. Good quality *in vitro* plants could be grown under aseptic conditions, but a lot of losses occurred during acclimation where several tree species survived satisfactorily [5,6]. Nodules appearing on newly formed roots remained sterile in peat-perlite substrates which induced us to look for an adequate inoculation method with *Rhizobium* strains.

Vesicular arbuscular mycorrhizal fungi (AMF) are beneficial for improving nutrient uptake in almost 90% of higher plants. With black locust both endo- and ectomycorrhizal symbiosis can be found [7]. Inoculation with selected microsymbionts may improve the survival rate and growth vigour of micropropagated plants [8,9].

2. MATERIALS AND METHODS

The varieties propagated were collected for the initiation of cultures from adult trees, standing in a collection of varieties of the Forest Research Institute. Disinfected 10- to 15-mm long shoot tips were put directly on proliferation medium, consisting of modified major and minor elements of MS [10], where ammonium-nitrate was omitted and casein hydrolysate (250 mg/l) was added. The rate of multiplication varied from 2.5 to 3.6

A.C. Cassells (ed.), Pathogen and Microbial Contamination Management in Micropropagation, 351–354.

depending on the variety. Shoots about 10 mm long were used for root induction in a high concentration IBA medium and rooting in a half-strength hormone-free MS medium containing charcoal.

The rooted plantlets were acclimated in a mixture of 3:1 peat/perlite under high relative humidity in glasshouse conditions.

Rhizobium bacteria were isolated from root nodules of some black locust varieties by the Vincent method [11]. After some *in vitro* cleaning procedure 35 strains (from the 100 original ones) selected for further investigations and for checking their azote-fixing capacity in black locust seedlings. *Rhizobium* strains were further selected with regard to their infectivity and/or efficacy and kept on YEM solid medium at 5°C in a refrigerator.

Intact spores of vesicular arbuscular micorrhizal fungi belonging to the *Glomus* species were extracted by the wet sieving method of Gerdemann and Nicolson [12]. Sterilized spores (30 of each plant) were used for inoculation of maize (*Zea mays* L.) and green pea (*Pisum sativum* L.) grown in gamma-irradiated soil. After 3 months of growth endomycorrhizal colonization was controlled by the trypan-blue method of Kormanik *et al.* [13]. Soil containing infected roots was used for experimental mycorrhization. *Glomus* sp. strain was placed as a thin layer beneath the plantlet roots.

A 1 cm^3 *Rhizobium* suspension ($2-6 \times 10^8$ CFU/cm^3) was mounted on top of soil at each plantlet using ten strains in four replicates.

In vitro propagated plantlets were inoculated with both of the micro-organisms at the time of transplantation into the glasshouse and the *Rhizobium* inoculation was repeated 2 months later.

Survival rate, root nodulation, acetylene fixation activity (ARA; [14]), growth vigour and plant height were measured after 2.5 and 18 months of transplantation.

3. RESULTS AND DISCUSSION

Survival rate of micropropagated black locust has always increased following inoculation with *Rhizobium* strains, originating from five different varieties of *Robinia pseudoacacia* L. (Table 1). The increase of survival rate is highly dependent on the strain origin. The most effective *Rhizobium* sp. strains were isolated from the 'Bajti' black locust variety.

The starter positive effect of *Rhizobium* inoculation in most cases lasted over the later growing period, but other strains, not effective at first — like 'G 6/1' — were also able to develop a beneficial growth stimulation later.

Table 1. Effects of *Rhizobium* and endomycorrhizal (AMF) inoculation on the vegetative growth and acetylene fixation activity of micropropagated 'Rózsaszín AC' black locust variety

Treatment AMF + *Rhizobium* strains	No. of inoculated plantlets	Survival rate (%)	2.5-Month-old plantlets			1.5-Year-old plantlets	
			Height (mm)	Mass (G/PL)	ARA (C_2H_4/PL/h)	Shoot growth (mm)	Height (mm)
Control	240	21.0	415	20.88	0	1460	580
AMF (GL)	80	20.0	180	10.28	0	0	0
AMF+B 5/1	80	17.5	180	10.28	0	800	400
B 5/1	80	40.0	463	24.30	0.22	2600	880
AMF+R 3/1	80	37.0	180	5.80	0	3220	920
R 3/1	80	40.0	380	36.90	1.81	3040	990
B 2/1	80	35.0	620	28.17	0.99	730	580
B 11/1	80	52.0	555	29.03	5.85	1620	710
G 6/1	80	28.8	440	20.55	5.67	3430	1080
J 4/1	80	51.3	580	36.79	0.85	2630	920

No positive correlation whatsoever was found between the acetylene fixation activity (ARA) and the green yield or the plant height after 1.5 years of growth, suggesting the necessity of a multilevel *in vitro* strain selection. In this particular study, 100 *Rhizobium* strains have been isolated, and only a few proved to be especially effective. Strain 'R 3/1', originating from the host 'Rózsaszín AC', black locust variety proved to be no more efficient than strains of other varieties.

AMF inoculation, on the contrary, did not have any positive effect on the survival rate of micropropagated black locust plantlets. The growth rate of the AMF inoculated plantlets was significantly lower than that of the control, or those inoculated only with *Rhizobia*. It may be supposed that some AM isolates act as parasitic fungi at the very beginning of growth. The same result was found by Safir [15].

Growth was not only retarded by the inoculation of an AMF isolate alone but also when it was used together with *Rhizobia* at the time of transplantation. It has been found that micropropagated *R. pseudoacacia* trees inoculated only with a *Glomus* sp., without *Rhizobia*, could not survive the first year of growth.

The latter effect of micro-organisms highly depends on the strains used. The results of 1.5 years of growth show that, in the case of the same *Glomus* sp., inoculation could increase the total shoot growth and plant height in combination with 'R 3/1' *Rhizobium* strain, compared to the only 'R 3/1' inoculation, or could retard the plant development, when, e.g., combined with 'B 5/1' *Rhizobium* strain.

In most cases a positive correlation of inoculation and plant vigour was registered but further investigations are necessary to find the most effective candidates for developing a beneficial tripartite symbiosis with the requested black locust variety, selected for the particular environmental conditions. It can be supposed that growth can be further enhanced by the use of an adequate AMF strain and a carefully determined time of infection.

ACKNOWLEDGEMENTS

The authors are grateful to the National Committee of Technical Development, Hungary for financial support.

REFERENCES

1. Keresztesi, B. (1984) Az akác, Akadémia Kiadó, Budapest, p. 20–27.

2. Brown, C.L. (1980) in Tissue Culture in Forestry (Bonga J.M. and Durzan D.T., eds.) pp. 137–145. Martinus Nijhoff, The Hague, Boston, London.

3. Balla, I. and Vertesy, J. (1985) in In Vitro Problems Related to Mass Propagation of Horticultural Plants, Symposium, Book of Abstracts II, p. 37. Int. Soc. of Hort. Sci, Gembloux.

4. Barghchi, M. (1987) Plant Sci. 53, 183–189.

5. Han, K.H. and Keathley, D.E. (1989) Robinia 112–114.

6. Balla, I., Koves-Pechy, K. and Vertesy, J. (1990) In: Proc. of 7th Congr. on Plant Tissue and Cell Micropropagation, Amsterdam, p. 23.

7. Harley, J.L. and Harley, E.L. (1987) New Phytol. 105, 1–102.

8. Calvet, C., Pera, J., Estaun, V. and Camprubi, A. (1989) Agronomie 9, 181–185.

9. Guillemin, J.P., Gianinazzi, S. and Gianinazzi-Pearson, V. (1991) Fruits 46, 355–358.

10. Murashige, T. and Skoog, F. (1962) Physiol. Plant.15, 473.

11. Vincent, J.M. (1970) A manual for the practical study of root-nodule bacteria. J.B.P Handbook, No. 15, Blackwell Sci. Publ. H., Oxford, Edinburgh.

12. Gerdemann, J.W. and Nicolson, T.H. (1963) Trans. Br. Mycol. Soc. 46, 235–244.

13. Komnanik, P.P., Bryan, C.W. and Schultz, R.C. (1980) Can. J. Microbiol. 26, 536–538.

14. Hardy, R.W.F., Bums, R.C., Hosten, R.D. (1973) Soil Biol. Biochem. 5, 47–81.

15. Safir, G.R. (1987) in Ecophysiology of VA Mycorrhizal Plants. (Safir G.R., ed.) p. 224, CRC Press, Inc. Boca Raton, FL.

A REVIEW OF TISSUE PROLIFERATION OF *RHODODENDRON*

RICHARD H. ZIMMERMAN

US Department of Agriculture, Agricultural Research Service, Fruit Laboratory, 10300 Baltimore Avenue, Beltsville, MD 20705-2350, USA

1. INTRODUCTION

Tissue proliferation (TP), a disorder first found in the mid-1980s on micropropagated rhododendrons, is characterized by the formation of gall-like or tumor-like growths on the lower portion of the stem above or below the soil line. The superficial similarity in appearance of the gall-like growths of TP to crown gall, incited by the bacterium *Agrobacterium tumefaciens*, has created considerable concern among growers and nursery inspectors. Significant economic loss has occurred when nurserymen destroyed plants showing the disorder rather than selling ones that they thought might be diseased.

In the late 1980s, a series of reports [1–5] were published detailing variation in micropropagated plants of *Rhododendron* as they grew in the nursery. These variations included changes in flower form and color, leaf shape and variegation, plant form and growth habit (particularly dwarfed shoots).

The first published report specifically about the disorder now known as tissue proliferation appeared in 1992 [6] and additional descriptions of this disorder and its effect on rhododendrons soon appeared [7–12]. By the time these first reports appeared, research on the problem had already begun with the initial emphasis on attempting to isolate pathogenic strains of *Agrobacterium* from the gall-like tissue of plants exhibiting the disorder. In addition, nurseries were surveyed to determine if there were any clear links to particular cultivars, micropropagation laboratories, nursery practices, herbicides, fertilizers, or other factors.

Several regional meetings were held among growers, laboratory operators, extension personnel and research scientists to exchange information in order to understand the magnitude of the problem and to seek solutions. These culminated in a workshop in early 1993 bringing together scientists from across the US who were already working on the problem or with experience in areas thought to be related to the problem [13,14]. This

A.C. Cassells (ed.), Pathogen and Microbial Contamination Management in Micropropagation, 355–362.
© *1997 Kluwer Academic Publishers. Printed in the Netherlands.*

workshop served to clarify the problem and to define areas in which research was feasible and needed. Another workshop of the same group of scientists held in early 1996 summarized progress to date on the various research programs underway. The current status of the problem in the nursery industry was also summarized as was the response of commercial micropropagation laboratories to the situation.

2. DESCRIPTION OF TISSUE PROLIFERATION

Tissue proliferation is characterized by gall-like growths on the trunk or lower stems of the plant above, at, or below the soil line, that usually appear after the plants are 1–3 years old [9,15]. Similar spherical growths on roots, usually at the tips, near the soil surface have been noted [16], but it is still uncertain whether these are characteristic of TP. The surface appearance of the growths resembles callus tissue, but it may be very hard. The growths can sometimes be manually broken off the stem with little effort. Shoots, probably adventitious, may grow from the gall-like growths; these are sometimes numerous but usually small in size with reduced leaf size. These shoots have short internodes and may soon die to be replaced by new shoots [7]. Internally, the growths have vascular tissue, although it may be distorted. Stem girdling can occur and girdled stems may break off. Poor root and shoot growth may also be related to girdling.

3. SCOPE OF THE PROBLEM

More than 35 cultivars of *Rhododendron* as well as several of *Kalmia latifolia* have been found with TP. Although it was originally identified only on elepidote cultivars, TP has also been found on several lepidote types and one deciduous azalea cultivar (Table 1).

Tissue proliferation was first seen on micropropagated plants and was originally thought to be limited to such plants. However, it has now been found on cutting-propagated [15,16] and seedling [15] plants, although the frequency is much lower than on tissue-cultured plants.

Plants with TP have been reported in at least 15 states of the US as well as Canada and the UK [15] (E. Dutky, personal communication; K. Mudge, personal communication). The US states are Washington, Oregon and California in the west with the remaining states all east of the Mississippi River stretching from Michigan south to Georgia and east to Massachusetts, thus covering the major part of the country in which nurseries produce *Rhododendron* and *Kalmia*.

Table 1. *Rhododendron* cultivars on which tissue proliferation has been observed. Lepidote cultivars are indicated by an asterisk

'Aloha'	'Chionoides'	'Montego'
*'Aglo'	'Cunningham's White'	'Mrs. T.H. Lowinsky'
'Anna Kruschke'	'Dopey'	'Nova Zembla'
'Anna Rose Whitney'	'Elisabeth Hobbie'	'Rio'
*'April Rain'	'Grace Seabrook'	'Rocket'
'Bambino'	'Hallelujah'	'Rutgers'
'Bessie Howells'	'Henry's Red'	'Scintillation'
*'Black Satin'	'Holden'	'Solidarity'
'Boule de Neige'	'Hong Kong'	'Spellbinder'
'Bravo'	'Lee's Dark Purple'	'The Honorable Jean Marie de Montague'
'Calsap'	'Lodestar'	'Vulcan'
'Catawbiense Album'	'Marlene Peste'	'Yaku Prince'
'Centennial Celebration'	'Molly Fordham'	

4. POTENTIAL CAUSES OF TISSUE PROLIFERATION

4.1. Pathogens

Crown gall disease is reported to occur on rhododendron [17], although this occurrence has been questioned [18]. Since the gall-like growths of TP often closely resemble crown gall, initial research efforts were concentrated on determining whether the disorder was, in fact, crown gall. *Agrobacterium* was isolated from gall-like growths on rhododendron, but these isolates did not induce galls on test plants or upon reinoculation in rhododendron [12, 14, 15]. These isolates also did not react to T-DNA probes used for detecting pathogenic strains of *Agrobacterium*. Furthermore, TP galls have vascular differentiation, which crown gall tissue lacks (S. James, unpublished).

Inoculation tests on tomato and rhododendron, using more than 100 strains of potentially pathogenic bacteria isolated from TP galls, resulted in no gall formation or TP on tomato and rhododendron test plants [18]. Similarly, inoculation of rhododendron with TP pieces or extracts produced no TP in the test plants [18]. Inoculation of 'Scintillation' rhododendron with several virulent *Agrobacterium* strains also did not result in any gall formation after more than 1 year [18]. Similar negative results were obtained after stab inoculations of nine other cultivars with *A. tumefaciens, A. rhizogenes, A. radiobacter* or *Nocardia vaccinii*, which infects *Vaccinium* species (Ericaceae) forming galls and adventitious shoots at or below the soil surface [19] similar to the symptoms seen in tissue proliferation on rhododendron (M. Brand, personal communication). Another study with *N. vaccinii* did not result in gall formation on plants of 'Nova Zembla' [15].

Other biotic agents known or thought to cause gall formation in other plants, including *Erwinia herbicola*, have not been identified in affected tissues of rhododendron [15]. Although it now appears unlikely that TP is a disease, the possibility of an unknown pathogen not yet isolated and identified cannot be completely ruled out.

4.2. Genetic variation

Propensity to form TP galls varies with cultivar. Some cultivars show a high proportion of affected plants, whereas most cultivars now being micropropagated seem not to show the disorder at all. Clearly the majority of susceptible cultivars are elepidotes (Table 1).

The occurrence of TP on micropropagated plants led to speculation that affected plants have undergone a genetic change. A mutation is considered unlikely and no evidence of a gross chromosomal rearrangement was detected using RAPD markers [20]. The fact that TP occurs on seedlings, on cutting propagated plants and on micropropagated plants of so many different rhododendron cultivars with such widely varying parentage also argues against a genetic (chromosomal) change being the cause. It is unlikely that the same mutation would occur repeatedly in such a broad spectrum of plants.

Epigenetic variation, a change in the timing of gene expression, is a common occurrence in plants. Typical examples are the distinct differences in leaf shape and plant form associated with normal phase change, or with rejuvenation seen in tissue-cultured plants as changes in leaf shape and greater ease of rooting. Other epigenetic changes that seem to occur frequently in tissue-cultured plants include increased branch angle and formation of more branches (i.e. release of axillary buds from correlative inhibition).

Habituation is another example of epigenetic variation that has been found with 'Montego' rhododendron on which TP-like galls will form *in vitro*. Cytokinin-habituated cultures of this cultivar have proliferated readily on cytokinin-free medium for more than 5 years. More recently, shoot cultures established from cuttings taken from the original plant of this cultivar showed no evidence of TP *in vitro* on medium containing 10 μM 6-(γ,γ-dimethylallylamine) purine (2iP) after several years [21,22]. Shoots from TP-negative cultures were then transferred to medium containing either 10 or 50 μM 2iP and maintained for three 5-week culture periods. TP symptoms developed with either 10 or 50 μM 2iP, but the frequency and extent of TP symptoms was considerably lower on 10 μM [22]. Cultures of 'Montego' established from nursery plants exhibiting TP symptoms adapted to *in vitro* culture much more rapidly and showed TP symptoms *in vitro* even when grown on cytokinin-free medium [21,22]. Induction of TP symptoms *in vitro* requires formation of adventitious buds, but adventitious bud formation does not necessarily result in TP symptoms (M. Brand, personal communication). It should be noted that other TP-prone cultivars have not shown similar habituation *in vitro*.

4.3. Lignotubers

Another phenomenon that has been considered as having a possible relationship to TP is the inherent propensity for some ericaceous species to form lignotubers or burls. Lignotubers (also called basal burls [23]) are woody outgrowths at the stem base that contain numerous dormant buds and develop as part of the normal ontogeny of the plant [24], although basal swelling of the stem rather than an outgrowth may be present (K. Mudge, personal communication). In nature, these buds develop into shoots after injury to the plant, e.g. destruction of the shoot system by fire. Lignotuber-forming species include *Rhododendron griersonianum, R. maximum, R. occidentale* and *R. ponticum* [24,25] as well as *Kalmia latifolia* [23,24]. Thus, the possibility exists that TP is related to lignotuber formation. Although tissue-cultured plants of *Kalmia latifolia* produced basal burls, these had only one-third as many bud clusters after 7 years as did the basal burls on 4-year-old nursery-grown seedlings [23], so the tissue-cultured plants do appear to differ from seedling material in this regard. Also, plants of several cultivars showing TP symptoms did not produce shoots from the galls even when the plants were cut back to the galls [25,26] (K. Mudge, personal communication). At present, it seems unlikely that lignotuber formation is the explanation for TP, although propensity for lignotuber formation may have some bearing on the development of TP.

4.4. Nursery and landscape cultural practices

TP has been more evident on container-grown plants than on field-grown ones, either because of higher frequency or more developed symptoms. Container production uses porous, lightweight growing mixes that require more water and fertilizer, resulting in more vigorous growth. This, in turn, requires more frequent and heavier pruning or use of growth regulators to suppress excessive growth. The containers are set above ground, exposing the root systems to greater temperature fluctuations than those of field-grown plants. All of these factors impose more stress on the container-grown plants, which might induce the development of TP if other predisposing factors are present.

A nursery survey showed that cutting propagated rhododendrons developed TP less often than micropropagated plants of the same cultivar [16]. When cuttings were taken from tissue-cultured stock plants, TP occurred on up to 100% of the plants [16]. In another study, cutting propagated plants produced from TP-positive micropropagated stock plants developed TP symptoms only with 'Montego' (M. Brand, personal communication).

When micropropagated plants of *Rhododendron* 'Solidarity' were grown in 20-cm diameter ("2 gallon") nursery containers for 19 months at two fertility levels, 62% of the plants receiving the higher fertilizer rate had TP, whereas only 23% of those with the lower rate did [15]. The fertilizer was a controlled release formulation applied annually in the spring with the high rate slightly above the manufacturer's recommendation and the low rate half of the high rate. Plants evaluated after two growing seasons showed a significant correlation between median plant size and number of plants with TP.

Five cultivars of rhododendrons, with or without TP, were transplanted from containers to a field where they were grown in soil for 3 years under conditions similar to a landscape planting [16]. Since the site was more exposed than a normal landscape planting of rhododendrons, plant mortality was higher than normal, particularly during the first growing season. Plants with TP galls existing at time of field planting had significantly higher mortality after the first season than those without galls [16]. In a separate study, herbicide (oryzalin) application did not increase TP symptoms on 'Montego', but doubling the rate of herbicide showed some tendency to increase the number of plants with galls [16].

In another study, two rhododendron cultivars ('Montego' and 'Lee's Dark Purple'), each with or without TP, were compared by growing in the field at two locations at two levels of intensity of maintenance to determine treatment effects on susceptibility to *Phytophthora* infection and black weevil feeding injury. High maintenance plots received herbicide (isoxaben and oryzalin), insecticide (fluvalinate) and fertilizer treatments whereas low maintenance plots received none of these. No difference attributable to TP status was found in susceptibility to *Phytophthora* or in attraction to and damage by black vine weevils [18].

One puzzling aspect of the TP problem has been the persistent reports that a laboratory could deliver micropropagated plants from the same production lot to two different nurseries with one nursery then having severe problems with TP on these plants and the other nursery having none [14,25]. This situation hints at nursery practices as having a role in TP development, but definitive studies have not yet been completed.

5. SOLUTIONS TO THE PROBLEM

Control of the TP problem will require adjustments by both laboratories and nurseries. Many changes have already been implemented that seem to be reducing the frequency and severity of TP in nurseries.

Firstly, micropropagators need to follow laboratory procedures that are advisable for any commercially produced crop [27]. These include regular reinitiation of cultures from non-tissue-cultured, correctly identified stock plants, use of minimal growth regulator (particularly cytokinin) concentrations in the medium, avoidance of adventitious shoot production, reculturing only of stem tips from axillary shoots, rooting only of shoot tips and not of the bases, and cold storage of cultures to reduce the number of transfers.

Secondly, the cultivars to be produced in tissue culture must be selected carefully. Those that are very prone to developing TP need to be treated very carefully in the production system or may need to be eliminated from the micropropagation system. Laboratories have already started this process. When the cultivars in question are very important economically, a return to conventional cutting propagation may be necessary.

Thirdly, nursery production practices need to be re-evaluated, particularly when container production is used. It may become advisable to modify the growing mix to lessen the fertilizer requirements and improve the water-holding capacity. Growth retardant and herbicide use needs to be studied carefully to determine if lower rates or less frequent application is feasible.

6. CONCLUSIONS

Tissue proliferation is a disorder primarily of micropropagated rhododendrons that was discovered in the 1980s. It has caused significant economic loss to nurserymen and commercial micropropagation laboratories because its similarity in appearance to crown gall disease and possible reduction in plant quality led to the destruction of many plants. Although the cause(s) of TP is still unknown, greater awareness of the symptoms by growers, nursery inspectors, extension personnel and laboratory operators has moderated the crisis atmosphere that existed several years ago. Ongoing research may provide information on the cause and control of TP. In the meantime, continuing efforts by laboratory operators and nurserymen to modify past procedures seem to be enabling the growers to circumvent the problem to a large extent.

REFERENCES

1. Mezitt, R.W. (1987) Comb. Prcc. Int. Plant Prop. Soc. 37, 403–407.

2. Cross, J. (1987) Comb. Proc. Int. Plant Prop. Soc. 37, 407–410.

3. Mezitt, R.W. (1988) Comb. Proc. Int. Plant Prop. Soc. 38, 566–570.

4. Anonymous (1989) Am. Nurseryman 169(6), 15,17.

5. Knuttel, A.J. (1989) Am. Nurseryman 170(11), 43,45–49.

6. Anonymous (1992) Am. Nurseryman 175(3), 15,18.

7. Brand, M.H. (1992) Am. Nurseryman 175(5), 60–62,64–65.

8. Anonymous (1992) Am. Nurseryman 176(1), 17,20.

9. Lamondia, J.A., Rathier, T.M., Smith, V.L., Likens, T.M. and Brand, M.H. (1992) Yankee Nursery Q. 2(2), 1–3.

10. Wells, J. (1992) Am. Nurseryman 176(4), 8.

11. Rostan, T.J. (1992) Am. Nurseryman 176(10), 23,28.

12. Brand, M. and Kiyomoto, R. (1992) Comb. Proc. Int. Plant Prop. Soc. 42, 530–534.

13. Rostan, T. (1993) Am. Nurseryman 177(7), 15–16.

14. Linderman, R.G. (1993) Am. Nurseryman 178(5), 56–67.

15. McCulloch, S.M. and Britt, J.L. (1997) HortScience 32, in press.

16. Mudge, K.W., Lardner, J.P., Mahoney, H.K. and Good, G.L. (1997) HortScience 32, in press.

17. Moore, L.W. (1986) in Compendium of Rhododendron and Azalea Diseases (Coyier D.L. and Roane M.K., eds) pp. 29–30, APS Press, St. Paul, MN, USA.

18. LaMondia, J.A., Smith, V.L. and Rathier, T.M. (1997) HortScience 32, in press.

19. Demaree, J.B. and Smith, N.R. (1952) Phytopathology 42, 249–252.

20. Rowland, L.J., Levi, A. and Zimmerman, R.H. (1997) HortScience 32, in press.

21. Kiyomoto, R.K. and Brand, M.H. (1994) HortScience 29, 516 (abstract).

22. Brand, M.H. and Kiyomoto, R. (1997) HortScience 32, in press.

23. Del Tredici, P. (1993) Comb. Proc. Int. Plant Prop. Soc. 42, 476–482.

24. James, S. (1984) Bot. Rev. 50, 225–266.

25. Maynard, B.K. (1995) Comb. Proc. Int. Plant Prop. Soc. 45, in press.

26. Brand, M.H. and Kiyomoto, R. (1993) Yankee Nursery Q. 3(4), 5–6.

27. Brand, M.H. (1992) Am. Nurseryman 175(5), 66–71.

PROBLEMS WITH PLANT HEALTH OF *IN VITRO* PROPAGATED *ANTHURIUM* SPP. AND *PHALAENOPSIS* HYBRIDS

GISELA GRUNEWALDT-STÖCKER

Institut für Pflanzenkrankheiten und Pflanzenschutz, Universität Hannover, Herrenhäuser Straße 2, D-30419 Hannover, Germany

1. INTRODUCTION

Productive plants are the backbone of a successful horticultural company. Therefore, the business manager has to direct special attention to maintaining plant health in its comprehensive meaning. Since about 1989 the quality of young *Anthurium* and *Phalaenopsis* plants of several specialist German nurseries has deteriorated to such an extent that the existence of some companies has been threatened. Tests revealed that plant pathogens such as root-rot fungi and bacteria or pests, e.g. phytophagous mites, played only marginal roles and seemed to be of secondary importance as causal agents. There were many discussions on mistakes in cultivation and on unknown abiotic and biotic factors which might account for the loss of vitality; however, diverse manipulations of plant protection and plant nutrition showed no lasting improvement in plant growth and health. Moreover, these failures led to the assumption that earlier abiotic stress during the *in vitro* propagation might have caused the problems.

As the damages and economic losses increased beyond the scope of a single company, the Ministerium für Umwelt, Raumordnung und Landwirtschaft in Düsseldorf supported a case study by the author in 1995 [1]. The aim was to collect basic facts on the phenomenon of loss of vitality in *Anthurium scherzerianum* and *Anthurium andreanum* and in *Phalaenopsis* hybrids in order to specify central questions relevant to the practical situation, on which a causal analysis should be concentrated. Such an analysis in a scientific, experimental sense has not yet been carried out and it seems rather difficult in *Anthurium* and *Phalaenopsis* because of the long-lasting culture with many factors successively influencing plant health.

This subject is presented as one example of a complex task in phytomedicine: to define and characterise the healthy plant, even in the absence of pathogens and pests.

A.C. Cassells (ed.), Pathogen and Microbial Contamination Management in Micropropagation, 363–370.
© *1997 Kluwer Academic Publishers. Printed in the Netherlands.*

2. CASE STUDY

With information gathered in 11 laboratories and 14 nurseries in Germany, Belgium and The Netherlands on the damage phenomenon in *Anthurium* and *Phalaenopsis*, a factual basis was established for the evaluation of potential causes. As the problem is located in the economic domain of ornamental plant production with enormous competition, the economic interests of involved as well as non-involved companies limited adequate cooperation in clarification of the essential details in retrospective study of the damage period of more than 5 years. In particular, the in vitro growth media and methods actually used for producing either diseased or healthy plantlets were not disclosed.

With regard to the scientific literature and with the support of qualified scientific experts, the actual symptoms, the occurrence and the extent of the damage were investigated. The subsequent assessment of potential causes was guided by working hypotheses: these were divided into abiotic and biotic factors for *in vitro* and *in vivo* domains.

3. RESULTS

3.1. Symptoms of damaged greenhouse plants

The main symptoms of *in vitro* propagated young plants, beginning in community boxes and later on in multipot plates or single pots, are summarized as follows.

Anthurium: dry wilt during acclimatisation;
wilting after transplanting, stem and root rot;
retarded growth, growth stop on shoot and roots;
hardening of the roots, anomalous root morphology;
long time suppression of new growth,
only limited recovery after months of loss of vitality.

Phalaenopsis: reduced leaf size and leaf thickening, leaf deformation;
disturbance of apical dominance;
total growth stop on shoot and roots;
interruption of root elongation and of new root formation;
early rot of *in vitro* formed roots;
partial root rot and root constrictions;
wilting of young and older plants;
only limited recovery after months of loss of vitality.

There was no convincing evidence that these symptoms were caused by plant pathogens or pests.

3.2. Occurrence and extent of damage

From the inquiry into the circumstances of loss of vitality in several nurseries the following facts became apparent:

• The losses concern different plant species propagated by means of different *in vitro* propagation methods.

• They occurred in *Phalaenopsis* hybrids of different origins and from different plant breeders, especially after 'meristem' cloning, but also in seedlings. Similarly, *A. scherzerianum* clones of different breeders were damaged. Recently, *A. andreanum* and *Spathiphyllum floribundum* were included in the problem.

• Symptoms and damage are generally not estimated as clone-specific changes *in sensu* somaclonal variation [2,3]. There exist differences in the extent of damage in relation to plant genotype and the duration of its *in vitro* propagation (the longer the period, the greater the loss of vitality that can be demonstrated for *Anthurium* varieties). There exist differences in the degree of damage between single plants of a plant set, but almost all plants are far from healthy looking.

• Over the years the losses increased continuously within a nursery: retarded or completely stopped development of the young plants, impeded set sales on time, made parts of the set unsaleable and afforded additional nursery costs for the remaining slow-growing plants. Losses ranged from 30 to 100% per set.

• The problematic lots of *Anthurium* and *Phalaenopsis* were supplied by different *in vitro* laboratories.

• The losses showed up under different greenhouse conditions and were not limited to the first *in vivo* phase of young plants.

As a general conclusion, loss of vitality and disfunction of young plants was not a problem in companies that propagate own varieties in their laboratory, incorporated in the complete production cycle inclusive of greenhouse cultivation to mature plants. Therefore, many arguments indicate that plant propagation based on divided labour between laboratory and greenhouse of different companies is a fundamental part of the described problems (Fig. 1). Here, economic competition forces minimization of micropropagation costs, e.g. by increasing yield of plantlets per explant, a fact that might result in a loss of plant health.

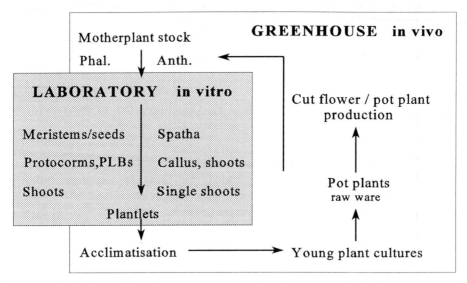

Figure 1. Production of *in vitro* propagated *Anthurium* spp. and *Phalaenopsis* hybrids based on divided labour in laboratory and greenhouse level in different companies

3.3. Evaluation of potential causes of damage

The overall results of the assessment of potential causes of damage are summarized in Table 1. It was obvious that one single damage factor could not be the reason for the trouble over several years. According to the working hypotheses, the results now enable one to focus on some essential points for future research on loss of vitality.

Biotic damage factors during the *in vivo* phase were of minor importance for the problem in *A. scherzerianum* as well as in *Phalaenopsis* hybrids; symptoms, occurrence and extent of the damage point towards primary causes other than microbial pathogens and pests, which could not be detected in a sufficient frequency and distribution.

Biotic damage factors *in vitro* such as exogenous microbial contaminants and saprophagous mites can be ruled out as relevant causes; although they may occur, they are known to cause trouble and are discarded in good laboratory practice [4,5]. In fact, we do not know enough about the effects of endophytes and also latent viruses on the productivity of young transplants to estimate them in the context of the actual problem. As *in vitro* propagated *Anthurium* and *Phalaenopsis* plants were frequently introduced to renew the motherplant stocks, an influence of the explant-donor plants on the successful growth of the clonal progenies might, in principle, be possible; however, this hypothesis has not yet been tested.

Table 1. Survey on the assessment of possible causes for the loss of plants health in micropropagated *Anthurium* spp. and *Phalaenopsis* hybrids

Potential causes of damage	*Anthurium*	*Phalaenopsis*	
		Clones	Seedlings
Biotic factors *in vivo*			
- microbial pathogens	Se	Se	Se
- viruses	-	?	?
- pests	-	Se	Se
Biotic factors *in vitro*			
- exogenous contaminants	-	-	-
- endophytes	?	?	?
- viruses	-	?	?
- basic plant material	?	?	?
Abiotic factors *in vivo*			
- basic parameters - (climate, substrate, nutrition)	Se	Se	Se
- plant protect. compound	Pr	Pr	Pr
Abiotic factors *in vitro*			
- chem., physic. parameters - (pH, nutrients, H$_2$O-potential)	Pr	Pr	Pr
- phytohormones	Pr	Pr	?

Pr: possible primary cause; Se: possible secondary cause; ?: not to estimate, but possible; -: without importance.

As the basic parameters of crop culture are so heterogeneous between nurseries and over the damage period, abiotic factors *in vivo* defy any damage analysis as long as there are no serious and long-lasting mistakes in the culture schedule to detect. They can be ruled out as the primary cause; however, less favourable growth conditions will secondarily contribute to and intensify already established damage in plant vitality. Plant protection chemicals that were applied very early and frequently in the nursery culture seem to be important for the induction of growth disturbances, especially if they were absorbed by *in vitro* grown plant parts. With regard to the morphological and physiological deviations of *in vitro* grown transplants in comparison with conventionally propagated seedlings or cuttings, interactions with systemically acting plant protection compounds are rather probable. Such an initial effect would explain the damage in different crop plants with different propagation procedures and in different sets in various nurseries. Also, the varying degree of damage and recovery of single plants within a set could become explicable. This effect of plant protection chemicals, however, can only be discussed as an initial damaging event in combination with the predisposition of the plants, depending on the conditions of the preceding *in vitro* phase.

Abiotic factors *in vitro*, such as organic and inorganic components of media or such as physical parameters within the culture vessels, may have long-lasting negative effects on plant growth and development; most of the knowledge on this part of the assessment was contributed by scientists and unfortunately not by commercial growers. Thus, we have to assume, but with good probability, that in the actual damage cases the effects of exogenously applied synthetic or natural phytohormones, cell division stimulating substances, have already caused a predisposition or a physiological impairment of the plantlets, even if their outer appearance satisfied quality demands.

In addition, plantlets were frequently stored in *in vitro* vessels under reduced growth conditions (light, temperature) for weeks or months either in normal storage rooms or in greenhouse compartments. This period may have had uncontrollable consequences on growth and development of the plantlets *ex vitro* during acclimatisation and later.

As the primary causes for the loss of plant health, *in vitro* factors such as phytohormones (cytokinins) and also other abiotic parameters come into consideration in that they can create the aforementioned damage on root and shoot development alone or in interaction with systemic pesticides and biotic factors (e.g. root-rot pathogens, larvae of fungus gnats) in the nursery.

4. DEFICITS IN KNOWLEDGE AND REQUIREMENTS IN FUTURE RESEARCH

From the main results of the assessment the following deficits in knowledge and actual necessities for research become apparent:

• Especially with regard to propagation systems based on divided labour between laboratory and nursery companies there is a basic demand for sufficient plant health of *in vitro* plants to ensure successful acclimatisation and later crop culture. The term 'healthy plant' includes more than to be free from pathogens and pests as well as apathogenic microbial contaminants (endophytes); it should be defined as sufficient in size and with normal differentiation as is necessary for speedy and productive growth *in vivo*.

• The *in vitro* propagation in general is not the problem, but problems emerging in some companies have to be solved. They might be set up by economic pressure against one's better judgement; but more likely, deficits in knowledge about longer-lasting effects of methodical changes in the propagation schedule on plant health have to be compensated for. Therefore, parameters to estimate the productivity of *in vitro* plantlets must be developed in order to identify physiological deviations from a normal healthy, efficient estate.

• There are still large gaps in understanding control mechanisms and effects of phytohormones. To ensure *in vitro* mass propagation in the long-term, it will be necessary to work out tolerance thresholds for the applied cytokinins as well as to conduct investigations on accumulation and side-effects of phytohormones and their metabolites according to plant species.

• Further, with regard to physiological differences between *in vitro* and conventionally propagated plants [6], it is necessary to investigate how susceptible *ex vitro* plants behave towards plant protection chemicals applied in different combinations, at different growth stages and with different frequencies. The interactive effects of selected fungicides and cytokinins on shoot and root formation of *Cordyline* spp. and *Spathiphyllum floribundum* [7–9] are the first references for this aspect with high practical relevance.

• For progressive development of propagation methods in clone selections not only should applicability of *in vitro* culture techniques be considered, but also the selection of tolerance towards stress situations *in vivo* deserves attention. Unfortunately, selection parameters suited for this purpose are still lacking.

• Investigations on the importance of endophytes for plant health during the *in vitro* and *in vivo* stage are necessary for micropropagated ornamentals in general. Endophytes do occur in *Anthurium* spp. and may be a limiting factor in starting mass propagation of new clones. They can , in principle, be detected; however, the considerable expenditure involved is only justifiable if elimination measures follow [10].

The assessment of possible causes for actual problems in plant health in micropropagated *Anthurium* spp. and *Phalaenopsis* hybrids revealed specific and general deficits in knowledge, both in practice and in science. For the benefit of ornamental growers and to stabilize confidence in the unrenouncable in vitro mass propagation, they should be dealt with soon.

5. REFERENCES

1. Grunewaldt-Stöcker, G.(1996) Systemanalyse über Vitalitätsverluste an in vitro-vermehrten Zier-pflanzen: Bestandsaufnahme und Bewertung möglicher Ursachen bei Anthurien und Orchideen. Abschlußbericht zum Forschungsvorhaben; Ministerium für Umwelt, Raumordnung und Land-wirtschaft des Landes Nordrhein-Westfalen, Düsseldorf, 108 p.

2. Geier, T. (1987) Acta Hortic. 212, 439–441.

3. Geier, T. (1990) in Handbook of Plant Cell Culture. Vol. 5: Ornamental Species (Ammirato P.V., Evans D.A., Sharp W.R. and Bajaj Y.P.S., eds) pp. 228–252. McGraw-Hill Publishing Comp., New York, Toronto.

4. Blake, J. (1988) Acta Hortic. 225, 163–166.

5. Debergh, P.C. and Vanderschaeghe, A.(1988) Acta Hortic. 225, 77–81.

6. Ziv, M. (1991) in Micropropagation. Technology and Application (Debergh P.C. and Zimmerman R.H., eds) pp. 45–69, Kluwer Academic Publishers, Dordrecht, The Netherlands.

7. Debergh, P.C., De Coster, G. and Steurbaut, W.(1993) In Vitro Cell Dev. Biol. 29 P, 89–91.

8. Werbrouck, S.P.O. and Debergh, P.C. (1995) J. Plant Growth Regul. 14, 105–107.

9. Werbrouck, S.P.O., Van Der Jeugt, B., Dewitte, W:, Prinsen, E., Van Onckelen, H.A. and Debergh, P.C. (1995) Plant Cell Rep. 14, 662–665.

10. Cassells, A.C. (1992) in Techniques for Rapid Detection and Diagnosis in Plant Pathology (Duncan J.M. and Torrance C. eds) pp. 179–192, Blackwell Scientific Publishers, Oxford.

Developments in Plant Pathology

1. R. Johnson and G.J. Jellis (eds.): *Breeding for Disease Resistance*. 1993
    ISBN 0-7923-1607-X
2. B. Fritig and M. Legrand (eds.): *Mechanisms of Plant Defense Responses*. 1993
    ISBN 0-7923-2154-5
3. C.I. Kado and J.H. Crosa (eds.): *Molecular Mechanisms of Bacterial Virulence*. 1994
    ISBN 0-7923-1901-X
4. R. Hammerschmidt and J. Kuć (eds.), *Induced Resistance to Disease in Plants*. 1995
    ISBN 0-7923-3215-6
5. C. Oropeza, F.W. Howard, G. R. Ashburner (eds.): *Lethal Yellowing: Research and Practical Aspects*. 1995　　　　　　　　　　　　ISBN 0-7923-3723-9
6. W. Decraemer: *The Family Trichodoridae: Stubby Root and Virus Vector Nematodes*. 1995　　　　　　　　　　　　　　　　　ISBN 0-7923-3773-5
7. M. Nicole and V. Gianinazzi-Pearson (eds.): *Histology, Ultrastructure and Molecular Cytology of Plant-Microorganism Interaction*. 1996　　ISBN 0-7923-3886-3
8. D.F. Jensen, H.-B. Jansson and A. Tronsmo (eds.): *Monitoring Antagonistic Fungi Deliberately Released into the Environment*. 1996　　ISBN 0-7923-4077-9
9. K. Rudolph, T.J. Burr, J.W. Mansfield, D. Stead, A. Vivian and J. von Kietzell (eds.): *Pseudomonas Syringae Pathovars and Related Pathogens*. 1997
    ISBN 0-7923-4601-7
10. C. Fenoll, F.M.W. Grundler and S.A. Ohl (eds.): *Cellular and Molecular Aspects of Plant-Nematode Interactions*. 1997　　　　　ISBN 0-7923-4637-8
11. H.-W. Dehne, G. Adam, M. Diekmann, J. Frahm, A. Mauler-Machnik and P. van Halteren (eds.): *Diagnosis and Identification of Plant Pathogens*. 1997
    ISBN 0-7923-4771-4
12. A.C. Cassells (ed.): *Pathogen and Microbial Contamination Management in Micropropagation*. 1997　　　　　　　　　　　　ISBN 0-7923-4784-6

KLUWER ACADEMIC PUBLISHERS – DORDRECHT / BOSTON / LONDON

DATE DUE

DEMCO, INC. 38-2971